T0201145

TEMPORARY STRUCTURE DESIGN

CHRIS SOUDER

WILEY

I would like to dedicate this book to Stuart (Bart) Bartholomew.

I only knew Bart for 14 years, but in this short time he had more influence on me than most. Not only did Bart design the temporary structure class that this book was designed for, he was instrumental in my decision to change to the teaching profession. Countless breakfast and lunch meetings had me listening in amazement to the years of construction, teaching, and consulting experiences. Bart, you are a model of integrity, honesty, and ethical behavior.

CONTENTS

ABOUT THE AUTHOR

Chris Souder graduated with an undergraduate degree in construction management in 1988 before going to work for Kiewit Pacific Co. in northern California. Chris had a successful 16-year career with Kiewit and was involved with many projects in the heavy civil arena. Chris held positions from field engineer to project manager to lead estimator. Some of the projects Chris was involved with were the Woodland WWTP expansion in Woodland, California, Highway 85 Bridge construction for CalTrans in San Jose, California, WWTP Expansion and new facilities for the City of Roseville at its Booth Rd. and Pleasant Grove Plants, Highway 101 Retrofit work for CalTrans in San Francisco, California, new Highway 880 construction of bridge structures for CalTrans in Oakland, California, following the 1989 Loma Prieta earthquake, Water storage facilities for the City of Sacramento, new bridge and 2 miles of road construction including a pump station in Oroville, California, an expansion of the Sacramento River WTP facility for the City of Sacramento, and various estimating assignments for both heavy highway and water treatment facilities throughout northern California. These projects as a whole had total revenues in excess of $420 million.

Chris then pursued an Interdisciplinary Master's degree in construction planning at California State University, Chico, while teaching full time in the construction management program. Today, Chris teaches temporary structures and scheduling and project controls to fourth-year students at Chico State while maintaining a continuous portfolio of consulting projects and industry trainings ranging from cost estimating, temporary structures design, and scheduling services. While teaching, Chris received the terminal degree in construction management by completing his M.S. in construction planning at Chico State. This education, combined with 16 years of heavy civil industry experience makes Chris a most effective type of professor in the construction management discipline.

PREFACE

Temporary structure design is not taken lightly by the owner, engineer, or contractor. It has and should always be a practice that is performed by a licensed engineer in its specific discipline. However, the construction manager should be versed in the design procedures to a point where he can request a particular design or review a concept or submittal with the ability to understand the basic components of the design.

In 1989, the fourth edition of *Simplified Mechanics and Strength of Materials* was written. This book is an example of the present book's goal. I was inspired by the simplicity that Parker and Ambrose displayed in their text. I truly believe that this subject can be well understood by the construction manager without the ultimate goal of becoming a licensed engineer. However, if that is the goal of the student, this text will prepare you to take the next step in engineering pursuing your goal to be licensed.

There is a need for this topic in a construction management (CM) degree, both undergraduate and graduate, civil engineering (CE), both under graduate and graduate, or in industry that is simplified enough that the student, intern, or engineer can simply follow the major concepts without sacrificing key engineering principles. Different universities approach the temporary structures topic in several ways. Some, like Chico State, make it the culminating experience following statics and mechanics. This text will compliment a similar program. Others teach a "structure" class that gives the students a basic understanding of how structures are designed. The latter focuses more on permanent design. Many civil engineering students graduate and go on to work for state agencies or heavy civil contractors. Both of these careers rely heavily on the design of temporary structures. With the state agency, one will be reviewing and inspecting temporary structures. With the contractor, one will be involved with helping design and building temporary structures. These two paths are very rewarding for a CM or CE undergraduate or graduate student.

I also wanted students of temporary structures to be able to comprehend the more complicated analysis that come with more difficult loading conditions without the need for a complete understanding or need for indeterminate structure analysis. I want the student to be aware of the available software on the market today that can simplify even the most complicated loading condition.

I also thought it was important that the student or engineer of this subject be able to understand and perform simple cost estimates of the designs that are explained in each chapter. Most chapters have brief explanations of cost analyses so the educated decisions can be made during the design phase.

ACKNOWLEDGMENTS

To Jessica, Jason and Devon, thank you for putting up with me through this life of construction. To my brothers Greg and Mike, Greg for encouraging me to go into a construction management program and Mike for his constant support. To my parents, Dick and Ines Souder for your constant support, even if you really don't know what I do for a living.

To Robert Towne and Ruth Younger for your hours of assistance to the manuscript.

To Valentina Pozin and Don Hamann for your sound engineering advise and "hand holding."

To the following industry folks and colleagues:

Jim Dick
Steve Floyd
Howard Mattfield
Jim Cole
Dan Collins
Jon Re
Dave Mitchell
Clint Cole
Ken Riley
Dave Jack
Dan Griffin
Shawn Drobny
Brad Kaufman
Bill Cooke

Matt Halleen
Dan Munson
Dave Hazen
Rovane Younger
Bruce Yoakum
Lee Cushman

for your contributions, advice, and case study assistance.
To the following companies:

Kiewit
Traylor Bros.
Golden State Bridge
Cushman Contractors Inc.
Flatiron Corporation
Pankow Builders, LTD

for your case study contributions.
I thank you all.

TEMPORARY STRUCTURE DESIGN

CHAPTER 1

STATICS REVIEW

1.1 STATICS REVIEW

In construction management and civil engineering programs, students are required to take statics and strength of material classes in preparation for their successor. The successor might be a generic "structures" course, a temporary structure course, or maybe no successor course at all. Whichever direction the curriculum goes, the basics of statics and strength of materials is the common denominator.

This book has been written under the assumption that the student has a background in statics and strength of materials and these skills only need to be refined. Temporary structures utilizes many of the less complicated aspects of statics and strength of materials, so even if the student did not master the two prerequisites, he should still be successful in the subject matter of this book. In addition, temporary structure design is a very practical subject, and the student should be energized to see that the challenges that this book covers are real construction situations that the student will experience for his or her entire career.

1.2 UNITS OF MEASURE

At the time of this writing, local and state projects in the United States continue to use the English "Imperial" unit system (feet, pounds, etc.). While most of Europe and the rest of the world use the metric system, the United States has resisted this movement. Even the California Department of Transportation, which had converted current and future projects to the Imperial system of measures late in the 20th century, has gone back to using the Imperial system in the early 21st century. Since England

1

TABLE 1.1   Units of Measure

Unit Name	Unit of Measure
Length	
Foot	ft (′)
Inches	in (″)
Area	
Square feet	SF, ft^2
Square inches	in^2
Volume	
Cubic feet	CF, ft^3
Cubic inches	in^3
Force and Pressure	
Pound	lb, #
Kip	k (1000 lb)
Pounds per ft	lb/ft
Kips per ft	k/ft
Pounds per SF	lb/SF, psf
Pounds per linear foot	lb/ft
Kips per linear foot	kpf
Kips per SF	k/SF, ksf
Pounds per CF	lb/CF, pcf
Moment	
Foot-pounds	ft-lb
Inch-pounds	in-lb
Foot-kips	ft-k
Inch-kips	in-k
Stress	
Pounds per ft^2	psf, lb/ft^2
Pounds per in^2	psi, lb/in^2
Kips per ft^2	ksf, k/ft^2
Kips per in^2	ksi, k/in^2
Temperature	
Degree Fahrenheit	°F

has also gone to the metric system, their "English" Imperial system is now referred to as the U.S. units. Because this text has been written for students in the United States, examples will be given in U.S. units only. Table 1.1 shows most of the common units of measure used in this book.

1.2.1 Common Units of Measure

With any engineering subject, the use of variables to represent different engineering values is standard. These symbols derive either from the Greek alphabet or English letters. Regardless, a great number of symbols are necessary to represent the various engineering concepts. Table 1.2 shows the notations and symbols most used in this book.

TABLE 1.2 Notation and Symbols

Subject	Symbol (Variable)	Description
Properties	S	Section modulus
	I	Moment of inertia
	E	Modulus of elasticity
	A	Cross-sectional area
	r	Radius of gyration
	R	Radius of a circle
	e	Eccentricity
	a	Moment arm distance
	b	Beam width
	c	Distance from centroid to top or bottom edge
	d	Depth of beam
	D	Diameter of a circle
	g	Acceleration of gravity
	h	Height or depth
	K	Distance from top of beam to tangent of web
	e	Effective length of a column or strut, distance from top of beam to web tangent
Stress	f_b	Bending stress
	f_v	Shear stress
	f_v'	Twice shear stress used for short-term shear loading
	$f_{c\parallel}$	Normal compression stress parallel to the grain of the wood
	f_c	Normal compression stress perpendicular to the grain of the wood
Soil Mechanics	c	Cohesion (psf)
	ϕ	Angle of internal friction (degrees)
	β	Passive slip plane angle (degrees)
	α	Active slip plane angle (degrees)
	μ	Coefficient of friction
	γ	Unit weight (pcf)
	π	$\pi = 3.1416$
	k_a	Active soil coefficient
	k_p	Passive soil coefficient
	T	Temperature

1.3 STATICS

Statics is the study of an object that is not moving, hence static or equilibrium. A force is a motion or change of motion in a body. A common force that is produced on Earth naturally is gravity. Gravity is the tendency of the weight of a body to be attracted to the center of the Earth. The mass of some unit weight is placed in motion by gravity or other means. The force of the mass originates at the center of gravity of the body in question. Thus, there is direction and a known weight. Another way to describe a force is something that has magnitude and direction.

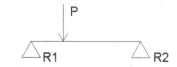

FIGURE 1.1 Force not centered on beam.

Active forces are those forces created by the magnitudes and directions (P) such as the resultant force of a load of concrete. Reactions (R) are also forces, but they are the by-product of the sum of the resultant forces, such as two beams supporting another beam on each of its ends. For instance, Figure 1.1 shows an active P (10 k) force acting downward on the beam, 4 ft from the left side. The reactions are the forces at the supports pushing back (upward). The reactions between the left and right side are not the same because the active force is not centered on the beam.

Forces can be in line with the longitudinal axis of a member or act eccentrically. Axial, compression, and tension forces typically act in line with the linear axis of the member. Eccentric loading introduces forces coming from different directions from the normal forces. The forces are not always vertical or horizontal but sometimes act in directions that are not normal to either axis. Standard structural members are designed to best transfer forces along their linear axes; however, we are not always able to comply with this optimum design.

For the purpose of temporary structures, forces are either uniform (linear) or concentrated. Concentrated loads are also known as point loads and are represented by force arrows and force value in pounds or kips as shown above. Linear loads come in different forms. They can be triangular, trapezoidal, or rectangular depending on the material creating the forces. These forces are represented by a sequence of arrows and a value in pounds per linear foot (lb/ft) or kips per linear foot (kips/ft) as shown in Figure 1.2. Different types of forces created in temporary structures will be discussed in Chapter 3.

1.3.1 Centroids/Center of Gravity

When moment of inertia is introduced to the statics student, it is typically discussed after center of gravity is understood. The student should spend time understanding where and why each object has a center of gravity. In this book, the center of gravity (COG) and "centroid" or "centroidal axis" is used synonymously.

The center of gravity of a square, circle, or rectangle is in the center of each dimension of a material of uniform density. The vertical is referred to as the y axis and the horizontal is referred to the x axis. The center of gravity of circles (pipes) and rectangles is illustrated in Figure 1.3.

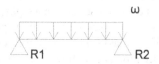

FIGURE 1.2 Uniform load on beam.

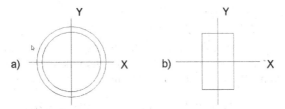

FIGURE 1.3   Center of gravity of pipes and rectangles.

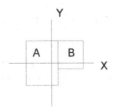

FIGURE 1.4   Center of gravity of a composite shape.

Shapes that are not symmetrical have centers of gravity that are also not symmetrical. For instance, if the shape in Figure 1.4 is considered, where would the center of gravity be in the x and in the y axis?

Example 1.1   If A has a base of $4''$ and a height of $6''$, and B has a base of $4''$ and a height of $3''$ and the top of A and B are flush, determine the composite's COG in the x and y directions.

Step 1: Calculate the area of A and B and determine each shape's individual center of gravity:

$$A = 4'' \times 6'' = 24 \text{ in}^2$$

A's COG is half the base and half the height, $2''$ and $3''$:

$$B = 4'' \times 3'' = 12 \text{ in}^2$$

B's COG is half the base and half the height, $2''$ and $1.5''$. The sum of the two areas is $24 + 12 = 36 \text{ in}^2$.

Step 2: Determine the distance of the x axis to the bottom of the composite by adding the moments of each about the bottom plane. Figure 1.5 illustrates this dimension.

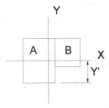

FIGURE 1.5   Center of gravity determining y'.

A's COG is $\frac{1}{2}$ (6″) = 3″ from the bottom.
B's COG is 3″ + $\frac{1}{2}$ (3″) = 4.5″ from the bottom.

Now considering their areas, total the areas about the bottom and make them equal to the combined area and its unknown distance to the new centroid (y'). This will be how far upward the horizontal centroid is:

$$[(24 \text{ in}^2)(3'')] + \left[(12 \text{ in}^2)(4.5'')\right] = [(36 \text{ in}^2)(y')]$$

Solve for y':

$$y' = \frac{(24 \text{ in}^2)(3'') + (12 \text{ in}^2)(4.5'')}{36 \text{ in}^2} \qquad y' = 3.5''$$

Step 3: Determine the distance of y' from the left side of the composite by summing the moments of each about the left side. Figure 1.6 illustrates this dimension.

A's COG is $\frac{1}{2}$ (4″) = 2″ from the left.
B's COG is 4″ + $\frac{1}{2}$ (4″) = 6″ from the left.

Once again considering their areas, sum the areas about the left side and make them equal to the combined area and its unknown distance to the new centroid (x'). This will be how far to the right the vertical centroid is:

$$[(24 \text{ in}^2)(2'')] + [(12 \text{ in}^2)(6'')] = [(36 \text{ in}^2)(x')]$$

Solve for x':

$$x' = \frac{(24 \text{ in}^2)(2'') + (12 \text{ in}^2)(6'')}{36 \text{ in}^2} \qquad x' = 3.33''$$

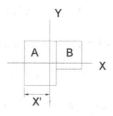

FIGURE 1.6 Center of gravity determining *X*.

Figure 1.7 shows the solution graphically.

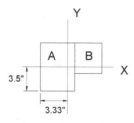

FIGURE 1.7 Center of gravity determining solution.

Figure 1.8 shows some sample calculations for some common shapes.

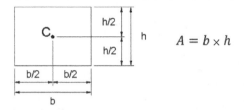

$$A = b \times h$$

$$A = \frac{\pi d^2}{4}$$

FIGURE 1.8 Centroids of area of common shapes.

1.3.2 Properties of Sections

Cross-Sectional Area The cross-sectional area of a beam consists of the depth times the width or the combination of these if there is an irregular shape. For a rectangular or square section, it is defined as

$$A = b \times d$$

where b = beam width
$\quad\quad\quad d$ = beam depth

Centroidal Axis The centroid of a beam section is important as one begins to study moment of inertia. The centroid is the center of gravity of the beam and can be measured about the x or the y axis. The distance from the centroid to the outermost surface of a beam is important because this is the location where the highest stress occurs. When a beam is placed in a bending condition, its stress is measured by the beams section modulus value. The section modulus is a derivative of the beams moment of inertia and the distance from the centroid to the outermost fiber (surface).

Moment of Inertia Moment of inertia is introduced in statics following the understanding of centroids and center of gravity. These basic concepts are key to understanding how beam sections are influenced by how far their sections extend beyond the centroidal axis. Moment of inertia of beams increases as the beams cross-sectional properties are farther from the centroidal axis. Moment of inertia is used to determine deflection in beams and is paramount to the value of section modulus, a very important component of a beam's bending resistance (to be covered in the next section).

The moment of inertia for a rectangle or square section can be solved by the following simple formula that is also referred to as the first moment:

$$I = \frac{bd^3}{12}$$

where b = beam width (in)
  d = beam depth (in)

For composite sections, the moment of inertia is determined by the sum of the first and second moments. The first moment is the moment of inertia of the individual components of the composite (web and flanges). A composite section consists of different shapes (usually rectangles) that are put together to create one common unit made of the same material in most cases. The top and bottom horizontal components are referred to as flanges and the vertical components are referred to as webs or webbing. The second moment takes into account the distance of the composite components to the centroid. This is referred to by engineers as the parallel axis theorem. Large moments of inertia are developed when the individual components become farther from the centroid. Therefore, a tall beam would have a higher moment of inertia as the top and bottom flange get farther from the centroid. The moment of inertia of a composite beam can be solved with the following formula referred to by engineers as the parallel axis theorem:

$$I = \Sigma I + \Sigma A d^2$$

where I = first moment (in^4)
  A = area of an individual component (flange or web)
  d = distance from the centroid to the center of the individual component

Example 1.2   Determine the moment of inertia about the horizontal axis for the composite beam section shown in Figure 1.9.
  Beam properties:

The thickness of the flange and the web are both 2″.
The flange is 6″ wide (*A*).
The web is 8″ tall, not including the flange thickness (*B*).

FIGURE 1.9   Section drawing of composite section.

Step 1: Determine the centroid (COG) about the horizontal axis.
  To do this, total the moments of each section *A* and *B* about the bottom of the composite. Refer to Figure 1.10.

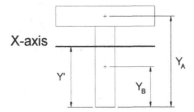

FIGURE 1.10   Section drawing of composite section.

Area of $A = 2 \times 6 = 12$ in^2
Area of $B = 2 \times 8 = 16$ in^2

Distance from bottom to center of A:

$$Y_A = 8 + 1 = 9''$$

Distance from bottom to center of B:

$$Y_B = \tfrac{8}{2} = 4''$$

$$(28)(y') = (12)(9) + (16)(4); \; y' = 6.14''$$

Step 2: Determine the first and second moments using Table 1.3. Calculate d, which is the distance from the section to the centroid. Table 1.3 summarizes the different values that make up the moment of inertia. It is recommended that one always use a similar table to organize the information to avoid mistakes.

$$\sum I + \sum Ad^2 = 260.75 \text{ in}^4$$

The first and second moments add up to 260.75 in^4.

TABLE 1.3   Moment of Inertia Summary

Section	Area	d (in)	Ad^2 (in^4)	I (in^4)	$I + Ad_2$ (in^4)
A	12	$9 - 6.14 = 2.86$	98.15	4	102.15
B	16	$6.14 - 4 = 2.14$	73.27	85.33	158.60

Section Modulus   Section modulus is a derivative of moment of inertia. It increases or decreases based on the distance from its centroid to its outermost fiber, either top or bottom of the beam section. Section modulus is used when determining bending stress of a beam. Section modulus can be derived by the following formula or obtained from a beam section (see Appendix 1). This appendix comes from values calculated in the *Manual of Steel Construction* (MSC). The authors of the MSC have calculated all the beam properties for all available beams manufactured in the United States and some other countries. For wood sections, the National Design Specification for Wood Construction (NDS) provides values for the most common wood sections provided. Appendix 6 illustrates these values, which are rectangular

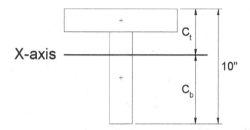

X-axis

c_t

10"

c_b

FIGURE 1.11   Distance to extreme fibers.

and square sections. If the moment of inertia and the centroid are known, then the section modulus can be derived from both. Figure 1.11 illustrates the distance c from either top or bottom of the composite.

$$S = \frac{I}{c}$$

where I is the moment of inertia and c is the distance from the centroid to the outermost fiber of the beam. If the composite beam in Example 1.1 was used as a simply supported beam, the top would be in compression and the bottom would be in tension. The beam height is 10 in, the center of gravity is 6.14″ from the bottom. Therefore, the center of gravity is $(10 - 6.14) = 3.86″$ from the top.

Based on the section modulus formula above:

$$S_{\text{bottom}} = \frac{260.75 \text{ in}^4}{6.14″} = 42.50 \text{ in}^3$$

$$S_{\text{top}} = \frac{260.75 \text{ in}^4}{10″ - 6.14″} = 67.55 \text{ in}^3$$

For a rectangular or square section, since the centroid is in the center of the section, the section modulus is as follows:

$$S = \frac{bd^2}{6}$$

where b = beam width
       d = beam depth

Example 1.3   Determine the section modulus of the section shown where the base dimension is 3″ and the depth of the section is 9″. Refer to Figure 1.12 for details.

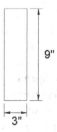

9"

3"

FIGURE 1.12   Vertical rectangular section.

Since $S = bd^2/6$, therefore, $S = (3)(9)^2/6 = 40.5$ in^3.
And since $I = bd^3/12$, therefore, $I = (3)(9)^3/12 = 182.3$ in^4.

If this section is turned 90° and its base becomes 9″ and depth becomes 3″, this would dramatically change the beam's section modulus (and moment of inertia). Let's recalculate the section modulus of the beam in the weak direction. Refer to Figure 1.13 for details.

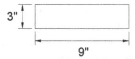

3″

9″

FIGURE 1.13   Horizontal rectangular section.

Since $S = bd^2/6$, therefore, $S = (9)(3)^2/6 = 13.5$ in^3.
And since $I = bd^3/12$, therefore, $I = (9)(3)^3/12 = 20.25$ in^4.

The bending resistance (S) between the strong axis and the weak axis is three times higher. The deflection and buckling resistance (I) between the strong and the weak axes is nine times higher.

Modulus of Elasticity   Modulus of elasticity is a measurement of stiffness for a particular material. Every material has a different value for modulus of elasticity. The most common materials in temporary structures will be discussed in succeeding chapters. Modulus of elasticity is used in temporary structures to determine deflection in beams and allowable buckling stress in columns. This will be discussed in more detail in the following chapters.

Radius of Gyration   Radius of gyration is a distance from the centroid toward the outer portions of the beam or column that resists buckling. Rectangles (lumber and tube steel) have a strong and a weak radius of gyration value when the x and y axes are different dimensions. Squares (lumber and tube steel) have the same radius of gyration in both the x and y axis Circles have a radial radius of gyration; it's the same for 360°. Finally, angle iron and similar sections have three axes of radius of gyration, and the weak axis for buckling is the z axis. This z-axis radius of gyration for angle sections should be used for buckling calculations if the designer wants to use the worst-case scenario. Figure 1.14 shows these three axes. The radius of gyration is a

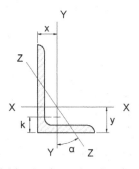

FIGURE 1.14   Angle section showing three axes.

derivative of the cross-sectional area (A) of a section and its moment of inertia (I). Radius of gyration is calculated as follows:

$$r = \sqrt{I/A}$$

where I = moment of inertia
  A = cross-sectional area

Notice that the units work out to inches from the center of the cross section outward. There is also a different radius of gyration for each axis, the strong and weak. Usually when column buckling is a concern, the weak r value is used because that is the direction that buckling will occur.

Example 1.4 What is the radius of gyration of a 4″ diameter standard pipe with the following properties?

$$A = 3.17 \text{ in}^2 \quad \text{and} \quad I = 7.23 \text{ in}^4$$

$$\sqrt{7.23 \text{ in}^4/3.17 \text{ in}^2} = 1.51 \text{ in}$$

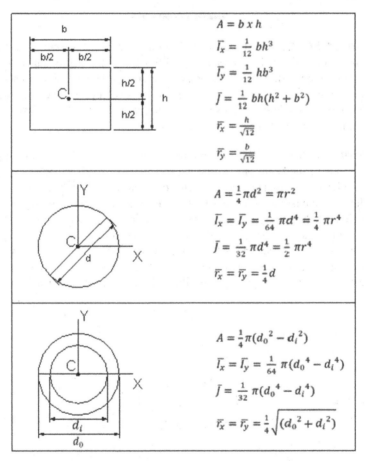

$$A = b \times h$$
$$\bar{I}_x = \frac{1}{12} bh^3$$
$$\bar{I}_y = \frac{1}{12} hb^3$$
$$\bar{J} = \frac{1}{12} bh(h^2 + b^2)$$
$$\bar{r}_x = \frac{h}{\sqrt{12}}$$
$$\bar{r}_y = \frac{b}{\sqrt{12}}$$

$$A = \frac{1}{4} \pi d^2 = \pi r^2$$
$$\bar{I}_x = \bar{I}_y = \frac{1}{64} \pi d^4 = \frac{1}{4} \pi r^4$$
$$\bar{J} = \frac{1}{32} \pi d^4 = \frac{1}{2} \pi r^4$$
$$\bar{r}_x = \bar{r}_y = \frac{1}{4} d$$

$$A = \frac{1}{4} \pi (d_0{}^2 - d_i{}^2)$$
$$\bar{I}_x = \bar{I}_y = \frac{1}{64} \pi (d_0{}^4 - d_i{}^4)$$
$$\bar{J} = \frac{1}{32} \pi (d_0{}^4 - d_i{}^4)$$
$$\bar{r}_x = \bar{r}_y = \frac{1}{4} \sqrt{(d_0{}^2 + d_i{}^2)}$$

FIGURE 1.15 Common shape properties.

The radius of gyration for this $4''$ pipe determines how much buckling resistance the pipe would have if used as a column or strut.

Figure 1.15 shows common shapes used in temporary structures and illustrates the important properties.

Resultants The resultant of a force system is the system converted to a single force. The location of the resultant is determined by the shape of the system. Figure 1.16 shows examples of some simple resultants from a few different force systems.

The resultant of the rectangular force is located in the center.

The resultant of the triangular force is located one third the distance from the left.

The resultant of the trapezoidal force is located a distance (x) somewhere between the triangular resultant and the rectangular resultant.

Statics students typically spend a great deal of time on force vectors, force triangles, and truss analysis. This text does not review these concepts as they are not used with basic temporary structure analyses. This will become evident in Chapter 3 when the different forces in temporary structures are introduced. In statics, it was also very important to understand the difference between different types of supports: roller, fixed, hinged, and the like. Temporary structure supports are hinged the majority of the time. The rest of the time, the supports or connections are fixed by welding or bolting. The reason this is not common is because temporary structures are "temporary," and any effort to fix connections cost money in labor and materials both while installing and while removing them.

Free-body diagrams (FBDs) are drawings that summarize the forces (concentrated), their directions, and the beam geometry and conditions. As learned in statics, the free-body diagram can be the whole structure or part of the structure. This text tends to draw FBDs of the whole structure. Those teaching temporary structures should always insist on a properly drawn FBD accompanying all temporary structures computations. Mistakes most often occur when students have not taken time to create a complete FBD of the structure. A typical FBD is illustrated in Figure 1.17.

Moments Created by Forces A moment is a tendency of a force to create rotation. The point at which the moment is rotating is the center of the moment.

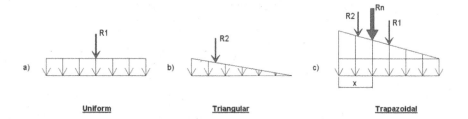

FIGURE 1.16 Resultant force location.

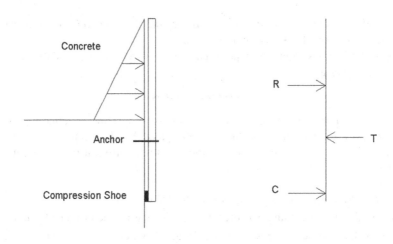

Concrete

Anchor

Compression Shoe

R

T

C

<u>Structure Diagram</u>   <u>Free Body Diagram</u>

FIGURE 1.17   Free-body diagram example.

F1   F2

FIGURE 1.18   Static equilibrium.

The perpendicular distance between the force line of action and the center of the moment is the moment arm. Therefore, a moment is the force times the moment arm distance. If force is represented in pounds and the moment arm is represented by feet, then the moment units are in foot-pounds (ft-lb).

Laws of Equilibrium   This text focuses on static equilibrium. Static equilibrium is when a body is at rest and all forces cancel all other forces out. To achieve static equilibrium, three laws form the cornerstone to temporary and permanent design. These are shown in Figure 1.18.

All forces in the horizontal direction must equal zero, $\Sigma Fx = 0$.

All forces in the vertical direction must equal zero, $\Sigma Fy = 0$.

All moments must equal zero, $\Sigma M = 0$.

Example 1.5   Determine the X and Y forces and the moment created in the simple structure at support A with the forces shown in Figure 1.19.

5 k

4'

3 k

5'

A

FIGURE 1.19   Example 3 sketch.

$$\Sigma Fy = 0 \quad -5\,k + F_{VA} = 0 \quad F_{VA} = 5\,k$$

$$\Sigma Fx = 0 \quad +3\,k - F_{HA} = 0 \quad F_{HA} = -3\,k$$

$$\Sigma M = 0 \quad +(5\,k \times 4\ ft) - (3\,k \times 5\ ft) = 0 \quad MA = +5\ ft\text{-}k$$

F_{VA} = vertical force at point A

F_{HA} = horizontal force at point A

M_A = moment at point A

ΣF_X = sum of forces horizontally

ΣF_Y = sum of forces vertically

Determining Reactions If we go back to Figure 1.1 and ask what are the reaction forces at the left support and the right support, this would be possible by adding the moments about each end of the beam.

Example 1.6 Determine the left and right reactions of the following beam where $P = 5\,k$, the beam is 15 ft long, and the concentrated force is located 4 ft from the left side as shown in Figure 1.20.

FIGURE 1.20 Simple beam reaction example.

To determine the reaction forces at R_1 and R_2, total the moments about one of the two reactions. Let's first add up the moments about R_1.

$$\Sigma MR_1 = 0 \quad -[(5\,k)(4')] + [(R_2)(15')] = 0$$

therefore, $R_2 = 1.33$ k.

Now for the other reaction, we can use one of two methods. We can either total the moments about R_2, in order to get R_1 (similar to what we just did), or we can add up the forces in the vertical direction:

$$\Sigma MR_1 = 0 \quad +[(5\,k)(11')] + [(R_{1V})(15')] = 0$$

therefore, $R_{1V} = 3.67$ k.

$$\Sigma F_V = 0 \quad -5\,k + 1.33\,k + R_{2V} = 0$$

therefore, $R_{2V} = 3.67$ k.

FIGURE 1.21   Simple beam resultant force solution.

It is recommended that the moment method by itself be used or that both methods be used as a double check. The vertical force method can produce errors if the R_1 calculation was incorrect. The engineer should also do a visual check. This means, do the results make sense? For example, if we reversed the reactions by mistake and did a visual check, we would realize that the larger reaction force is farther from the 5-k force and that would not be logical. The larger reaction force has to be on the side that the 5-k force is closest to. Figure 1.21 illustrates the resultant forces of Example 1.6.

Example 1.7   Now let's look at a more complicated example. Figure 1.22 contains multiple loads and load types.

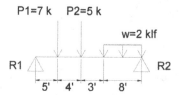

FIGURE 1.22   Simple beam with various loads.

This beam is 20 ft long; the 7-k (P_1) load is 5 ft from the left side; the 5-k (P_2) load is 9 ft from the left side; and the uniform load is 8 ft long, starts 12 ft from the left, and extends to the far right of the beam.

Step 1: Since there is a uniform load, first it must be converted to a concentrated load.

$$P_3 = wL$$

where P_3 = force in klf
       w = uniform load
       L = length of the beam

$$P_3 = 2 \text{ klf} \times 8 \text{ ft} = 16 \text{ k}$$

Step 2: Place the P_3 load of 16 k at the center of the uniform load and draw a new FBD. Double check that the dimensions equal 20 ft. Draw an FBD similar to Figure 1.23.

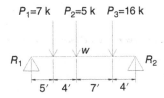

FIGURE 1.23   Simple beam various loads resultant forces.

Step 3: Add the moments about the left reaction (R_1) in order to calculate R_2.

$$\Sigma MR_1 = 0 \quad -[(7\,k)(5')] - [(5\,k)(9')] - [(16\,k)(16')] + [(R_2)(20')] = 0$$

therefore, $R_2 = 16.8$ k.

Step 4: Determine R_1 using both methods shown above in Example 1.1.

$$\Sigma MR_2 = 0 \quad +[(7\,k)(15')] + [(5\,k)(11')] + [(16\,k)(4')] - [(R_1)(20')] = 0$$

therefore, $R_1 = 11.2$ k.

$$\Sigma F_V = 0 \quad -7\,k - 5\,k - 16\,k + 16.8\,k + R_{1V} = 0$$

therefore, $R_{1V} = 11.2\,k$.

The visual test is not as easy this time, but with the 16-k load close to the right support, it would be expected that the right side would have a larger reaction.

CHAPTER 2

STRENGTH OF MATERIALS REVIEW

2.1 STRESS

When forces act on a structure, internal stresses are created that resist the forces over a unit area. Materials experience several different types of stress depending on the forces and loading conditions. The three most common types of stress are normal, bending, and shear. In these three types, compression and tension play a large part on how the stress occurs, especially in normal and bending stress. Stress is measured in force over an area, such as pounds per square inch (psi), kips per square inch (ksi) or pounds per square foot (psf). Generally, psi and ksi refer to steel and wood forces, and psf designates soil and concrete pressure. Strength of a particular material is measured by how that material can resist forces. When these forces are measured over a unit area or section property, stress is the result.

2.1.1 Normal Stress

Normal stress is when a force is directly applied in compression or tension to a unit area. The total force divided by the smallest cross-sectional area equals the highest stress on the cross-sectional area. The higher the force and/or smaller the area will result in a higher stress. On the other hand, a lower force and/or larger area will produce a lower stress. Normal stress is most commonly displayed using the following formula, which can be modified to solve for any of the three variables as shown in the other two formulas:

$$f = \frac{P}{A}$$
$$P = fA$$
$$A = \frac{P}{f}$$

where f = stress
$\quad\quad$ P = axial force
$\quad\quad$ A = cross-sectional area

2.1.2 Bending Stress

Bending stress occurs in beams that span from one support to another (or multiple supports) and are loaded either uniformly, with concentrated loads, or both. When a beam is loaded, it begins to deflect. Deflection (to be covered later in this chapter) is acceptable as long as the beam is not overstressed and as long as the maximum deflection requirement is not exceeded. The following formula calculates basic bending stress. Section modulus is derived from moment of inertia, which was discussed in Chapter 1.

$$f_b = \frac{M}{S}$$

where f_b = bending stress in the most extreme fiber of the member
$\quad\quad$ M = maximum moment in the beam under load considering the span
$\quad\quad\quad$ length, the load, and the type of supports
$\quad\quad$ S = section modulus of the beam section

2.1.3 Shear Stress

Shear stress occurs mostly where there are concentrated forces and reactions. The force at these locations is trying to fail the member, like scissors cutting paper. One side of the shear area is acting in the opposite direction of the force on the other side of the shear area. The cross-sectional area of the member at this location is the property resisting this failure. The type of material is the other key component to resisting shear. Steel, wood, aluminum, and the like are going to have different shear resisting capabilities. Each of these will be discussed in this book under different applications.

$\quad$ For wood or any other rectangular sections, shear is calculated with the following equation:

$$f_v = \frac{1.5\,V}{A}$$

where V = maximum shear in beam
$\quad\quad$ A = area of cross section

$\quad$ For steel sections, shear takes into account only the web sectional area and is calculated with the following equation:

$$f_v = \frac{V_{max}}{t_w d}$$

where V = maximum shear in beam
$\quad\quad$ t_w = web thickness
$\quad\quad$ d = depth of beam

2.1.4 Horizontal Shear Stress

Horizontal shear occurs in the connections between flanges and webs of beams and between the laminates in plywood. When a beam is placed in a bending condition, it deflects the beam's internal parts move horizontally. Without a solid connection between the internal parts, these parts will slide horizontally. In order to resist this sliding, one can use welding, bolting, gluing, and the like. In the case of a wide flange beam, the flange and web are cast together monolithically. Therefore, the horizontal shear resistance is substantial without additional connectors. The student of strength and materials will figure out the weld size and length or the number and size of bolts used in a composite connection. The National Design Standard (NDS) has established horizontal shear stress values for plywood. This chapter will review the basic concept behind horizontal shear. Horizontal shear is calculated by the following formula:

$$v = \frac{VQ}{Ib}$$

where V = maximum shear in the beam, usually at the reaction point (pounds or kips)
  $Q = Ad$ (in^3)
  d = distance from the centroid to the center of the portion of beam in question (in)
  A = cross-sectional area of the top flange (in^2)
  I = moment of inertia (in^3)
  b = width of the connection point from the flange to the web (in)

Example 1.2 is a good example to use to illustrate horizontal shear stress where the top flange meets the web. If a wood flange was attached to a wood web, could the horizontal shear be determined between the flange and the web? If this beam had to span 20 ft with 200 pounds per linear foot (plf) throughout the length, what would be the horizontal shear stress between the top and bottom flange? Figure 2.1 illustrates this example, and Figure 2.2 shows the center of gravity.
  Beam properties:

The thickness of the flange and the web are both 2″. The flange is 6″ wide (A). The web is 8″ tall, not including the flange thickness (B).

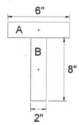

FIGURE 2.1   Composite T section.

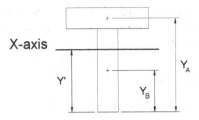

FIGURE 2.2   Center of gravity.

Cross-sectional areas of each member:

Area of $A = 2 \times 6 = 12$ in^2.
Area of $B = 2 \times 8 = 16$ in^2.
Distance from bottom to center of $A = 8 + 1 = 9''$.
Distance from bottom to center of $B = \frac{1}{2}$ of $8'' = 4''$

$$(28)(y') = (12)(9) + (16)(4);$$

$$y' = 6.14''.$$

In Chapter 1, the moment of inertia was calculated as $I = 260.75$ in^4.
If this beam had to span 20 feet with 200 plf throughout the length, then the shear at the ends would be

$$V = \frac{wl}{2}$$

$$V = (200 \text{ lb/ft})(20 \text{ ft})/2 = 2000 \text{ lb}$$

Area of the top flange is $a = 12$ in^2
d = distance from the center of the top flange to the centroid of the composite as calculated in Chapter 1 as $6.14''$ from the bottom. From the top, the centroid would be $10'' - 6.14 = 3.86''$. Since the top flange is $2''$ thick, half of $2''$ would be the center of the flange, therefore, $3.86'' - 1'' = 2.86''$ which is d.
The second moment is the product of the area and d.

$$Q = Ad \quad Q = (12 \text{ in}^2)(2.86 \text{ in}) = 34.32 \text{ in}^3$$

b = thickness of the web where it connects to the top flange, $b = 2$ in.
Since all the values are now known, the horizontal shear can be determined.

$$v = \frac{VQ}{Ib}$$

$$v = \frac{(2000 \text{ lb})(34.32 \text{ in}^3)}{(260.75 \text{ in}^4)(2'')} = 131.6 \text{ psi}$$

A quick check should be made with the units. The pounds remain and three of the 5-in units cancel, leaving inches squared in the denominator. Therefore, pounds per square inch remain, which are the units of horizontal shear stress. If this shear stress was compared to most allowable shear stresses on any specie of wood, this would be considered high. The question remains if the flange and web can be attached with a reasonable number of nails or an adhesive to resist the horizontal shear stress while not damaging the web or flange.

2.1.5 Modulus of Elasticity

Modulus of elasticity for a material is the ratio between unit stress (psi) and the unit deformation (in/in). The unit of measure remains pounds per square inch (psi). The unit is symbolized with E and is an indication of stiffness of a material. Modulus of elasticity is used in determining deflection in beams and buckling in columns.

For steel, E is approximately 29,000,000 psi. For wood, E is between 1,300,000 and 1,900,000 psi, depending on the specie of wood. For concrete, E ranges from 2,000,000 to 5,000,000 psi depending on the mix design. This text does not study concrete design.

2.2 BENDING MOMENTS

2.2.1 Maximum Bending Moments

Shear and moment diagrams are used when these values cannot be calculated from our simple, standard formulas such as $wL^2/8$, $PL/4$, or Pa and when our support arrangement is something other than a simple support. Construction management and engineering students spend a great deal of time learning how different beams react under certain loads. Rather than detailing all aspects of beam design, this book will review the overall concepts. Later, shear and moment diagrams will be discussed in detail. Figure 2.3 shows the three most common and simple maximum moment formulas used: from left to right, uniform load, concentrated load in the center of span, and concentrated load off center of span, respectively. If these arrangements are encountered, shear and moment diagrams are not necessary.

Chapter 3 will discuss different loading conditions and where the loads might come from in a temporary structure condition. For the purpose of this book, the loads on beams are classified as either concentrated or distributed. In temporary structure design, both are used. However, the distributed load is probably more common. Steel beams that are supported by columns typically support other beams. If the beams being supported are close together, then this may be analyzed as a distributed load. If these same beams are spaced more than one fourth the length, they may be considered as individual concentrated loads. Moments are created in beams when they are subjected to bending.

The load on an individual beam is the concentration of forces that are within the beam's tributary width. The tributary width of a beam is the distance that extends

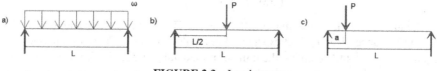

FIGURE 2.3   Load types.

FIGURE 2.4   Tributary width on a beam.

halfway to the beam on each side, which is also equal to the spacing of the supporting beams. If beams are spaced at 4 ft on center, one beam sees the forces from 2 ft each side or 4 ft of tributary width. Figure 2.4 illustrates the tributary width on one beam, spaced equally.

2.2.2 Maximum Shear

Beams are typically supported by columns or other beams. These points of support are reaction points and points at which high shear value occurs. At the reaction points, the beam has a concentrated shear force where the reaction force is directed upward and the beam load is directed downward at the same point. A shear diagram is used to show the values of shear at any given point on a beam. The shear value can be calculated at any point along a beam from one end to the other and is dependent on the types of loads, the support locations, and types of supports. In this book, most supports will be free or hinged. It is not common to fix (weld or bolt) connections of beams and columns in temporary structures because it adds unnecessary costs to the installation and removal of the elements.

The magnitude of the shear at any point of a beam is equal to the sum of all the forces to one side (usually to the left). These forces have positive and negative values. Forces directed upward (reactions) are positive and forces directed downward (concentrated loads and distributed loads) are positive values. The difference between the two is the amount of shear at a particular section.

2.2.3 Law of Superposition

Law of superposition comes in handy when more than one common force is applied to a simple span, but it is not quite complicated enough for a shear and moment diagram. For instance, if a simply supported beam was loaded with a uniform load and a single or multiple concentrated load, the maximum moment could be calculated by adding the moment caused by the uniform load to the moment caused by the concentrated load. The combination of concentrated loads either fall into the condition of Figure 2.3(b) or 2.3(c). Figure 2.5 illustrates three concentrated loads on a simply supported beam.

The law of superposition says that the moment of inertia for the beam and forces shown can be calculated by adding the maximum moment formula of the P_1a to the maximum moment formula $P_1L/4$. This, of course, only works if the two end forces are the same distance from the supports and the center force is in the middle of the beam:

$$M_{\max} = P_1a + \frac{P_1L}{4}$$

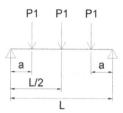

FIGURE 2.5   Law of superposition.

2.3   MATERIALS

This book will focus on the two most common materials used in construction: steel and wood. Steel is a manufactured product that results from its chemical composition being cast during a heating process. For steel, its carbon content is the component that has the greatest effect on its strength. The steel type used in most temporary structures is ASTM A36, the number 36 representing the steel's yield stress point during American Society for Testing and Materials (ASTM) testing. Unless specifically noted, this type will be used for examples in this book.

2.3.1   Factors of Safety

Factors of safety (FOS) are reductions applied to material stress values that lower the amount of stress a material can see in its useful life. For this book, the useful life of a material has been defined as several hours to several months and it is considered temporary. Factors of safety can be applied to yield stresses or ultimate stresses of a particular material. For the purpose of discussion, mild A36 steel will be used. There are three points of stress we will use to understand factor of safety: (1) the ultimate (F_{ult}), (2) the yield (F_y), and (3) the allowable (F_{all}). The yield stress of a material is typically close to the plastic limit. The yield is the point at which a material, such as steel, begins to deform, but when the load is released will go back to its original shape. The ultimate stress of a material is achieved when a material fails. If we use mild A36 steel as an example, we can understand these points of stress.

2.3.2   Grades of Steel

Steel comes in many grades for both structural shapes and bolts. In construction, the most common steel is A36 (grade 36). The 36 represents the steels yield strength of $F_y = 36$ ksi. The number system used in grade of steel is named after the ASTM testing methods. A36 steel is considered mild steel because of its carbon content. As the carbon content changes, the steel strength and corrosion resistance changes. Other steel grades used in construction are A328, A572 (grade 50, 50 ksi), A572 (grade 55, 55 ksi), and A572 (grade 60, 60 ksi). There are many other grades available. As the grade increases, typically so does the price per pound. It is becoming more popular for an engineer to specify a higher grade steel in order to maximize the design and, hopefully, still be economical.

An easy and common sample would be a $1'' \times 1''$ steel bar placed in a laboratory testing machine. The specimen would be pulled in tension and the forces would

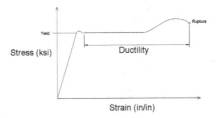

FIGURE 2.6   Stress–strain graph.

be recorded. The steel would immediately begin to undergo stress (pounds per square inch, psi) as the force is increased. This applied force will continue to increase along with the stress as shown in Figure 2.6. This is a linear function until a certain point. When the force reaches approximately 32,000 lb, the steel sample begins to deform. This deformation is in the form of elongation. If the force is removed at this point, the sample would go back to its original length. This is the plastic limit. When the force is increased beyond 36,000 lb, the elongation continues. Beyond this point, if the load is removed, the sample would maintain a portion, if not all, of the deformation. As the force is increased even further, the sample continues to elongate. When the force reaches approximately 58,000–70,000 lb, it will see its ultimate stress and fail. Ductility is a material's ability to deform in tension and stretch to a longer length.

Some key data points during this test of stress are:

$$\text{Yield stress } (F_y) \; 36{,}000 \text{ lb}/1 \text{ in}^2 = 36{,}000 \text{ psi}$$
$$\text{Ultimate stress } (F_u) \; 58{,}000 \text{ lb}/1 \text{ in}^2 = 58{,}000 \text{ psi}$$

The value derived from these data points is that allowable stress (F_b) 21,600–24,120 psi (depends on 0.6–0.67 FOS).

Generally, a factor of safety is applied to the yield stress. In other words, the allowable stress (F_b) is calculated by factoring down the yield stress. If an engineer wanted to apply a 1.5 : 1 factor of safety, he would divide 36,000 psi by 1.5. In this case the allowable stress (the maximum amount of stress the steel could withstand without failing) would be 24,000 psi. If the engineer followed the recommendation of the American Institute of Steel Construction (AISC) and applied a 60% factor of safety, the allowable stress would be 36,000 psi × 0.60, or 21,600 psi. A 67% factor of safety would multiply 36 ksi by 0.67, or 24.12 ksi. This example for steel would represent allowable "normal stresses." Other types of stress such as shear and normal stress would have a lower factor of safety. Table 2.1 shows the most common types of stress for materials used as temporary structures.

2.3.3 Compact Beam

The factor of safety can be adjusted to $0.66F_y$ if it meets a compact section requirement. To see if a beam meets this requirement, determine its ratio of d/t. This is the ratio between the beam's depth and minimum thickness of its vertical component (web):

$$\frac{d}{t} < \frac{190}{\sqrt{F_y}} \quad ; \quad \frac{190}{\sqrt{36 \text{ ksi}}} = 31.67$$

TABLE 2.1 Allowable Stresses for A36 Steel

Stress Type	Factor of Safety	Allowable Stress (psi)
F_b (bending)	0.6–$0.67F_y$	21,600–24,120
F_v (shear)	$0.40F_y$	14,400
F_c (compression)	$0.90F_y$	32,400
F_t (tension)	$0.60F_y$	22,000
	$0.50F_u$	29,000
	$0.33F_u$	19,000
E Modulus of elasticity	29,000,000–30,000,000 psi	

TABLE 2.2 Compact Beam Sample Calculations

Beam	d (in)	T (in)	d/t (ratio)	Conclusion
C6 × 8.2 channel	6.0	0.20	30.0	30 < 31.67, compact
W12 × 58 WF beam	12.19	0.36	33.86	33.86 > 31.67, not compact
TS8 × 8 × 1/2	8.0	0.5	16.0	16 < 31.67, compact

where d = depth
       t = web (vertical) thickness
       F_y = yield stress or 36 ksi

Table 2.2 illustrates calculations that check whether beams meet the compact requirements.

When a beam is considered compact, $F_b = 0.67F_y$.

2.3.4 Wood

Plywood   Plywood has a strong and weak direction, depending on the direction of the supporting members (studs or joists). This will be discussed in detail in Chapter 9, but Table 2.3 has been provided to show the allowable stresses of standard grade BB-OES plywood used. It is common for contractors to use other grades of plywood, typically higher grades such as high density overlay (HDO) or medium density overlay (MDO). The manufacturer of these grades of plywood will supply a property data sheet including allowable stresses, deflection, spans, and so forth.

Dimensional Lumber   Dimensional lumber is the most popular material used for formwork and other lightweight temporary structure systems. Its versatility, weight,

TABLE 2.3 Plywood Allowable Stress Ranges

Strong (Face Grain Parallel to Span)	
Bending stress (F_b)	1500–2000 psi
Shear stress [$F_v(I_b/Q)$]	55–60 psi
Modulus of elasticity (E)	1,500,000–1,600,000 psi

TABLE 2.4 Dimensional Lumber Allowable Stress Ranges

Strong (Face Grain Parallel to Span)

Bending stress (F_b)	700–2400 psi
Shear stress (F_v)	85–145 psi
Adjusted shear stress (F_v')	170–190 psi
Compression, parallel to grain (F_{cll})	400–730 psi
Compression, perpendicular to grain (F_c)	625–1650 psi
Modulus of elasticity (E)	1,200,000–1,900,000 psi

TABLE 2.5 Load Factor Uses

Factor	Permanent (P) or Temporary (T)
Duration (permanent to impact)	T
Wet use (moisture)	P, T
Flat use (loaded in weak direction)	P
Size (timber vs. lumber)	P
Temperature (heat and cold)	P
Beam stability (lateral stability)	P
Shear	T
Incising (pressure treatment)	P
Repetitive member (multiple uses)	P, T
Column stability (slenderness)	P
Buckling stiffness (slenderness)	P

cost, and ease of use make it a highly preferred material for temporary structures. Table 2.4 shows the ranges of strengths in dimensional lumber.

Load factors are multipliers that adjust the allowable stresses either up or down, depending on the factor. For permanent design, there are approximately 10 factors. Temporary structures use only a few of these, such as load duration, wet use, shear, and temperature. The range depends on the type of use and severity of each. For instance, temperature factor will depend on how extreme the temperature goes up and down. The wet use factor will depend on the variability of moisture. The factors only affect specific stresses. One factor may affect bending stress and modulus of elasticity, whereas another factor may affect shear and compression parallel to grain. For example, compression perpendicular to grain requires only the wet use and temperature factor. Some of these available factors are shown in Table 2.5.

2.4 DEFLECTION

Deflection occurs in all loaded beams. It is defined as the measurement between the straight, undeflected beam and the deflected beam, typically in inches. Some beams, if their spans are long enough, deflect under their own weight as well. In this case, during the beam analysis, the weight of the beam should be added as an additional uniform load.

Deflection is compared to engineering standards such as L/240, L/360, L/480, and so forth. These standards determine the maximum deflection allowed per span of beam. For example if a 32-ft beam span had to stay within an L/360 standard, the maximum deflection would be

$$32 \text{ ft} \times 12''/\text{ft}/360 = 1.0667 \text{ in}$$

Note that L is converted to inches in the numerator.

The amount of deflection seems relatively high; however, for a beam to deflect 1.0667 in. in 32 ft is negligible when it comes to temporary structures if the movement in this beam does not affect any other part of the system. In permanent construction, the excessive deflection of a stud or joist may cause the stucco or drywall to crack; therefore, the deflection limits will take this into consideration and be more stringent, such as L/480. Also, the stress on this same beam with this same deflection would typically be low. Generally, in temporary beam design, the stress on the steel dictates over the amount of deflection within a standard factor of safety of 1.5 or 2.0 : 1.0. Deflection tends to govern only when there is a light load over a very long span.

When beams are designed to resist bending stress and they have an appreciable amount of deflection, the designer would typically calculate the amount of deflection and add camber strips to the top of the beam to offset the deflection anticipated. This way, when the beam deflects, the top of the beam becomes a straight line again, thus not leaving undulations in the finish product. Chapter 11 discusses the use of camber strips in bridge construction. A very common deflection calculation for a uniform load on a simply supported beam is

$$\frac{5wl^4}{384EI}$$

In temporary structure design, L/360 is very common. A lower tolerance could be encountered in a support of excavation system where horizontal movement that causes some settlement adjacent to the excavation is acceptable due to the lack of structures nearby. However, in another case, the requirements for a temporary structure may be more stringent. For example, the engineer designing excavation support within the city streets of San Francisco may call for an L/1000 tolerance on the shoring design so the displacement of the adjacent roads would be immeasurable. The shoring system selected, deep soil mixing with wide flange beams embedded or sheet piling with waler support beams, is designed for almost no lateral movement.

A similar tolerance was required on the SR 520 Evergreen Floating Bridge Pontoons in the State of Washington. The formwork that was used had steel plate sheeting instead of plywood in order to meet the engineer's deflection requirements. Deflection requirements such as these examples were not frequently encountered until more recently.

2.5   SHEAR AND MOMENT DIAGRAMS

Shear and moment diagrams are a staple to engineering temporary structures design when the loading and support conditions are less common. As mentioned before, there are three simply supported loading conditions where the maximum moment can be determined through a simple formula. In addition, if more than one of these three conditions are combined, the law of superposition can be used. As a rule of thumb while creating shear and moment diagrams, clockwise rotation will be a negative moment and counterclockwise rotation will be a positive moment. Positive shear will be upward and negative shear will be downward.

When the standard formulas cannot be used, two options can be used. The first is to create a shear and moment diagram, which we will do in this section. The second would be to use beam software that calculates these values from some simple input data provided by the temporary support designer. This software will be covered later in this chapter and will be used for structures classified as indeterminate structures, those having more than two unknown reactions.

In statics, strength of materials, or mechanics classes, students spend a great deal of time mastering shear and moment diagrams. This text will refresh the user on the basics of shear and moment diagrams without repeating the methods for exact values at specific locations along the beam. The methods used here will only derive the maximum and minimum values for shear and moment. Basic shear and moment diagrams done manually include the following:

1. Two supports (two unknowns). More than two unknowns are indeterminate structures, which are covered later in this chapter.
2. Concentrated loads, uniform loads, or varying uniform loads.

The method used in this text builds the moment diagrams from the areas of the shear diagrams. It is very important to build the shear diagrams correctly so the moment diagrams are accurate. Basically, the sum of the areas to the left of the shear diagram at any point is equal to the moment in the beam at that point. Following are three examples of building shear and moment diagrams. The first will be a very simple condition showing how $M = wl^2/8$ is derived; the second is a relatively simple example with two loads; and the third will be intentionally complicated (although not very realistic) in order to assure that the concepts are mastered.

Example 2.1   Draw a shear and moment diagram for the following loading condition: simply supported beam (l), uniform load (w) the same length as the beam.

Step 1: Determine the (R)esultant of the uniform load and its location along the beam.
$$R = wl, \text{ distance } \frac{1}{2} \text{ centered on the beam}$$

Step 2: Determine the shear/reactions at the left and right side supports:
$$R_l = R_r = wl/2$$

Step 3: Draw the shear diagram with these reactions. The diagram has a positive shear at the left equal to the reaction at the left ($+wl/2$). The shear value decreases as the shear decreases from left to right. The amount of decrease is equal to the value of w per linear measure from left to right. When the uniform load w gets to the right support, the shear value should be $-wl/2$ and come back to a zero value.

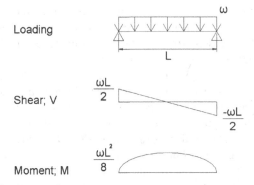

Loading

Shear; V

Moment; M

FIGURE 2.7   Shear and moment diagram for Example 2.1.

Step 4: Draw the moment diagram by determining the points of change from the shear diagram and the area of the shear at different locations along the beam. Since this example is just a single uniform load, there are no abrupt changes in shear except at the end reactions, and it only crosses zero shear once. The zero shear locations are an indication of a maximum point of moment, either positive or negative. Figure 2.7 represents the shear and moment diagram for Example 2.1.

It should be pointed out that the shear diagram is broken into two halves, both of which are triangular shapes. The maximum moment is in the center of the beam (where shear value crosses a zero), and its value is equal to the area of the shear diagram from the center to the left support:

$$M = \text{Area} = \frac{bh}{2} = \frac{(L/2)(wL/2)}{2} = \frac{wL^2}{8}$$

This same process can be used to prove $m = PL/4$ and Pa. Have your instructor illustrate these in class or try this during a study session.

Example 2.2   Draw a shear and moment diagram for the following loading condition: simply supported beam $L = 20'$, uniform load of 5 kips per linear foot (klf) the same length as the beam length and a concentrated load of 25 k located 8 ft from the left support. This is represented in Figure 2.8.

Step 1: Determine the resultant (R) of the uniform load and its location along the beam.

5 klf × 20 ft = 100 k, located in the center of the 20′ span. Add this to the 25-k concentrated load, and we have a total resultant downward force of 125 k. This is the value that the reactions should equal in the next step.

Step 2: Determine the shear/reactions at the left and right side supports by adding the moments about one side or the other. Let's use the left side.

$$\Sigma M_{\text{left}} = 0 \quad -(25 \text{ k} \times 8 \text{ ft}) - (100 \text{ k} \times 10 \text{ ft}) + (R_r \times 20 \text{ ft}) = 0$$

therefore $R_r = 60$ k.

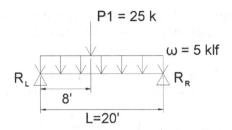

FIGURE 2.8 Loading diagram for Example 2.2.

Now add the vertical forces, $+60 \text{ k} - 125 \text{ k} + R_1 = 0$; therefore, $R_1 = 65$ k.

Step 3: Draw the shear diagram with these reactions. The diagram has a positive shear at the left equal to the reaction at the left $+65$ k. The shear value decreases as the shear decreases from left to right. The amount of decrease is equal to the value of w per linear measure from left to right. The amount of shear loss due to the uniform load is 5 klf × 8 ft = 40 k. Now the shear is at a positive 25 k (65 k − 40 k = +25 k). When the uniform load (5 klf) gets to the 25-k concentrated load (8 ft from the left support), there is an abrupt change in shear equal to the 25-k load. Here the shear diagram drops straight down 25 k, the shear value is now 25 k less or 0 k (25 k − 25 k = 0 k). From here the shear diagram continues the downward slope due to the uniform load of 5 klf. There are 12 ft to go, therefore 5 klf × 12 ft = 60 k. Subtract this 60 k from the 0-k shear and this equals −60 k. Now, finally go up the amount of the reaction at the right side (60 k), and this is the end of the shear diagram. We are at 0 k of shear at the right support, the same place we started at on the left side. Figure 2.9 shows the diagram for this example.

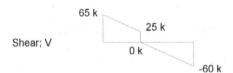

FIGURE 2.9 Shear and moment diagram for Example 2.2.

Step 4: Draw the moment diagram by determining the points of change from the shear diagram and the area of the shear at different locations along the beam. The most significant location is at the 25-k concentrated load 8 ft from the left side. Also notice the shear diagrams are triangles and trapezoids. Therefore to calculate the areas under the shear diagrams, one must separate the triangle from the rectangle. Zero shear is always an indication of a maximum point of moment. The most important values when determining the moments are the maximum values, so the shape of the parabolic curve is not as important as the value at the extreme points. After all, to determine the minimum beam size in beam design, one must know the *maximum moment*. In this case the maximum moment is equal to the area within the shear diagram. The positive shear and the negative shear will equal each other in order for the shear diagram to equal zero value at each end.

The shear diagram is a trapezoid (triangle and a rectangle); therefore, the area of both need to be calculated in order to come up with the total moment.

$$\text{Top triangle } (A_1): \quad A_1 = \frac{40\,\text{k} \times 8'}{2} = 160 \text{ ft-k}$$

$$\text{Bottom rectangle } (A_2): \quad A_2 = 25\,\text{k} \times 8' = 200 \text{ ft-k}$$

$$\text{Total area}: \quad M_{\text{max}} = 360 \text{ ft-k} = \text{max moment}$$

Figure 2.10 illustrates this calculation.

Moment; M

FIGURE 2.10   Maximum moment.

Example 2.3   Draw a shear and moment diagram for the following varying linear load with a max of 500 plf on a 12-ft beam with supports at the left and 4 ft from the right side. Figure 2.11 depicts the diagram for Example 2.3.

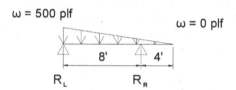

FIGURE 2.11   Loading diagram for Example 2.3.

Today, this condition would be input into a beam software program and the results would be seconds away. However, in order to get confident in producing an accurate shear and moment diagram, this solution will be derived manually.

Step 1: Calculate and locate the resultant of the triangular linear load.
$(500 - 0)(12')/2 = 3000$ lb located one third from the left and two thirds from the right.

Step 2: Determine the reactions at both supports, by summing the moments about one side or both:

$$R_l = 1500 \text{ lb}$$

$$R_r = 1500 \text{ lb}$$

It is a coincidence that they are equal. This is only because of the triangular linear load over a beam with a second support two thirds to the left.

Step 3: Draw an FBD using the beam length, support locations, and the resultant as a concentrated force. Figure 2.12 shows the free-body diagram for Example 2.3.

$R_L = 1'500$ lb $R_R = 1'500$ lb

FIGURE 2.12   Free-body diagram for Example 2.3.

Step 4: Draw the shear diagram and moment diagrams. This is shown in Figure 2.13.

Shear; V

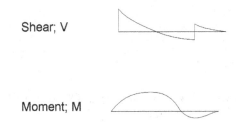

Moment; M

FIGURE 2.13   Shear and moment diagram for Example 2.3.

Example 2.4   Beam Software Programs
There are an unlimited number of beam software programs available on the market today. Some are much easier than others. It is recommended that the most basic and easy-to-use software available be used because even the most elementary program typically does more than the temporary structure designer needs. The program should be able to do the following:

1. Accept multiple supports, hinged or fixed.
2. Accept uniform, concentrated, or varying uniform loads that represent all construction loads.
3. Accept beam properties such as moment of inertial values.

The software should be able to produce the following:

1. Shear diagram with support reactions
2. Moment diagram indicating positive and negative moments and identifying which one has the highest magnitude
3. Deflection diagram (optional)

Consider an example of a useful diagram produced by simple beam software. The diagram is duplicating a typical formwork waler as the beam. The supports are

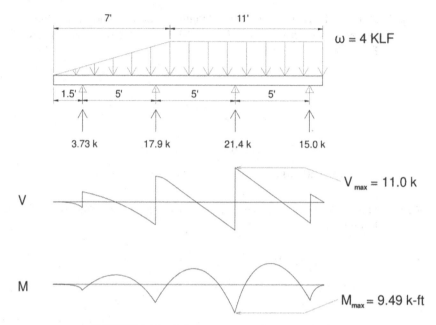

FIGURE 2.14   Software-generated S and M diagram.

the ties; and since there are more than two, this is already a complicated shear and moment diagram because it is indeterminate. Finally, the loading is indicative of a concrete pressure diagram, which is typically rectangular until it gets close to the top and then tapers down to 0 KLF. The program calculates the reactions at each tie (support), the shear diagram, the moment diagram, and a deflection diagram. The deflection is only accurate if a beam was selected while setting up the beam information. However, usually the purpose of the diagram is to determine the maximum moment before selecting the beam. Therefore, sometimes the deflection is not accurate until the beam is selected, the steel type is selected, and the moment of inertia has been input into the program. Figure 2.14 shows a typical software-generated diagram.

2.6   BEAM DESIGN

As mentioned earlier when discussing steel beams, the maximum moment of a beam under a particular load and support arrangement is the most important information in order to determine the beam size necessary to *resist the bending stress in the beam.* For wood beams, this information with the *Shear resistance* is combined to design the beam. The information from the shear diagram is helpful for this check. These two checks, however, do not guarantee the beam will not have stability problems. Stability will be covered later in two ways. Early on, a ratio between the depth and the flange width will be used as a quick way to check stability. The second method, to be covered later in this book, will be a more accurate measure of a beam's stability using the ratio d/A_f.

For this chapter, beam design will only have to comply with bending stress and shear stress. It will become very apparent that shear in steel beams is rarely, if ever, an issue. For wood, however, shear can definitely be the governing criteria of the beam selection. Once again, here are the bending moment and shear formulas for both steel and wood.

Steel:
$$f_b = \frac{M}{S}$$

where S = section modulus of the beams horizontal axis
M = maximum moment

$$f_v = \frac{V}{dt_w}$$

where V = maximum shear in beam
t_w = web thickness
d = depth

Wood:
$$f_b = \frac{M}{S}$$

where S = section modulus of the beams horizontal axis
M = maximum moment

$$f_v = \frac{1.5V}{A}$$

where V = maximum shear in beam
A = cross-sectional area of the beam ($A = bd$)
b = base dimension of wood beam
d = depth of wood beam

In Example 2.2, a beam is supporting a concentrated load and a uniform load. These values will be used for the steel beam design. The maximum shear and moment were as follows:

$$V_{max} = 65 \text{ k}$$

$$M_{max} = 360 \text{ ft-k } (360{,}000 \text{ ft-lb})$$

Steel Beam Design Before a beam can be designed, the allowable bending and shear stress should be determined. Using AISC standard values for stable beams, allowable bending stress (F_b) should not exceed $0.667F_y = 24$ ksi and allowable shear stress (F_v) should not exceed $0.40F_y = 14.4$ ksi. The steel beam will be designed for bending stress and then checked for shear stress.

Step 1: Bending Stress—Determine the minimum section modulus using the allowable bending stress and maximum moment in the beam from above.

$$S_{min} = 360 \text{ ft-k} \times 12''/\text{ft}/24 \text{ ksi} = 180 \text{ in}^3$$

List wide flange beams between 12″ and 24″ with a minimum section modulus of 180 in³. Table 2.6 shows the available beams.

TABLE 2.6 Available WF Beams Meeting
Required Section Modulus Requirement

Beam	S_x (in^3)
W12 × 136	186
W14 × 120	190
W16 ×	N/A
W18 × 97	188
W21 × 93	192
W24 × 84	196

Looking at this list of beams, we can only make a decision based on the weight of the beam and the fact that they all have an acceptable section modulus. Based on weight (which is how steel is purchased, by the pound), the W24 would be the most economical beam. Without any other criteria to select the beam by, such as stability, weight will be the determining factor. Later in this book, stability of beams will be addressed.

Depending on the number of beams a project would need in this example, cost savings could amount to a great deal of money when comparing two of the beams above. For instance, assume that the project required 40 of these 15-ft beams and the staff was deciding between the W24 × 84 and the W18 × 97. The weight difference per linear foot between the two beams is 13 lb (97 − 84 = 13). The total linear footage of beams required is 40 each × 15 ft = 600 lf. At 13 lb/lf, the difference in cost would be 600 lf × 13 lb/lf = 7800 lb of steel. If the cost of steel was $0.65/lb, this would equal $5070.00.

Wood Beam Design

Example 2.5 Example 2.4 has loads and a span that generate a much higher moment than a normal size wood beam would be able to handle. Therefore, another beam condition will be developed in order to explain wood design. Figure 2.15 illustrates a 15-ft span simply supported beam with a 2000 lb/lf load from left to right (similar to Figure 2.14).

Using the simple moment formula for a simply supported beam with a uniform load:

$$V_{max} = \frac{(2000)(15)}{2} = 15,000 \text{ lb}$$

$$M_{max} = \frac{(2000)(15)^2}{8} = 56,250 \text{ ft-lb}$$

Before a wood beam can be selected, the specie of wood should be known or at least the allowable stress values. In this example, it will be assumed that the allowable shear stress of the wood is $F_v = 140$ psi, and the allowable bending stress will be $F_b = 1500$ psi. Using the formulas above, the beam size can be determined using both bending and shear. The beam should be sized for bending and then checked for shear.

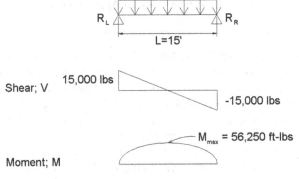

FIGURE 2.15   Example 2.5 shear and moment diagram.

Step 1: Determine beam size based on bending:

$$1500 \text{ psi} = \frac{(56{,}250 \text{ ft-lb})(12''/\text{ft})}{S_x}$$

where $S_x = 450 \text{ in}^3$.

Table 2.7 suggests that the 6×24 ($S_x = 506.2 \text{ in}^3$) is the most economical because it has a minimum section modulus greater than 450 in^3 while maintaining the lowest cross-sectional area. Cross-sectional area determines the cost relating to board feet. Therefore, if the board has less board feet (bf) per linear feet (lf), then it will be cheaper.

Step 2: Shear Stress—The shear stress should be checked before determining this to be the best beam. The maximum allowable shear stress (F_v) from above is 140 psi.

$$140 \text{ psi} = \frac{1.5 \,(15{,}000 \text{ lb})}{A} \quad \text{and } A = \frac{1.5(15{,}000)}{140 \text{ psi}} = 160.7 \text{ in}^2$$

The area of a 6×24 is 129.3 in^2, therefore $A_{\text{all}} = 160.7 > 129.3$ (Not Good).

The area of the 6×24 is *not* sufficient. Another way to check this would be to solve for f_v and compare to 140 psi:

$$f_v = \frac{1.5 \,(15{,}000 \text{ lb})}{129.3 \text{ in}^2} \quad f_v = 174 \text{ psi} > 140 \text{ psi} \quad \text{(Not Good)}$$

Since this beam fails the shear test, the next beam that met the S_x test ($S_x \geq 450 \text{ in}^3$) will have to be tried.

$$f_v = \frac{1.5(15{,}000 \text{ lb})}{166.3 \text{ in}^2} \quad f_v = 135.3 \text{ psi} < 140 \text{ psi} \quad \text{(OK)}$$

The 10×18 works for bending and shear. A table containing values for wood properties can be found in Appendix 6.

Cost Comparison for Wood Beams   The cost of wood can also be made part of the wood example. How does the cost compare between the 6×24 and a 10×18, which had an adequate section modulus of 484.9 in^3 (greater than 450 in^3)? The 10×18 has a cross-sectional area of 166.3 in^2 and the 6×24 has a cross-sectional area of 129.3 in^2. For this example, the price of timber is assumed to be \$1.20/bf. Table 2.7 provides a cost comparison of two sizes of lumber.

Similar to the example for the steel beam design, the cost savings can be very significant. For example, the 600 linear feet used earlier would equate to

$$600 \text{ lf} \times (\$18.00 - \$14.40) = \$2160.00 \text{ cost difference}$$

Buckling Stress   When explaining buckling stress, the terms "post," "column," and "strut" are used interchangeably. When posts, columns, and struts are supporting an axial force, the tendency is to buckle in their weak direction. The weak direction is the direction parallel to the smaller dimension of the post. Buckling is basically like deflection in a beam but is caused by an axial force rather than a bending moment. If a post is square or if a pipe is used as a post, the post can buckle any direction of its 360° (the square in 90° directions). If a rectangular post is used, then the post will deflect in the weak direction as shown, unless appropriately braced. Figure 2.16 provides a sketch of column buckling.

Buckling stress is the amount of stress a post, column, or strut can withstand considering (1) its weakest radius of gyration, r_y, and (2) its largest unsupported length, l_u. Normal stress of an axial loaded post, when buckling is not considered, is P/A. For A36 steel, the allowable would be approximately 27,000 psi. For wood posts, the allowable would be in the range of 625–1650 psi. However, when buckling is considered, taking into account r_y and l_u, the allowable stresses (buckling stress) can drop as low as 3000 psi (estimate) for A36 steel and 150 psi (estimate) for different species of wood. These numbers are ballpark figures in order to make the point. These changes

TABLE 2.7   Cost Comparison of Lumber

Beam	Area (in^2)	bf per lf	Cost per bf (\$)	Cost per lf (\$)
6×24	129.3	12	1.2	14.40
10×18	166.3	15	1.2	18.00

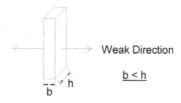

Weak Direction

$b \leq h$

FIGURE 2.16   Column buckling sketch.

in allowable stress values can be considered adjustments to the *P/A*, axial normal stress value.

Slenderness ratio (SR) is the ratio between the length of the post in inches and the weakest radius of gyration or the smallest dimension of a rectangle, depending on whether it is steel or wood. The ratio of an 8-ft long 2×4 S4S would be l/d ($8' \times 12''/\text{ft})/1.5'' = 64$. This is a ratio because the inches cancel so there are no units. The ratio, if we lengthened the beam to 12 ft, would be 96 (50% higher). The higher the slenderness ratio, the more susceptible the section (2×4 in this case) is to buckling. Instead of *P/A* being 1250 psi, the allowable buckling *P/A* would be below 100 psi. This is a considerable difference when a beam is being used as a post, column, or strut.

In strengths of materials, the student studied the Johnson and Euler theories of column buckling. These two theories distinguished between long and short columns, depending on their slenderness ratios being above or below 120. This text uses one method to calculate either scenario. A table will also be introduced in Appendix 4 for trial-and-error column design calculations.

End connections of a column are crucial to the proper design. The effective length is the adjusted length after taking into account the effective length factor, which is determined by how the ends of the column are attached. The ends are typically pinned in temporary structures; however, other options are fixed (welded or bolted), free like a flag pole, and fixed but free to move laterally. Table 2.8 shows the multiplying factor used to adjust the effective length for the slenderness ratio calculation.

Two formula methods for columns will be used: one for steel and one for wood members. The slenderness ratio will be used for both; however, steel will use the weakest radius of gyration, while wood will use the least dimension in cross section.

Steel:

$$F_{bs} = \frac{\pi^2 E}{[(kl_u/r)]^2 \times \text{FOS}} = \frac{P}{A}$$

(units in psi or ksi)

where F_{bs} = allowable buckling stress (psi)
       E = modulus of elasticity for steel, usually 29,000,000 psi
       k = connection-type multiplier, depending on fixed, hinged, etc. (no units)
       l_u = unsupported length of column; if there are supports, then the longest distance between supports (in)
       r = weakest radius of gyration (in)
    P/A = normal stress and is set equal to allowable buckling stress in order to determine
     P_{all} = (or area allowed)

Refer to Appendix 2 for pipe properties.

TABLE 2.8   End Connection Factors

Connection Type	Factor
Pinned × Pinned	1.0
Fixed × Pinned	0.7
Fixed × Fixed	0.5
Fixed × Free	2.0

Example 2.6   Buckling of a Steel Column

Let's look at a 16-ft-long $3\frac{1}{2}''$ standard steel pipe with an FOS of 1.5 : 1.0. To determine the allowable axial load this pipe can support, follow these steps.

Step 1: Look up the cross-sectional area (A) and the radius of gyration (r). There is only one value for r because it is a pipe.

$$A = 2.68 \text{ in}^2$$

$$r = 1.34 \text{ in}$$

Step 2: Calculate $F_{bs} = \pi^2(29,000,000)/\{[(1.0)(16 \times 12''/\text{ft})/1.34'')]^2 \times 1.5\} = 9294$ psi.

Step 3: Set $F_{bs} = P_{all}/A$ and solve for $P_{all} = 9294$ psi $\times 2.68$ in$^2 = 24,909$ lb.

This 16-ft-long, $3\frac{1}{2}''$ standard pipe can support 24,909 lb before failing by buckling. If buckling was not an issue, then $P_{all} = 27,000$ psi $\times 2.68$ in$^2 = 72,260$ lb. This is a 47,451-lb reduction because of buckling.

Wood Columns:

$$F_{bs} = 0.30E/(l_u/d)^2$$

where F_{bs} = allowable buckling stress (psi)
$\quad\quad E$ = modulus of elasticity for the wood specie (psi)
$\quad\quad l_u$ = unsupported length of column if there are supports, then the longest distance between supports (in)
$\quad\quad d$ = least dimension of the cross section

Table 2.9 explains how d is the least dimension of a cross section. A square cross section has the same dimension for d, whereas, a rectangular cross section only has one d, the smaller one.

TABLE 2.9 d Dimensions for Various Lumber Sections

Size	d Dimension (in)
2×4	1.5
4×4	3.5
4×6	3.5
4×8	3.5
6×6	5.5

The FOS safety for wood is built into the modulus of elasticity value. P/A is normal stress and is set equal to allowable buckling stress in order to determine P_{all} or area allowed.

Example 2.7   Buckling of a Wood Column

Let's look at a 12-ft-long, 4×4 with $E = 1,700,000$ psi. To determine the allowable axial load this 4×4 can support, follow these steps.

Step 1: Look up the cross-sectional area (A) and the radius of gyration (r). There is only one value for r because it is a pipe.

$$A = 3.5 \times 3.5 = 12.25 \text{ in}^2$$

$$d = 3.5 \text{ in}$$

Step 2: Calculate $F_{bs} = 0.30 \ (1{,}700{,}000 \text{ psi})/[(12 \text{ ft} \times 12''/\text{ft})/(3.5)]^2 = 301.3$ psi.

Step 3: $P_{all} = 301.3 \text{ psi} \times 12.25 \text{ in}^2 = 3690.3 \text{ lb}.$

This 12-ft-long, 4×4 can support 3690.3 lb before failing in buckling. If buckling was not an issue, then $P_{all} = (650 \text{ psi} - 1650 \text{ psi}) \times 12.25 \text{ in}^2 = 7963 \text{ lb to } 20{,}213 \text{ lb}$ depending on the specie of wood. This is a 4000 lb – 17,000 lb reduction because of buckling.

Example 2.8   Steel Column Design Using HP Shapes

Determine the allowable load (P_{all}) of a HP 14×89 that is 32 ft long. Appendix 3 will be used for this example.

Step 1: Look up the properties of a HP14 $\times$ 89 in Appendix 1.

$$r_y = 3.53 \text{ in}$$

$$A = 26.1 \text{ in}^2$$

Step 2: Calculate the slenderness ratio (SR).

$$\text{SR} = \frac{32' \times 12''/\text{ft}}{3.53 \text{ in}} = 108.8$$

Step 3: Look up the allowable buckling stress of SR = 108.8 in Appendix 4 for 36 ksi steel.

$$F_{bs} = 11.91 \text{ ksi}$$

Step 4: Calculate the P_{all} of the HP 14×89

$$P_{all} = 11.91 \text{ ksi} \times 26.1 \text{ in}^2 = 310.9 \text{ k}$$

Normal stress on steel would allow $F_{bs} = 27{,}000$ psi; however, because of buckling, the allowable stress was lowered to 11.91 ksi.

2.6.1 Combined Stress

Combined stress occurs when a material section has a condition causing normal and bending stress at the same time. One example would be a pipe full of water and attached to the side of a post or column. The pipe and water weight (P) create two stresses: axial (normal) stress from the weight of the pipe over the cross-sectional area of the post and a bending stress from the weight of the pipe (P/A), and e eccentricity distance away from the centerline of the post creating a bending moment and stress on the section modulus of the post Pe/S. See Figure 2.16.

Combined stress $= P/A + Pe/S$, units in psi or ksi

Using this same scenario, given the information to follow, determine the combined stress in psi.

Given: 4×4 S4S wood posts at 12 ft on center

4″ STD pipe full of water

Assume the pipe is adequately attached to the side of the post

Step 1: Determine the weight of the pipe and the water inside the pipe. The dimensions and properties of 4″ standard pipe are found in Appendix 2. It has an ID $= 4.026″$ and an OD $= 4.50″$. The inner diameter (ID) will be used for the water volume and the outer diameter (OD) will be used for the distance e eccentricity. The pipe itself weighs 10.79 plf.

Water weight: Either convert the unit weight of water to cubic inches or convert the inside pipe volume to cubic feet. Let's use the former method. Water weighs 62.4 pcf, therefore, $62.4 \text{ pcf}/1728 \text{ in}^3/\text{ft}^3 = 0.03611 \text{ lb/in}^3$ of water.

The inside volume of the pipe is the cross-sectional area $\times$ 12″ length of pipe (1 ft).

$$\pi(4.026″)^2/4 = 12.73 \text{ in}^2 \times 12 \text{ in} = 152.76 \text{ in}^3$$

Water weight in a 12″ piece of pipe $= 152.76 \text{ in}^3 \times 0.03611 \text{ lb/in}^3 = 5.51$ plf.

Add the weight of the pipe per 1 linear foot, 5.51 plf $+$ 10.79 plf $=$ 16.31 plf.

Step 2: Calculate the weight of 12 ft of pipe

$$16.31 \text{ plf} \times 12 \text{ ft} = 195.7 \text{ lb}$$

Step 3: Calculate the distance from the center of the pipe to the center of the post.

$$e = \tfrac{1}{2}(\text{pipe OD}) + \tfrac{1}{2}(\text{post width}), \ e = \tfrac{1}{2}(4.50″) + \tfrac{1}{2}(3.50″) = 4.00″$$

Step 4: Calculate the axial stress due to the weight of the pipe on one post loading the cross-sectional area of the 4×4 post.

$$\frac{P}{A} = \frac{195.7 \text{ lb}}{3.5″ \times 3.5″} = 15.98 \text{ psi}$$

Step 5: Calculate the bending stress on the post due to the weight of the pipe e distance from the post causing a bending stress. The section modulus of the 4×4 post is 7.146 in^3

$$\frac{Pe}{S} = \frac{195.7 \text{ lb} \times 4.0″}{7.146 \text{ in}^3} = 109.54 \text{ psi}$$

FIGURE 2.17 Combined stress.

Step 6: Combine the two stresses similar to Figure 2.16. Figure 2.17 depicts this combined stress:

$$\text{Combined stress} = 15.98 \text{ psi} + 109.54 = 125.52 \text{ psi}$$

This does not consider how the pipe is attached to the post. That would have to be designed separately.

Combined stress can also occur when a column or strut that are supporting an axial load encounter a concentrated or uniform lateral load against the weak or strong axis. For instance, when a steel beam is used as a strut in a cofferdam, its original purpose is to resist the axial force from the soil loading that is pushing from each side. However, during the excavation, what would happen if a cubic yard of soil weighing approximately 3800 lb fell on the beam. Not even considering the dynamic force of the 3800 lb force, the strut would have to support the axial force at the same time the lateral force of the dropped soil applied a bending moment to the beam. The sketch in Figure 2.18 is an FBD of how these two forces would affect the beam.

Similar to the example above, the axial stress must be added to the bending stress. The difference in this example is that the bending stress moment is represented as $P_1 L/4$. The axial force is still represented by P/A. Another way to explain the combined stress analysis in a case like this would be to analyze each stress as it relates to its allowable stress. In other words, the axial force is a buckling stress case and the bending moment is a bending stress case. Therefore, the allowable buckling stress needs to be known and the allowable bending stress should be determined (between $0.6F_y - 0.67F_y$).

$$\frac{P/A/\ \pi^2 E}{(kl/r)^2 \times \text{FOS}} + \frac{PL/4/S}{0.60 \times 36{,}000 \text{ psi}} < 1.0$$

The theory behind this method is that if the two individual stresses are within their own ranges, they will pass the stress test; or, if together, they do not exceed 100% of the combined allowable stresses, they should be within the limits.

FIGURE 2.18 Combined stress on a beam.

Example 2.9 Determine the combined stress of a 6″ standard pipe strut. The strut is 15 ft long, the axial force from the soil is 30 k and there is a cable in the center of the pipe holding a 1500-lb trash pump. The FBD shown in Figure 2.18 depicts this situation.

Since the strut is 15 ft long, the cable is connected 7.5 ft from each end. The properties needed for the pipe are

$$A = 5.58 \text{ in}^2$$

$$S = 8.5 \text{ in}^3$$

$$r = 2.25 \text{ in}$$

Determine the axial stress compared to its allowable buckling stress and then determine the bending stress compared to the allowable bending stress on A36 steel assuming $0.6F_y$:

$$\frac{30,000\#}{5.58 \text{ in}^2} + \frac{\dfrac{1500\# \, (15'')}{4}}{8.5 \text{ in}^3} \leq 1.0$$

$$\frac{\Pi^2 \, 29,000,000 \text{ psi}}{\left[\dfrac{(15' \times 12)}{2.25}\right]^2 \times 1.92} = 23.293 \text{ psi}$$

$$(0.60 \times 36,000 \text{ psi}) = 21,600 \text{ psi}$$

$$\frac{5376 \text{ psi}}{23,293 \text{ psi}} + \frac{7941 \text{ psi}}{21,600 \text{ psi}} = 0.60 \leq 1.0 \quad \text{(OK)}$$

In this example, the pipe would not fail in combined stress because, when combined, the axial stress and the bending stress do not exceed more than 100% of the combined allowable stresses.

CHAPTER 3

TYPES OF LOADS ON TEMPORARY STRUCTURES

3.1 SUPPORTS AND CONNECTIONS ON TEMPORARY STRUCTURES

There are a few luxuries we have as construction managers in temporary structures design. The first one is that the construction manager and designer have a great deal of flexibility in system selection during the design phase. There are typically many different solutions to solve the same problem. These solutions are generally compared to one another and costs are placed on each. The best solution usually is the one that costs the least, is most schedule efficient, and does not come with any safety and health risks to the workers, the general public, or property.

Another luxury we have as construction managers is the way we support the individual members of our temporary structures. In statics, it is very important how the ends of beams or columns are supported when determining the amount and types of stresses incurred from bending moments, shear, buckling, and deflection. In temporary structures, supports are almost always considered to be simply supported or hinged (Figure 3.1). This means that when a beam rests on its support (another beam), they are not welded, bolted, or otherwise joined/fixed together. One member is simply resting on another, and, when the loads are applied, the beam is free to rotate about the point where they make contact. When supports are bolted or welded (fixed connections at the ends), the beam's maximum moment can be reduced from $wl^2/8$ to $wl^2/12$, but the labor involved with making the connection becomes a large contributor to the overall costs on the temporary structure.

When beams are simply supported, their deflection pattern begins and ends above the supports. When this is the case, we have to use the more conservative moment and deflection formulas. They are more conservative because they produce more moment

FIGURE 3.1   End connections.

and more deflection. Two examples of these formulas are

$$\frac{wl^2}{8}$$

$$\frac{5wl^4}{384EI}$$

These two formulas will produce higher moments in the beam and higher deflection values than if the beam is supported by multiple supports or fixed at either or both ends. The following formulas are used to calculate maximum bending moment and maximum deflection in beams with three or more spans:

$$\frac{wl^2}{10}$$

$$\frac{wl^4}{1740EI}$$

Columns are somewhat similar. The distance needs to be analyzed between where they make contact with the ground and what they are supporting. These points of contact are said to be hinged at the top and bottom so that when the axial force is applied to the column and the column's reaction is to buckle, it does so at the points of contact. In this case the effective length is equal to the full length of the column. In this text, we will assume hinged connections for all contact points and support locations. In a later chapter, connections will be discussed briefly so the student can gain some understanding of how other types of connections affect temporary structures designs. Even though hinged connections are the focus of this textbook, the student should understand the different types of connections that affect stress in columns. In mechanics, the student learns the following end conditions and applies the applicable k value. As mentioned in Chapter 2, k is the multiplier that is used, depending on the connection.

Both ends hinged, $k = 1.0$.
One end hinged, one end fixed, $k = 0.70$.
Both ends fixed, $k = 0.50$.

Figure 3.2 illustrates what happens to l_e when lateral braces are added to a post. P_2 would be used if there were only a top and bottom support. However, if an

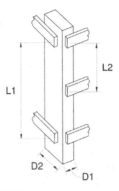

FIGURE 3.2   Sample post supports.

additional support was added in the center of the post, P_1 would represent the new effective length.

As a simple example, if a column is 12 ft tall, resting on a concrete slab and supporting the underside of a beam, the effective length (l_e) is equal to $12' \times 1.0 = 12'$. If we made a connection between the supporting slab and the post that was adequate for fixity, the effective length (l_e) would then be $12' \times 0.70 = 8.4'$.

Another consideration in supporting temporary structures is not only what type of support but how many supports are used under a continuous single member. In statics and strength of materials, this is the difference between simple moment calculations and an indeterminate structure that has more than two unknown reactions (supports). When a member spans over more than two supports, much time can be spent determining the reaction values. In this book, these two types of support systems are categorized into either a simply supported beam or a continuous beam over two or more supports. This concept will be discussed in more detail when beams, columns, and struts are analyzed in other chapters.

3.1.1 Forces and Loads on Temporary Structures

Axial   Axial forces are common in columns, struts, and tension members such as coil rods and anchors. In a simple case, the force is directed concentrically through the center of the supporting member. Other cases have the loads off center to the centerline of the axial member. In this case, combined stress should be considered.

Combined Loads   Combined stress is when there is an eccentric load causing axial (normal) stress and bending stress simultaneously. The two stresses are combined (added or subtracted) to result in a total stress at the extreme fiber. In temporary structures, the largest combined stress is the most important, as this is the worst-case situation. This concept was discussed in length in Chapter 2.

Compression   Compression has several considerations when it comes to wood members. Forces can be applied parallel to grain or perpendicular to grain as well as induce buckling stress. The first two stresses are basically P/A-type stresses. Buckling stress, however, is a reduced value to normal stress, which takes into account the length of the column or strut and its radius of gyration or least dimension in cross

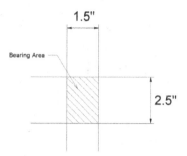

FIGURE 3.3 Compression zone.

section depending whether steel or wood is used. Figure 3.3 illustrates the area of the compression zone.

Compression in steel is treated the same as wood; however, the concern is not as relevant. Allowable stress on A36 steel can be as much as 75–90% of F_y, its yield, or 27–32.4 ksi.

Tension Tension forces can be produced by hanging supports in coil and DWIDAG-type rods or tieback, soil, and rock anchors using rods or high-strength strand.

Concentrated and Uniform Loads Concentrated loads occur when one beam element's load path travels through another beam element. The bottom of one beam's flange applies load to the top of another beam's flange. If the beam spacing is a quarter of the span or more, this load type can be considered a uniform load. If the spacing is less, one can consider this multiple concentrated loads.

Uniform and Varying Loads Uniform loads are probably the most common forces in temporary structures because they represent a continuous force that acts perpendicular to the supporting face. These forces come from water, concrete, wind, and soil. Some of these uniform forces are constant and some are varying.

3.1.2 Materials—How Different Materials Create Different Forces

Water Water can produce both uniform and varying forces. If the water pressure is static on a cofferdam, for example, then it is considered varying, with the highest force being at the bottom and reducing to zero at the top. This condition is shown in Figure 3.4. Water pressure, in this case, is the product of the depth and the unit weight of water.

Static Water Pressure Static water projects a triangular pressure diagram and acts perpendicular to the supporting face. The pressure is a direct result of the depth of water and the unit weight of water (always 62.4 pcf in this book). The maximum pressure (P_{max}) for a 25-ft deep (h) body of water in its static state is

$$P_{max} = \gamma_{water}h$$
$$P_{max} = 62.4 \text{ pcf} \times 25 \text{ ft} = 1560 \text{ psf}$$

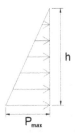

FIGURE 3.4   Water pressure diagram.

Flowing Water Flowing water will apply a uniform force to the side of the structure that is obstructing flow, such as a cofferdam or permanent pier or column. Figure 3.5 shows how high water levels can get in construction. The river shown in this figure reached its peak flood elevation just 4 months before construction of its replacement bridge began. This would be a 3-year project.

Figure 3.6 provides a simple water pressure diagram. To calculate flowing water force, the following formula is used from physics:

$$P = c_d \gamma \left[\left(\surd^2 \right) / (2g) \right] \quad \text{(units in psf)}$$

where P = pressure in psf
   C_d = drag force coefficient of the obstruction
   γ = unit weight of water (pcf)
   $\surd$ = velocity of flowing water (ft/s)
   g = acceleration of gravity at Earth's atmosphere = 32.2 ft/s^2

The drag force coefficient, C_d, takes into account the friction created by the water against the side of the obstruction as the water passes. The drag force will vary based

FIGURE 3.5   Flooding waters on a bridge pier.

FIGURE 3.6 Flowing water pressure diagram.

on the shape of the obstruction. The drag force can range from 1.2 for cylinders such as steel pipe to 2.0 for Z-shaped sheet piling with convolutions in the sheets created by the Z shape.

Example 3.1 Determine the pressure from a flowing river with the following characteristics. Then determine the resultant if the water is 20 ft deep.

Velocity of water = 12 ft/s
Unit weight of water = 62.4 pcf
Drag coefficient of 1.3

Step 1: Calculate the pressure in psf.

$$P = 1.3(62.4 \text{ pcf}) \left[\frac{12^2}{2(32.2 \text{ ft/s}^2)} \right] \quad P = 181 \text{ psf}$$

Step 2: Since the force is a rectangular load, the resultant is in the middle or the flowing water (see Figure 3.7). The resultant needs to be calculated by multiplying the pressure times the depth of the water:

$$R = P \times \text{Depth} \qquad R = 181 \text{ psf} \times 20 \text{ ft} = 3620 \text{ plf of flowing water}$$

Remember, the drag coefficient can vary from 1.2 to 2.0.

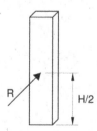

FIGURE 3.7 Resultant force location.

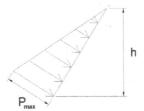

FIGURE 3.8 Triangular force on an incline.

Loads on Inclining/Sloped Surfaces Pressure from water is developed from the unit weight of water and the water depth, and the direction of the pressure acts perpendicular to the surface of the obstruction or dam. What happens when the surface of the obstruction or dam is sloped instead of vertical? Does the pressure increase? No, because we already determined that the pressure magnitude comes from the unit weight and depth of water. Does anything else change? Yes, the length of supporting surface becomes the sloped surface of the triangle. What does this mean?

The static water example above used a vertical surface as an obstruction in order to calculate the water pressure of 1560 psf. The resultant of this pressure over the 25-ft-tall dam is determined by calculating the area of the triangular pressure diagram (see Figure 3.8).

The resultant is 1560 psf $\times$ 25 ft$/2$ = 19,500 lb acting perpendicular against the surface one third of the distance from the bottom.

If the surface was sloped at a 45° angle, still maintaining the 25-ft water depth, the pressure would remain the same at the bottom. However, the pressure triangle would lengthen, thus increasing the resultant magnitude. This occurs because when the resultant is recalculated, the length of the pressure triangle is now 25 ft $\times$ 1.414 (the square root of 2) = 35.35 ft.

$$R = \frac{1560 \text{ psf} \times 35.35 \text{ ft}}{2} = 27,573 \text{ lb}$$

The resultant increased by a factor of 1.4, which happens to be the 45° angle square root of 2.

Concrete Concrete is a unique force in that it is liquid for a period of time but then begins to set and support itself eventually. Concrete can be supported with a vertical form in which the force acts laterally against the form face. This force is called hydrostatic pressure because it is present for a period of time until the concrete begins to set and support itself. This period of time is between 2 and 6 h, depending on mix design and admixtures used to retard or slow down the initial setting of the concrete.

Concrete can also be supported as an elevated slab element where its force is a gravity load acting downward to the ground. Both of these cases are discussed below.

Concrete supported as a gravity force—Concrete weighs approximately 150 pcf. When the concrete is reinforced, 5–10 lb can be added to the concrete weight. In this text, reinforced concrete weighs 160 pcf (150 + 10 = 160 pcf). The thickness of the concrete element is very important because in order to convert the load to psf, the unit

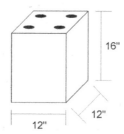

FIGURE 3.9 Reinforced concrete column.

weight must be multiplied to the thickness in feet (see Figure 3.9). Let's consider Example 3.2.

Example 3.2 Determine the maximum load in psf of 16-in-thick reinforced concrete.

$P = 160$ pcf $\times 16''/12''$/ft $= 213.3$ psf. The force is now in PSF after multiplying the unit weight in pcf times the thickness in feet. The sketch in Figure 3.9 depicts a $12'' \times 12'' \times 16''$ thick slab section.

In Chapter 10, a live load will be added to the dead load in order to come up with the actual load on the forms. The dead load takes into account the concrete and rebar weight and the live load is the estimated weight of the tools, equipment, and workers (see Figure 3.10).

Concrete as a hydrostatic force is probably the most common load in construction. If concrete was treated as a material that is always liquid, like water, the pressure would be 150 pcf $\times h$. However, concrete does not stay liquid for a long period of

FIGURE 3.10 Concrete placement on a bridge superstructure.

time; it begins to set and support itself after a period of time (typically 2–6 h). For this reason, special formulas are used to determine the maximum pressure. The pressure diagram for concrete is either trapezoidal or triangular. The trapezoidal load is more common. The triangular pressure diagram is used when the concrete is considered fully liquid. Both will be discussed in Chapter 8.

The following pressure diagrams in Figure 3.11 show what trapezoidal and triangular loads of concrete hydrostatic pressure would look like.

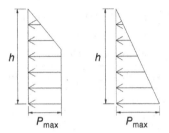

Hydrostatic (concrete)   Full Fluid Pressure

FIGURE 3.11   Hydrostatic and full liquid pressure.

Wind   Wind projects a horizontal uniform load from top to bottom until it reaches a certain height. This is indicative of a wind force on a column form or wall panel. Wind loads are calculated in pounds per square foot (psf) but can be converted from miles per hour (mph) of wind speed in a particular geographical zone. Table 3.1 shows how pressure (psf) corresponds to different wind speeds in mph. The table also relates this force to the force generated against a 4 × 8 sheet of plywood and against an

TABLE 3.1   Wind Speed Converted to Wind Pressure

mph	psf	Force on 4 × 8 Sheet of Plywood (lb)	Force on Average Size Person (lb)
10	0.27	8.6	2.2
20	1.08	34.6	8.6
30	2.43	78	19.4
40	4.32	138	35
50	6.8	216	54
60	9.7	311	78
70	13.2	423	106
80	17.3	553	138
90	21.9	700	175
100	27	864	216
110	33	1045	261
120	39	1244	311
130	46	1460	365
140	53	1693	423
150	61	1944	486

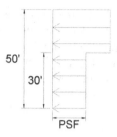

FIGURE 3.12 Wind pressure diagram.

TABLE 3.2 Wind Pressure at Different Height Zones

Height Zone (height above grade) (ft)	Wind Pressure (psf)
0–30	20
30–50	25
50–100	30
Over 100	35

average size human being's body. If someone has ever experienced lifting up a piece of plywood in the wind, he would understand the force being described.

When a project is constructed in a particular region, the designer and contractor can obtain wind force records common to the area. The designer of the permanent structure needs this information for the permanent design. Likewise, the contractor and his designer need this information for any temporary structure design. Wind force can affect forms for walls and columns, rebar for walls and columns, and any other structure above the ground that would be affected by wind forces.

State Department of Transportation and the AISC have developed best practices for column rebar guying. This will be covered in Chapter 12. However, the understanding of the wind forces created on a rebar cage that stands over 20 ft in height is very important. The best practices typically increase the wind speed (pressure) as the column gets higher. For instance, a 50-ft–high column designed for a 90-mph wind speed would be designed for 21.9 psf in the first 30 ft. The remaining 20 ft would then see a higher pressure. The increase in the upper section may be between 5 and 10 mph, thus 36–31 psf. The pressure diagram in this case would look like the Figure 3.12. Table 3.2 is an example of wind forces in California rounded to the nearest 5 PSF.

When the two charts are compared, it is obvious how the wind force used on these projects is the maximum wind speed that is recorded in the region over a long period of time. Even though reaching top speeds may not be likely, the designer must anticipate the worst-case scenario. As one can see, Table 3.2 uses wind pressure forces between 20 and 35 psf. Looking at Table 3.1, this is the equivalent of approximately 80–120 mph winds.

Example wind force will be illustrated here. Using the chart in Table 3.2, a wind load will be applied to a vertical wall.

Example 3.3 Calculate the resultant force of a 50-mph wind on a 20-ft-high × 15-ft-wide wall. Draw a pressure diagram like that depicted in Figure 3.13.

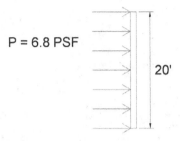

P = 6.8 PSF

20'

FIGURE 3.13 Wind pressure diagram.

Step 1: Determine the force from the wind as a 12″ strip load as the resultant:

$$R \ (12'' \ \text{strip}) = 6.8 \ \text{psf} \times 20 \ \text{ft} = 136 \ \text{lbs per linear foot (plf) of wall}$$

The resultant force of this 20-ft strip load is 10 ft from the bottom.

Step 2: Determine the resultant force of the 15-ft-wide wall.

$R(\text{wall}) = 136 \ \text{plf} \times 15 \ \text{ft} = 2040 \ \text{lb}$ located 10 ft from the bottom and 7.5 ft from the left and right. Draw a resultant diagram like that in Figure 3.14.

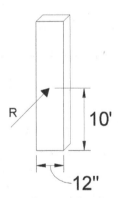

R

10'

12"

FIGURE 3.14 Resultant diagram.

Soil Pressure Soil pressure places a varying load laterally on a support of excavation (SOE). The forces from soil pressure have many variables, but the resulting pressure can be estimated using several different theories on soil pressure. This book will use the most common theories of soil mechanics to illustrate the pressure applied to an SOE from different types of soil. The theories of Coulomb and Rankine, which use the results of Mohr's circle, will be simplified to the extent that a great deal of time and effort are not spent explaining the origins of the theories. Like many examples in this book, the concepts can be understood without all the details that the original authors used in their calculations. This subject will be studied further in Chapter 5. Other effects on shoring, besides the soil pressure, are the water table and equipment surcharge loads.

FIGURE 3.15   Soldier beam and lagging system.

The water table elevation can affect one of four things:

Lighten the effective weight of a sandy, granular soil.

Increase the unit weight of a clayey soil.

Introduce water into the excavation through boiling or piping upward through the subgrade.

Add additional water load to the shoring design.

Most excavations have construction equipment working near or against the top of the shoring system. The weight of the equipment, as a gravity force, projects a lateral force on the shoring system. This force is added to the other lateral forces from the soil to achieve the maximum pressure the system will experience. Figure 3.15 shows a basic soldier beam and lagging support system. This type of system will be designed in Chapter 6.

Soil Supported as a Hydrostatic Force   Like concrete, soil that is supported laterally projects a horizontal force that is triangular, rectangular, curved and sometimes trapezoidal. The soil properties play a large role in the amount of pressure applied at the bottom of the pressure diagram. Later in this book, soil pressure formulas will be studied. For the purpose of this chapter, the understanding of the different pressure diagram shapes is the most important. Depending on the theory used, a soil pressure diagram can be triangular, rectangular, or trapezoidal. Some soil pressure examples are shown in Figure 3.16. These diagrams range from rectangular forces to triangular and forces in between. Soil pressure is probably the most difficult to represent of all temporary structure forces. Chapter 5 will explain the basics of the theories combined.

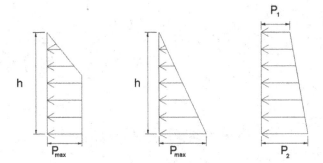

FIGURE 3.16   Different soil pressure diagrams.

Soil as a Vertical Force   Soil supported is a gravity force since soil weighs approximately 90–140 pcf. The thickness of the soil is very important because in order to convert the load to psf, the unit weight must be multiplied by the thickness in feet. For example, let's consider the following soil load.

Example 3.4   Determine the maximum load in psf of a 6-ft-thick soil column weighing 110 pcf.

110 pcf × 6 ft = 660 psf. The force is now in psf after multiplying the unit weight in pcf times the thickness in feet. Figure 3.17 depicts a 12″ × 12″ × 6 ft soil column.

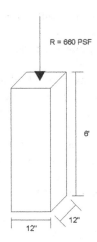

FIGURE 3.17   Soil as a column.

Supporting Existing Structures   Existing structures need to be supported during retrofit and expansion projects. It is very common to have to support bridge components during the construction of seismic retrofitting. Sometimes whole sections of permanent support systems such as piers and columns have to be held in place during the removal and reconstruction of the new supports. After the 1989 earthquake in the San Francisco Bay Area, billions of dollars were spent reinforcing, removing,

FIGURE 3.18   Temporary support of an existing structure.

and replacing bridge elements in order to withstand a reoccurring earthquake of similar or greater magnitude. Figure 3.18 shows a structure being supported at a water
treatment plant project that is being upgraded and increased in treatment capacity.

Oftentimes, buildings or water and wastewater facilities have to be expanded for
increased use beyond their original design use. Buildings need to handle more people,
or a water facility needs to treat and pump more gallons per day. If a new structure has
to be added next to an existing structure, it may be necessary to support the existing
structure from the risk of being undermined. Also, elevated slabs, beams, or columns
may have to be supported in order to lengthen or increase their cross-sectional areas.
When existing structures are involved, the unit weight of the materials being supported should be used. If the material type is questionable, use the more conservative
approach. When there is a combination of different materials, the unit weight of the
heaviest material should be used.

CHAPTER 4

SCAFFOLDING DESIGN

Scaffolding in construction is a common method of access to elevated areas of work. Scaffolding has been used for centuries in forms of earthen fills, timber, and bamboo and, in the most current century, steel and aluminum combined with wood members. Since scaffolding is so regulated in the United States and other progressive countries, the design of its components are typically standardized and performed only by the manufacturers of the systems themselves. When contractors are in need of a scaffolding system, they typically buy or rent preengineered systems.

This chapter does not imply that a contractor would have an engineer design custom scaffolding. Scaffolding is simply used in this book to illustrate simple engineering practices, load paths, and factors of safety.

4.1 REGULATORY

The Occupational Safety and Health Administration (OSHA) has the jurisdiction and sets the standards for scaffolding in the United States. Since scaffolding supports human beings in the workplace, higher standards are set to protect the health and safety of the workers and the general public. This chapter will discuss the safety factors that are required in scaffolding, cover a complete design of a scaffold, and discuss the different types of loads for which scaffolds are rated. The reader should consult OSHA documents for specifics on correct installation of a scaffold system.

4.2 TYPES OF SCAFFOLDING

The two types of scaffolding covered in this chapter are (1) tube and coupler and (2) premanufactured frames. The tube-and-coupler type will be used to teach scaffold

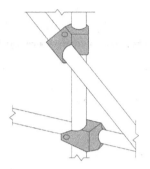

FIGURE 4.1 Tube and coupler.

design and the premanufactured frames will be discussed because they are the most popular type of scaffolding used today.

Tube-and-couple scaffolding is made up of individual pieces of pipes and couplers as shown in Figure 4.1. Wood or aluminum scaffold planks are added to complete the system. This scaffolding is good for irregular shapes of buildings and structures because each piece is selected and all are put together to maximize efficiency. Because we can design each component as we follow the load path, and the system is custom assembled, tube-and-couple scaffolding provides a perfect design example.

Premanufactured frames are the most common type of scaffold because the components are a fixed dimension and predesigned. As long as the pieces are assembled per the manufacturer's instructions and loaded within the load ranges, the scaffold is safe and follows OSHA regulations. Figures 4.2 and 4.3 illustrate some of the components of premanufactured frames.

The assembly shown in Figure 4.3 is connected to similar assemblies to make full systems as shown in Figure 4.4. These assemblies can be integrated into multistory systems for all types of construction.

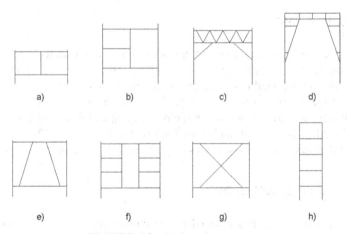

FIGURE 4.2 Various frame types.

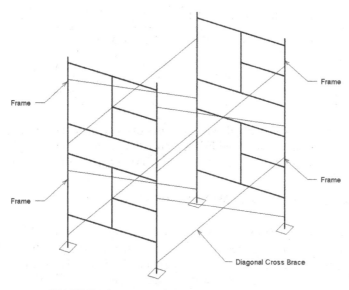

FIGURE 4.3   Assembly of frames and cross braces.

FIGURE 4.4   Scaffolding used for building construction.

4.3 LOADING ON SCAFFOLDING

Scaffold loading can be determined by a duty schedule based on type of work and use of scaffold, or it can be determined by the actual weights for which the scaffold will be used. The duty schedule is a scaffold rating that determines a design load in psf based on the type of construction for which the scaffold will be used. Table 4.1 describes

TABLE 4.1   Scaffold Duty Schedule of Loads

Duty Level	Design Load (psf)	Type of Work
Light duty	25	Painting, cleaning
Medium duty	50	Carpentry, light framing
Heavy duty	75	Concrete, steel, masonry, demolition

three different levels of scaffold use. The types of use dictate the typical personnel, materials, and tools that are used by the trades that perform the prescribed type of work. This book will use this information to determine the scaffold loading.

4.4 SCAFFOLDING FACTORS OF SAFETY

As mentioned above, scaffolding tends to require a larger factor of safety (FOS) than most other temporary structures. OSHA oversees the requirements for scaffold and imposes more stringent conditions than the standard 1.5 : 1 to 3 : 1 FOS. For scaffold in this chapter, the following FOS will be used for design. The temporary structure designer should consult specific local requirements before proceeding with scaffold design.

Scaffold structural steel and lumber members   4.0

Wire rope   4.0–6.0

4.5 SCAFFOLD COMPONENTS

4.5.1 Planking

Planking is the part of the scaffold with which the workers, materials, and tools make direct contact. The planking spans from bearer to bearer and transfers the loads from the type of use to the bearers. The two most common materials for planking are wood and aluminum. This book will use wood planking in its design examples. Aluminum planking is premanufactured and charts are available through the manufacturer, which that prescribe the usage limits and maximum allowable spans.

4.5.2 Bearers (Lateral Supports)

Bearers, typically pieces of pipe, support the planking and transfer the planking load to the posts with special connectors. The width of the bearers determines how much work space is available. The more work space, the more efficient the workers will be, but the more expensive is the scaffold in material and labor costs, both to install and to remove. The shorter the bearers, the lower the scaffold cost, but with less work space, which translates to lower worker efficiency.

4.5.3 Runners

Runners span from post/bearer to post/bearer. They traditionally do not carry any loads except to transfer lateral stability from post set to post set. In the design stage, the runners will not be designed for size.

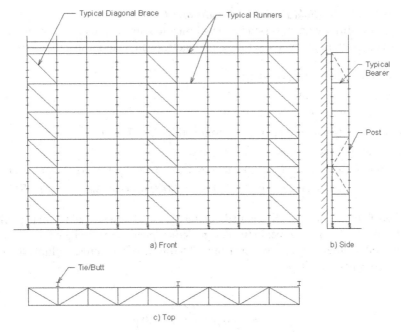

FIGURE 4.5 Typical scaffold section for design examples.

4.5.4 Posts

Posts transfer the load from the bearers and take all the concentrated loads from all levels to the ground or supporting surface. Post length is determined by how much room the contractor needs between the levels of the scaffold. This distance can also be dictated by the distance between the floors of the building so that there is a scaffold level at each floor level. For simplicity, posts are made up of the same pipe material as the bearers and runners. Figure 4.5 shows a typical scaffold section of seven levels. This will be the basis for our design example for tube-and-coupler scaffold.

4.5.5 OSHA

As mentioned earlier, this book is not designed to cover the rules and regulations set forth by both state and federal OSHA. The intent is strictly a design aid. For more information on OSHA rules and regulations, go to www.osha-safety-training.net.

4.6 SCAFFOLD DESIGN

Scaffold can be designed with two methods. The first method is when we know all the spacing of the different parts and we know the material type. From this information, the material size is determined. The second method is when the material size is known and the spacing of each component must be determined. In this book the first method mentioned will be used.

Example 4.1   Scaffold Design with Tube and Coupler

Design a scaffold with A36 standard steel pipe to carry the load of a medium-duty construction operation with the following features and material specifications:

Medium-duty loading = 50 psf
Three-level scaffold
Scaffold width front to back = 4 ft
Distance between posts along the front and back = 7 ft
Distance between levels = 9 ft
Planking: Douglas Fir Construction Grade, F_b = 900 psi (includes a 4 : 1 FOS)
Steel pipe A36 standard schedule 40, F_b = 9000 psi (includes a 4 : 1 FOS)

Step 1: Plank Design—Since medium duty is the required loading, 50 psf will be loaded on the planking to begin the design. When planking is designed, a 1-ft width is selected in order to keep the units from psf to plf. In other words, multiply the P_{max} by 1 ft so w units are in plf.

$$P_{max} = 50 \text{ psf}$$

$$w = 50 \text{ psf} \times 1 \text{ ft} = 50 \text{ plf}$$

Draw an FBD of one span of planking from bearer to bearer, as shown in Figure 4.6.

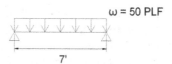

ω = 50 PLF

7'

FIGURE 4.6   Simple planking FBD.

The length of planking is equal to the distance between the bearers or the distance between the post sets along the front or back. The given information above allows for a 7-ft planking span. Since plank design is based on 1-ft increments, w = 50 plf.

Determine the minimum thickness of the planking by combining the most common bending stress formulas. These are

$$M = \frac{wL^2}{8}$$

$$f = \frac{M}{S}$$

$$S = \frac{bd^2}{6}$$

Since w and L are known, moment can be calculated by using the following formula:

$$M = \frac{(50)(7')2}{8} = 306.3 \text{ ft-lb}$$

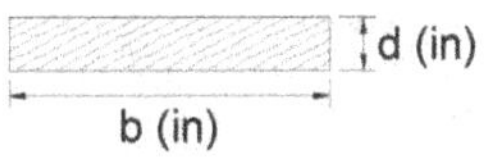

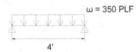

FIGURE 4.7 Cross section of a 12″ plank section.

Now we know that F_b is 900 psi based on the given information regarding the material. Also, since a 12-in (1-ft) strip of scaffold is the basis of design, shown in Figure 4.7, then b in the section modulus formula is set at 12 in. Therefore, the bending stress formula can be completed with only one unknown, d.

Very Important: In order to keep the units correct, the max moment must be multiplied by 12″/ft in order to work with the unit of inches in section modulus. It will look like this:

$$900 \text{ psi} = \frac{(306.3 \text{ ft-lb})(12''/\text{ft})}{(12'')(d)^2/6}$$

solve for d, where $d = 1.43$ in.

Conclusion: The plank thickness must be a minimum of 1.43″ thick, which would best be accommodated using a 2X nominal plank, which measures an actual 1.5″ in thickness, slightly greater than the required 1.43″.

Step 2: Determine the minimum size bearer using A36 standard steel pipe.

Draw an FBD of one bearer spanning from a front post to a back post. Figure 4.8 shows a sample FBD of this condition. Calculate the uniform load on the single bearer by multiplying the design load by the plank span. $w = 50$ psf × 7 ft = 350 plf.

ω = 350 PLF

4'

FIGURE 4.8 FBD of a single bearer.

Using the same bending stress formulas, determine the minimum standard steel pipe size for a bearer. First calculate the allowable bending stress of A36 steel with a 4 : 1 FOS. Since the steel pipe chart in Appendix 2 gives the section modulus for each size, calculate the minimum S for the pipe.

$$F_b = \frac{f_y}{4.0} \quad \frac{36 \text{ ksi}}{4.0} = 9 \text{ ksi } (9000 \text{ psi}) = F_b$$

$$M = \frac{wL^2}{8}$$

$$f = \frac{M}{S}$$

$M = (350)(4 \text{ ft})^2/8 = 700 \text{ ft-lb}$, therefore, $9000 \text{ psi} = (700)(12''/\text{ft})/S$, $S_{\min} = 0.933 \text{ in}^3$

Select the standard steel pipe with a section modulus (S) greater than or equal to 0.933 in^3.

From the pipe chart in Appendix 2, the best pipe would be the standard $2\frac{1}{2}$-in pipe.

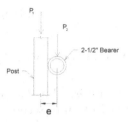

FIGURE 4.9 Combined stress in pipe posts.

Step 3: Determine the minimum size steel post to carry three levels of load while still staying within the combined stress limits. There are two modes of failure for a post, buckling failure and combined stress failure. Buckling can occur when the combined load of all the scaffold levels exceeds the buckling resistance of the pipe. The buckling resistance of a pipe is dependent on a higher radius of gyration (r) combined with a low unsupported length (L). Combined stress occurs when the axial stress on the pipe (P_1/A) is added to the effects of the bending stress caused by the eccentric location of the bearer as it is attached to the post (P_2e/S). This condition is shown in the Figure 4.9. Buckling typically fails before combined stress, so the post should be designed for buckling and then checked for combined stress.

The first task is to determine the load on the post for both conditions. P_1 is the accumulation of all the levels to the lowest post on the ground. P_2 is the load of one level to the bearer and onto the post. These loads are calculated as follows:

$$P_1 = 3 \text{ levels} \times \text{plank length} \times \tfrac{1}{2} \text{ bearer length} \times P_{max}$$

$$P_1 = (3)(7')(2')(50 \text{ psf}) = 2100 \text{ lb}$$

$$P_2 = P_1/3 \text{ levels or plank length} \times \tfrac{1}{2} \text{ bearer length} \times P_{max}$$

$$P_2 = 2100 \text{ lb}/3 = 700 \text{ lb or } (7')(2')(50 \text{ psf}) = 700 \text{ lb}$$

Buckling Stress for Steel Sections

Chapter 2 briefly described buckling in columns. In order to size the post based on buckling stress, the designer should select a standard piece of pipe from the chart for a trial-and-error process. Once the pipe is selected, record the values for r, A, and S (radius of gyration, cross-sectional area, and section modulus, respectively).

The pipe selected in this example will be an $1\tfrac{1}{2}$ standard steel pipe.

$$r = 0.623 \text{ in}$$

$$A = 0.799 \text{ in}^2$$

$$S = 0.326 \text{ in}^3$$

$$F_{bs} = \frac{\pi^2 E}{(kL/r)^2 \times \text{FOS}}$$

$$F_{bs} = \frac{P_{all}}{A}$$

where F_{bs} = allowable buckling stress in psi

E = modulus of elasticity of the steel, which is 29,000,000 psi unless noted otherwise (psi)

K = connection of the pipe to the bearer, which will always be a hinged connection (1.0 multiplier)

L = unsupported length of the pipe from the ground to level 1 (in)

r = radius of gyration of the pipe selected (in)

FOS = 4.0 factor of safety for scaffolding

π = 3.1416

$$F_{bs} = \frac{\pi^2 (29,000,000 \text{ psi})}{[(1.0)(9' \times 12''/\text{ft})/0.623'')]^2 \, 4.0}$$

F_{bs} = 2381 psi allowable buckling stress

This means that instead of an allowable stress of $F_y/4.0$, 36,000 psi/4.0 = 9000 psi, due to buckling only 2381 psi is allowed.

Now set $F_{bs} = P_{all}/A$, and solve for P_1 allowable.

$$2381 \text{ psi} = P_{all}/0.799 \text{ in}^2, \ P_{all} = 2381 \text{ psi} \times 0.799 \text{ in}^2, \ P_{all} = 1902 \text{ lb}$$

The actual load that was calculated earlier was 2100 lb. Since 1902 lb is less than 2100 lb, this pipe *DOES NOT WORK*.

If the two values were way off, then the next pipe selected would maybe be a couple sizes larger. However, since they were only 200 lb off, the 2″ standard pipe can be the next selected for the trial-and-error test.

2″ standard pipe:

$$r = 0.787 \text{ in}$$

$$A = 1.07 \text{ in}^2$$

$$S = 0.561 \text{ in}^3$$

Recalculate buckling stress; the only change is to radius of gyration; the length and other variables and constants are still the same.

$$F_{bs} = \frac{\pi^2 (29,000,000 \text{ psi})}{[(1.0)(9' \times 12''/\text{ft})/0.787'')]^2 \, 4.0}$$

F_{bs} = 3800 psi allowable buckling stress

$$3800 \text{ psi} = P_{all}/0.799 \text{ in}^2, \ P_{all} = 3800 \text{ psi} \times 1.07 \text{ in}^2, \ P_{all} = 4066 \text{ lb}$$

4066 lb > 2100 lb (OK)

The 2″ standard pipe is OK for buckling; now let's check this same pipe for combined stress.

Combined Stress: As mentioned earlier, combined stress is the combination of axial stress and bending stress at the same time. These two stresses act on the same pipe section, just below the connection of the bearer to the pipe post. In this example there are three levels of scaffold bearing down on the lowest level post and, at the same time, the first level load is bending the same pipe because the bearer is eccentrically loading the post at the same point. Combined stress is not calculated until the pipe has been selected for buckling. Combined stress is calculated as follows:

$$F_{CS} = \frac{P_1}{A} + \frac{P_2 e}{S}$$

This answer is in psi because normal stress and bending stress, both in psi, are being added together:

where P_1 = combined loads of all levels on a single lower post
A = cross-sectional area of the post
P_2 = load from a single level on the bearer that is eccentrically loading the lower post
E = modulus of elasticity of the steel pipe, typically 29,000,000 psi in this text
S = section modulus of the post

For buckling, the load for all three levels was needed. This value is still going to be used for P_1. P_2 is necessary for the eccentricity load on the post caused by the bearer.

$$P_2 = 700 \text{ lb}$$

Finally, the eccentricity distance is needed in order to calculate the bending stress.

$$e = \tfrac{1}{2}(\text{post OD}) + \tfrac{1}{2}(\text{bearer OD})$$

Post (2″ standard) OD = 2.375″

Bearer (2½″ standard) = 2.875″

$$e = \tfrac{1}{2}(2.375″) + \tfrac{1}{2}(2.875) = 2.625″$$

Using the section properties for the standard 2″ pipe from above and P_1, P_2, and e, calculate the combined stress on the post:

$$F_{cs} = \left(\frac{2100 \text{ lb}}{1.07 \text{ in}^2}\right) + \left[\frac{(700 \text{ lb})(2.625″)}{0.561}\right]$$

1963 psi + 3275 psi = 5238 psi

Allowable bending stress for any scaffold component is $F_y/4.0$ FOS = 36,000 psi/4 = 9000 psi.

Therefore, F_{cs} must be below 9000 psi; 5238 psi < 9000 psi   (OK)
Conclusion to scaffold design:

Planking: minimum 1½″ thick Douglas fin (DF) construction grade

Bearer: 2½″ standard steel pipe

Posts: 2″ standard steel pipe

It is the designer's option whether to size the posts differently for the second and third level. This takes more time and is more confusing to erect. However, if there is a lot of scaffolding, this could be more economical when purchasing materials.

4.6.1 Securing Scaffolding to the Structure

Securing scaffold to the structure is the responsibility of the contractor and the scaffolding engineer. When it is not specified, it is recommended to secure scaffolding every three levels. The attachment to the structure should be designed by a professional, registered engineer. The attachment should consider the scaffolding moving toward and away from the structure. Figure 4.10 shows what happens when scaffolding is not properly attached to the structure, considering wind and other forces besides the live load for which it was designed. Figure 4.11 indicates how a scaffold could be attached resisting both pushing toward the building and pulling from the building.

4.6.2 Hanging Scaffold

Hanging scaffolds are used more for cleaning and other exterior access to buildings when conventional lifts cannot be used due to reach or ground space. These types of scaffolds are not typical to construction activities, but this is a great way to illustrate and calculate the different components necessary to design a similar system. Later, in Chapter 11, falsework removal with winches is discussed. This same concept can be used when sizing the winches, wire rope, and counterweights on the winches. Figure 4.12 is a typical arrangement of how a scaffold may be suspended from a building or bridge. Two levels have been added to complicate the problem slightly.

FIGURE 4.10 Scaffold connection to building failure.

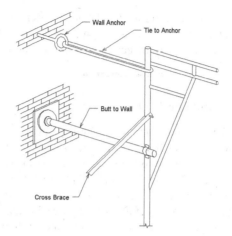

FIGURE 4.11   Scaffold connection to building example.

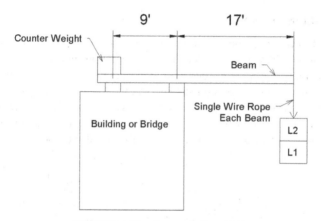

FIGURE 4.12   Suspended scaffold.

The hanging scaffolding in this example is two sets of supports on a bridge or building. Each support consists of a counterweight (CW), a beam, a front support, and a wire rope cable. It is assumed that the wire rope is connected sufficiently to the beam and the scaffold. The object is to determine the weight of the whole scaffold. Apply half of that weight to a set of supports and design the counterweight and the beam.

Example 4.2   Design a hanging scaffold system similar to Figure 4.12 and with the following characteristics:

A36 steel beams

Concrete counterweights (150 pcf)

IWRC 6 × 19 wire rope—improved plowed steel (IPS)

Scaffolding:

Light-duty Loading

32 ft long

5 ft wide

2 levels

Scaffold weighs 1200 lb

Factor of safety (scaffolding) 4.0 : 1.0

Factor of safety (wire rope) 6.0 : 1.0

Step 1: Determine the live load on the scaffold and add it to the 1200-lb scaffold weight.

Live load, $P = 2$ levels (32 ft)(5 ft)(25 psf) = 8000 lb

Dead load, 1200 lb

Total scaffold weight, 9200 lb

Weight on one support system 9200 lb/2 sides = 4600 lb

Step 2: Determine the size of the IWRC 6×19 wire rope necessary to carry half the scaffold. Table 4.2 has been provided to size the wire rope based on ultimate breaking strength. Tables 12.5 and 12.6 will be different so that students of temporary structures can experience different charts in the industry. The values in Table 4.2 are for this book only. Wire rope manufacturers have specific charts for their product.

TABLE 4.2 IWRC 6×19 Wire Rope Breaking Strength

Diameter (in)	IPS (tons)	XIP (tons)	XXIP (tons)
$\frac{1}{4}$	2.94	3.4	—
$\frac{3}{8}$	6.56	7.55	—
$\frac{1}{2}$	11.5	13.3	8.30
$\frac{9}{16}$	14.5	16.8	18.5
$\frac{5}{8}$	17.9	20.6	22.7
$\frac{3}{4}$	25.6	29.4	32.4
$\frac{7}{8}$	34.6	39.8	43.8
1	44.9	51.7	56.9
$1\frac{1}{8}$	56.5	65	71.5

Minimum breaking force in tons, 4600 lb $\times$ 6.0 = 27,600 lb or 13.8 tons.

Table 4.2 has three types of wire rope. The quality of the wire rope increases from left to right. It is recommended to use the lowest values in order to be safe. This is also because the designer is not 100% sure the field engineer will order the specified wire rope, if a higher quality is called out. According to Table 4.2, a minimum of $\frac{9}{16}''$ wire rope needs to be used in order to withstand the 13.8-ton breaking strength load with a 6 : 1 FOS.

FIGURE 4.13   FBD of single support beam.

Step 3: Determine the size of the counterweight. First draw an FBD of one support system as shown in Figure 4.13.

$$P = 4.6 \text{ k}$$

Sum moments about CW or R. Let's sum moments about R.

$$(CW)(9') = (4.6 \text{ k})(17'), \text{ CW} = (4.6 \times 17)/9, \text{ CW} = 8.7 \text{ k}$$

Sum forces in the vertical direction to determine R.

$$+R - 8.7 \text{ k} - 4.6 \text{ k} = 0 \qquad R = 13.3 \text{ k}$$

Remember, an FOS has not been applied to any of these forces.

Counterweight (CW) = 8.7 k, with an FOS of 4.0 : 1.0, multiply CW × 4.0, and this is what the counterweight must weigh.

$$8.7 \text{ k} \times 4.0 = 34,800 \text{ lb}$$

The unit weight of the concrete counterweight for this example is 150 pcf. In order to determine the amount of concrete required for this weight, solve for the volume in the following formula:

CW = 150 pcf × (bwh), where bwd is the volume in cubic feet

34,800 lb = 150 lb/ft^3 × volume, volume = 34,800 lb/150 pcf = 232 ft^3

Volume in cubic yards = 232/27 ft^3/yd^3 or 8.6 cubic yards of concrete

The size of a deadman (counterweight), similar to Figure 4.14, if it were a perfect cube, would be calculated by taking the cube root of 232 ft^3.

$$\sqrt[3]{232 \text{ ft}^3} = 6.14 \text{ ft } (6' - 2'')$$

Step 4: Beam Design—Select the most economical beam while maintaining at minimum of 1.25 : 1.0 beam stability ratio between the depth (d) and the flange width (b_f). Draw an FBD, as shown in Figure 4.14, indicating the forces on a single beam.

Note: A beam stability ratio is a quick way to make sure the beam is not too unstable and risking flange buckling failure. However, this does not mean the beam will not fail in lateral buckling. Later in the text, the student will learn how to check for beam stability.

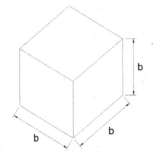

FIGURE 4.14   Counterweight.

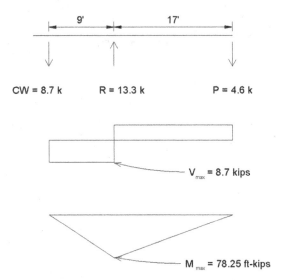

FIGURE 4.15   Shear and moment diagram.

In step 3, the reactions were calculated. Using the same FBD, a shear and moment diagram can be completed. The original forces (without an FOS) should be used for this process and then an FOS will be added later.

The shear and moment diagram gives us a maximum moment of 78.25 ft-k. Using A36 steel with a scaffold FOS of 4.0 : 1.0, the allowable bending stress on the beam is 36,000 psi/4.0 = 9000 psi. Using this steel, determine the most economical (lightest) beam maintaining the 1.25 stability ratio between the depth (d) and the flange width (b_f).

Determine the minimum section modulus by using the bending stress formula and solving for S_x.

$$S_x = (78.25 \text{ ft-k} \times 12''/\text{ft})/9 \text{ ksi} = 104.3 \text{ in}^3$$

List three wide flange beams with a minimum section modulus of 104.3 in³. Table 4.3 shows three beams that meet the minimum section modulus.

TABLE 4.3   Three Beams That Meet Minimum Section Modulus

Beam	S_x (in^3)	d (in)	B_f (in)	d/b_f (ratio)	Check (OK or NG)
W10 × 100	112	11.1	10.34	1.07	OK
W12 × 79	107	12.38	12.08	1.02	OK
W14 × 74	112	14.17	10.07	1.40	NG[a]

[a]NG = Does not meet all the requirements.

The W12 × 79 is lighter than the W10 × 100 while still maintaining a 1.25 ratio or lower. The W14 × 74 is the lightest and has the required section modulus, however, the beam is slightly too tall for its base and will probably have lateral stability issues.

Select the W12 × 79.

CHAPTER 5

SOIL PROPERTIES AND SOIL LOADING

Support of excavation in construction is used when excavations cannot legally or physically support themselves and protect the workers inside. Trench or excavation collapses are not the most frequent construction accident, but they occur in enough cases to be one of the most deadly.

The loading on support of excavation systems has been a subject of study for over a century, and many theories have been developed and accepted by engineers. This chapter will review some soil mechanic basics and discuss and implement some of the more common soil loading theories. A conservative approach is typically taken in the following examples due to the sensitivity nature of excavation support and safety.

5.1 SOIL PROPERTIES

Soils are classified into two main categories both for lab procedures and for temporary structure shoring design. The two classifications are coarse-grained soils (sands and gravels) and fine-grained soils (silts and clays). This chapter will mention several other characteristics that separate these two types of soils besides their particle size. However, particle size is the most obvious difference. In soils classification, one of the most elementary tests performed is the sieve analysis or gradation process. In this test, samples are put through a series of different size screens (sieves) to determine the amount of the sample that is retained or passes certain size screens. The smaller grained the material, the more screens the particles pass; and the larger grained the material, the more that is retained on the larger screens. The No. 200 sieve has openings measuring 0.075 mm or approximately 3/1000 of an inch. The amount of material, by weight, that passes this sieve (or is retained by this sieve), determines whether a material sample is a coarse-grained soil or a fine-grained soil.

Coarse-Grained Soils—Sands and gravels are defined as soils that have more than 50% of their particles retained on or above the No. 200 sieve (screen).

Fine-Grained Soils—Silts and clays are defined as soils that have more than 50% of their particles pass through the No. 200 sieve (screen). The Unified Soils Classification System is a commonly accepted testing system that includes information from a particle size analysis for both fine-grained and coarse-grained soils and a plasticity index test for fine-grained soils. A particle size analysis would place the samples in their correct main category (coarse grained or fine grained), and the coarse material would be separated between sands and gravels based on the amount of material passing or being retained by the No. 4 sieve. The No. 4 sieve measures 4.76 mm (0.187 in, which in fraction form is about (3/16)″). The difference between sands and gravels is at the 50% point of the No. 4 sieve. Less than 50% passing would be considered gravels and more than 50% passing would be considered sand. Beyond the No. 4 sieve criteria, the amount of fines (silts and clays) are measured in the coarse-grained materials to place them in an even more specific category.

Fine-grained soils are analyzed further for the plasticity index (PI). The PI is derived from the Atterberg limit test, which measures the plastic limit and liquid limit of a fine-grained soil in order to classify it as a clay or silt. The test sets limits of how much water makes the soil solid, semisolid, plastic, or liquid. The percentage difference in water between the plastic and liquid state is the PI. Table 5.1 shows how soils are classified based on both the sieve analysis and the plasticity index.

TABLE 5.1   Soil Classification Example

Level 1 Classification	Level 2 Classification	Fines	Soil Symbol	Soil Description
Coarse-grained soils	Gravels	Clean	GW	Well-graded gravel
			GP	Poorly graded gravel
		With fines	GM	Silty gravel
			GC	Clayey gravel
	Sands	Clean	SW	Well-graded sand
			SP	Poorly graded sand
		With fines	SM	Silty sand
			SC	Clayey sand
Fine-grained soils	Silts and clays $LL^a < 50\%$	Inorganic	CL	Lean clay
			ML	silt
		Organic	OL	Organic clay/silt
	Silts and clays LL^a 50% or more	Inorganic	CH	Fat clay
			MH	Elastic silt
		Organic	OH	Organic clay/silt
Highly organic	Dark in color		PT	Peat

^{a}LL = liquid limit.
ASTM Designation D-2487.

5.1.1   Standard Penetration Test and Log of Test Borings

Other tests will briefly be discussed later in this chapter, but the standard penetration test (STP) and the log of test borings are always available to the engineer for design and are typically included in the contract documents for construction and temporary support design.

Every engineer/owner has a particular procedure he or she likes to follow for the borings. State agencies tend to follow a fairly consistent format. The sample log in Figure 5.1 is from a CalTrans project on the Yerba Buena Island used for the demolition and construction of bridge structures after the new Bay Bridge was opened in 2013. Table 5.2 shows ranges for granular soil properties. State agency logs contain a great deal of useful information for both the engineer and the contractor. Here is some information that can be obtained by the log of test boring shown in Figure 5.1.

Bore hole number and date

Water table location/elevation and the date measured

Blow count: number of blows to produce 12 in of sampler penetration

Beginning elevation of bore (feet or meters above sea level)

Ending elevation of bore (feet or meters above sea level)

Size of sampler (diameter in inches or millimeters)

Description of material following the unified soil classification system

Dry unit weight of material (pcf)

Moisture content as a percentage

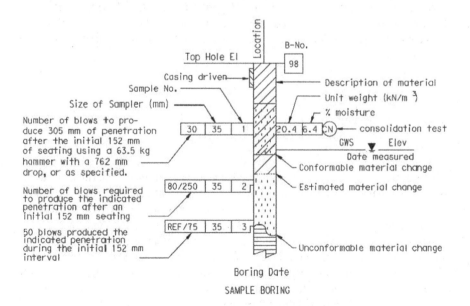

FIGURE 5.1   Log of test boring example.

TABLE 5.2 Properties of Granular Soils

Apparent Density	Relative Density (%)	SPT, N_{60} (blows/ft)	Friction Angle, ϕ (deg)	Unit Weight (pcf) Moist	Submerged
Very loose	0–15	$N_{60} < 5$	<28	<100	<60
Loose	16–35	$5 \leq N_{60} < 10$	28–30	95–125	55–65
Medium dense	36–65	$10 \leq N_{60} < 30$	31–36	110–130	60–70
Dense	66–85	$30 \leq N_{60} < 50$	37–41	110–140	65–85
Very dense	86–100	$N_{60} \geq 50$	>41	>130	>75

Data provided by CalTrans.

5.1.2 Unit Weights above and below the Water Table

Soil can have different properties depending if they are in-place (in situ) properties or below water table properties. This section discusses how the unit weights of different materials are affected by their relationship to the water table. More emphasis will be placed on unit weights below the water table since in-place (in situ) conditions are analyzed with more standard methods and deal with moisture content fluctuations. In this method, samples are dried in a kiln to remove the moisture before they are analyzed further.

When Soil is below the water table, its unit weight is altered by how it is affected when the water fills the voids. How the water affects the sample depends on whether the sample is a sand, gravel, silt or clay.

Void Ratio and Determining Unit Weights below the Water Table The void ratio is the ratio between the amounts of voids to the amount of solids in a sample by volume after the water has been extracted. Void ratio is calculated with the following formula:

$$e = \frac{V_v}{V_s}$$

where e = void ratio
  V_s = volume of the solids
  V_v = volume of the voids

This ratio varies between fine-grained and coarse-grained materials. It should also be understood how these two types of soils are affected by the intrusion of groundwater. For instance, coarse-grained materials typically have more voids due to their larger particle sizes. Therefore, when water is introduced into these voids, the particles, predominantly sands and gravels, tend to float or become buoyant. This water is known to be free draining. On the other hand, when water is introduced to a fine-grained material, the water tends to get trapped within the particles and is not free to travel through the material. This water is known to be non-free-draining.

An easy way to understand the difference between coarse-grained (CG) materials and fine-grained (FG) materials is illustrated in the Table 5.3.

TABLE 5.3 Coarse Grained versus Fine-Grained Materials[a]

Coarse Grained	Fine Grained
Sand and gravels	Silts and clays
Less than 50% passing No. 200 sieve	More than 50% passing No. 200 sieve
Free draining below water table	Non-free-draining below water table
Particle submerged below water table	Particle saturated below water table
High ϕ, low or zero C	Low or zero ϕ, high C
Strength through mechanical bonding	Strength through chemical bonding

[a] ϕ is the Internal friction angle (degrees). C is cohesion (psf).

The shear strength of a particular material is dependent on how the particles within the sample join together. In a coarse-grained (sand and gravel) material, the particles lock together mechanically with their angular shapes. The angle of internal friction, which is derived through lab testing, is typically between 20° and 45°. Anything below this range could be classified in the silts and clays category. Anything above this range would tend to have other mechanisms helping the angle of internal friction such as cementing capabilities. A fine-grained (silt and clay) soil relies on chemical bonding to join the particles together. Under a microscope, these particles would tend to lay flat and parallel to one another, as opposed to sands and gravels, which interlock in order to achieve shear strength. The silts and clays would not have a very measurable angle of internal friction, but its cohesiveness would be appreciable.

Typically during soil analysis, a 1-ft^3 sample is used in order to keep the ratios in relative order in comparison to other samples. A volume of 1 ft^3 one cubic foot also keeps the math straightforward in comparison. In the following examples, unit weights will be determined, given some initial soil information such as void ratio, specific gravity, and the unit weight of freshwater, which will always be 62.4 pcf in this text. (If salt water is being considered, unit weight is 64 pcf.)

Here are some common formulas that are used to determine unit weights based on a soils ratio of solids to voids and its specific gravity. Some of these variables were mentioned earlier.

$$e = \frac{V_v}{V_s}$$

$$1 \text{ ft}^3 = V_s + V_v$$

$$W_s = V_s \times SG \times \gamma_{\text{water}}$$

$$\text{Unit weight submerged (CG)} = W_s - (\gamma_{\text{water}} \times V_s)$$

$$\text{Unit weight saturated (FG)} = W_s + (\gamma_{\text{water}} \times V_v)$$

where W_s is the weight of the solids or the weight of a dry sample assuming the air has no weight; SG is the specific gravity of the material. Specific gravity is the relative density of a solid material as it relates to water. Specific gravity of water is 1.0; and γ_{water} is the unit weight of water. Freshwater weighs 62.4 pcf and salt water weighs 64 pcf.

Example 5.1   Unit Weight of a Coarse-Grained Material

Calculate the unit weight of a silty-sand below the water table that has the following properties:

Void ratio of 0.55

Specific gravity = 2.70

Step 1: Determine the volume of the solids (V_s). $V_s = 1/1 + e$

$V_s = 1/1.55 = 0.645$ (64.5 % of this sample is solid material)

Therefore:

$V_v = 1 - V_s$, or $V_v = 1 - 0.645 = 0.355$ (this value is not needed to complete solving this problem)

Step 2: Calculate the dry unit weight (W_s)

$W_s = \text{Water} \times \text{SG} \times V_s$, or $W_s = 62.4 \text{ pcf} \times 2.70 \times 0.645 = 108.7 \text{ pcf}$

Step 3: Calculate the *submerged* unit weight. Since this material is a coarse-grained material (sand or gravel), the particles become submerged under water and thus float in the water.

Step 4: Submerged unit weight $= W_s - (\text{water} \times V_s)$, or $W_s = 108.7 \text{ pcf}$
$- (62.4 \text{ pcf} \times 0.645) = 68.45 \text{ pcf} = 68.45 \text{ pcf}$

Notice how light the sample becomes when its particles are submerged in water. For the same reason a rock is lighter when it is held under water, a submerged sand/gravel becomes lighter.

Example 5.2   Unit Weight of a Fine-Grained Material

Calculate the unit weight of a sandy-clay below the water table that has the following properties:

Void ratio of 0.85

Specific gravity = 2.65

Step 1: Determine the volume of the solids (V_s). $V_s = 1/1 + e$ $V_s = 1/1.85 = 0.541$ (54.1% of this sample is solid material)

Therefore:

$V_v = 1 - V_v$, or $V_v = 1 - 0.541 = 0.459$ (this value is needed in step 4)

Step 2: Calculate the dry unit weight (W_s)

$W_s = \text{Water} \times \text{SG} \times V_s$, or $W_s = 62.4 \text{ pcf} \times 2.65 \times 0.541 = 89.5 \text{ pcf}$

Step 3: Calculate the *saturated* unit weight. Since this material is a fine-grained material (silts and clays), the sample becomes saturated under water, thus making the soil heavier.

Step 4: Saturated unit weight $= W_s + (\text{water} \times V_v)$, or $W_s = 89.5 \text{ pcf} + (62.4 \text{ pcf} \times 0.459) = 118.1 \text{ pcf}$

Notice how heavy the sample gets because of the water being trapped between the fine-grained particles. Anyone who has shoveled wet clay can attest to the weight of the material as well as the stickiness. However, while the unit weight of clay is heavier when saturated with water, the cohesiveness plays a role in allowing a clayey material to stand on its own when a trench is being excavated. This is just the opposite with a sand or gravel because sands and gravels tend to slope at the angle of repose.

5.1.3 Testing

There are many types of soil tests that can be performed for both permanent and temporary designs. The Atterberg limits test was mentioned earlier and is just one of many tests required for soil classification. Grain size tests were also mentioned as the premier test for defining coarse and fine-grained soils. In addition, triaxial shear develops angle of internal friction values, and the pocket penetrometer test is used in the field to determine in-place shear strength of soil so a competent person can classify the trenched material and what type of sloping is required. This type of field testing is discussed at the end of this chapter. These and other tests are listed in Table 5.4.

5.2 SOIL LOADING

5.2.1 Soil Mechanics

Soil can be considered a liquid pressure, but measures are taken to consider the manner in which the particles interact. With a course-grained material, the particles work together mechanically; and the higher the angle of internal friction, the more shear strength the soil will have. On the other hand, a fine-grained soil's particles bond chemically, which creates shear strength through cohesion. In addition, the same two soil classifications discussed above—fine grained and coarse grained—will be used when determining the proper formulas.

Two conditions will be discussed when soil loading is considered: active and passive pressure.

TABLE 5.4 In Situ Lab and Field Test Designations

Symbol	Test
AL	Atterberg limits
CA	Corrosivity
CN	Consolidation
MD	Maximum density
DS	Direct shear
R	R value
PP	Pocket penetrometer
SA	Grain size
SE	Sand equivalent
UC	Unconfined compression
UU	Unconsolidated, undrained triaxial
W	Wash analysis

5.2.2 Active Soil Pressure and Coefficient

Active soil pressure is when the soil loads the structure in question. Most shoring designs fall into this category because the purpose is to support an excavation or trench with adequate support that resists collapse and minimizes deflections and settlement.

The active soil coefficient is a multiplier used in the most common soil pressure formulas to introduce the angle of internal friction into the estimated soil pressure. The angle of internal friction is a very important element to soil pressure in Rankine and other theories. The angle of internal friction produces reliable information that is relevant to how soil loads a support system. This value, designated by the symbol ϕ can be obtained from soils reports and laboratory testing. For the active case, the active soil coefficient can be determined as follows:

$$K_a = \frac{1 - \sin \varphi}{1 + \sin \varphi}$$

This value will always be less than 1.000. For more accurate calculations, the soil coefficient values should be carried out to three decimal places (0.000). Anything less could result in pressures off by up to 20 psf.

Table 5.5 shows estimated unit weights (γ) and applicable active soil coefficients (K_a) for a variety of soil types based on their angle of internal frictions (ϕ). The angles decrease from gravels (highest) down to silts (lowest). These values can be found in soils reports and field testing reports, and some of these soil types and associated values are found in Table 5.5.

TABLE 5.5 Simplified Typical Soil Values

Soil Classification	ϕ Friction Angle of the Soil	Density or Consistency	γ Soil Unit Weight (pcf)	K_a Coefficient of Active Earth Pressure
Gravel,	41	Dense	130	0.21
gravel–sand	34	Medium dense	120	0.28
mixture,	29	Loose	90	0.35
coarse sand				
Medium sand	36	Dense	117	0.26
	31	Medium dense	110	0.32
	27	Loose	90	0.38
Fine sand	31	Dense	117	0.32
	27	Medium dense	100	0.38
	25	Loose	85	0.41
Fine silty sand,	29	Dense	117	0.35
sandy silt	27	Medium dense	100	0.38
	25	Loose	85	0.41
Silt	27	Dense	120	0.38
	25	Medium dense	110	0.41
	23	Loose	85	0.44

Passive Soil Pressure and Coefficient Passive soil pressure is when a structure loads the soil. For example, this could occur when a retaining wall being loaded from behind by soil subsequently pushes on the soil in front of the wall. Another case would be when an anchor plate or block is supporting a tieback system and the force from the tieback is pushing the anchor/plate against the soil (structure loading soil).

As with the active case, the passive has a soil coefficient that is derived from the angle of internal friction. The following value is the reciprocal of the active soil coefficient. Because the passive soil coefficient is the reciprocal, the value will always be greater than 1.000 and sometimes as high as 5.000.

$$K_p = \frac{1 + \sin \varphi}{1 - \sin \varphi}$$

5.2.3 Soil Pressure Theories

Several theories have been developed over the years for soil pressure analysis. Rankine, Boussinesq, Coulomb, Terzagie, and others have derived several theories that support how soil loads its support system. In the effort to simplify some of the temporary structure concepts in this text, the author has condensed the most common practices in order to give the student a solid, but simplified, understanding of soil loading. This is not to say that very detailed analyses should not be performed by a registered engineer, but the majority of our project cases can be solved using accepted theories. Test results on actual shoring systems have produced evidence that the most common soil pressure theories coincide with practical results within 20%, regardless of the method used. For this reason, the student should rely on future employers to dictate the methods used in determining soil pressures on structures. In addition, standard factors of safety should be used on a case-by-case basis.

Rankine Theory Rankine theory is a great method for a student of engineering because it covers all the soil properties, and one can make sense of how these properties apply to the amount of soil pressure generated by the conditions. This text will start with the Rankine theory because it employs all the soil property variables mentioned earlier: angle of internal friction, unit weight, and cohesion. Later in this chapter, other theories will be discussed that categorize cases into soil types and support conditions. Also, as mentioned before, Rankine theory can be used for both active and passive soil pressure cases.

Rankine Theory for Active Soil Pressure When soil is loading a structure in question, the pressure it develops based on Rankine at any depth of h is as follows:

$$P_a = \gamma h K_a - 2c\sqrt{K_a}$$

where P_a = active pressure at any given depth of h (psf)
γ = unit weight of the soil (pcf)
h = depth at which the pressure is being measured (ft)
K_a = Active soil coefficient; $K_a = 1 - \sin \varphi / 1 + \sin \varphi$
c = cohesion (psf)

The first part of the formula ($\gamma h K_a$) takes into account the weight of the soils and how deep the excavation or trench is. The soil coefficient is a multiplier in this case and greatly affects the outcome of the pressure value. As mentioned before, the active soil coefficient is always less than 1.000. Therefore, anything less than 1.000 is going to lower the value after taking into consideration the unit weight and the depth. The higher the degree of internal friction, the lower the value of K_a, thus the lower pressure would result. In contrast the lower the friction degree, the higher the coefficient value is closer to 1.000, thus the higher the pressure will be from the soil. This occurs because the smaller the friction value the less shear strength the soil has.

The second part of the formula ($-2c\sqrt{K_a}$) takes into account the amount of cohesion (clay) the soil contains. Notice the negative sign. This value is subtracted from the first part of the formula. Therefore, if there is cohesion in the soil, the soil pressure will be lower. The clay causing the cohesion helps strengthen the soil, thus reducing the pressure. If there is no cohesion to the soil, then this portion of the formula goes to zero, and the pressure only comes from the unit weight, the coefficient of friction, and the depth.

One thing to notice about this method is if the depth is 0 ft (ground level) and there is cohesion in the soil, the outcome pressure can be a negative pressure. This means that close to the top of the excavation, the soil is supporting itself. This should be made clearer in the examples to follow.

For an active soil case, when the soil unit weight is unknown, the *CalTrans Trenching and Shoring Manual* require the engineer to use 115 pcf.

The slip plane of the active case is the measured angle from the vertical of the structure to the plane of failure from the bottom of the excavation to the surface. This angle is calculated as follows:

$$\alpha = \text{Slip plane angle} = 45° - \frac{\varphi}{2}$$

where α = angle in degrees
   φ = angle of internal friction

Rankine Theory for Passive Soil Pressure   In the case where there is passive pressure and the structure is loading the soil, the formula is similar in its variables but very different in its outcome:

$$P_p = \gamma h K_p + 2c\sqrt{K_p}$$

where P_p = passive pressure at any given depth of h (psf)
   K_p = passive soil coefficient, $K_p = 1 + \sin \varphi / 1 - \sin \varphi$

The first and most obvious difference is that cohesion is added to the pressure due to depth, unit weight, and friction. When there is no cohesive value, this does not add any strength to the soil. However, a cohesive soil that does have cohesion will add a great deal of soil resistance. After all, passive pressure is basically soil resistance against the structure that is pushing against it. Passive soil pressure will always be a positive pressure because the clay strength is always added and not subtracted like active pressure.

The slip plane of the passive case is the measured angle from the vertical of the structure to the plane of failure from the bottom of the structure to the surface. This angle is calculated as follows:

$$\beta = \text{Slip plane angle} = 45° + \frac{\varphi}{2}$$

where β = angle in degrees
          φ = angle of internal friction
          α = angle in degrees

5.2.4 Soil Pressure Examples Using Rankine Theory

Rankine Theory for Active Soil Pressure (P_a) As already mentioned, the Rankine theory is a great basis for soil pressure education because it considers the unit weight of the soil, the soil coefficient, the depth of excavation, and how cohesive the soil might be. The following formula is based on the Rankine theory for active soil pressure (P_a).

$$P_a = \gamma h K_a - 2c\sqrt{K_a}$$

Figure 5.2 shows diagrams of active soil pressure for granular and cohesive soils.

Rankine Theory for Passive Soil Pressure (P_p) The following formula is used when calculating passive soil pressure using the Rankine theory:

$$P_p = \gamma h K_p + 2c\sqrt{K_p}$$

where $\gamma_{\text{in situ}}$ = unit weight of soil (pcf)
          h = depth of load from the surface
          K_p = passive soil coefficient

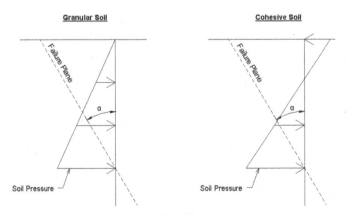

α = Slip plane angle = 45 degrees – $\varphi/2$

FIGURE 5.2 Active soil pressure diagrams for granular and cohesive soils.

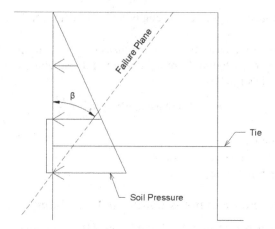

FIGURE 5.3   Passive soil pressure diagram.

Figure 5.3 provides the diagram of passive soil pressure.

$$K_p = \frac{1 + \sin \varphi}{1 - \sin \varphi}$$

where Φ = angle of internal friction:

$$\text{Slip plane} = \beta = 45° + \frac{\varphi}{2}$$

Example 5.3   Passive Soil Pressure Supporting a Steel Plate

We are supporting an SOE (support of excavation) with a tieback to a deadman, steel plate (see Figures 5.4 and 5.5). The following describes the soil properties and the

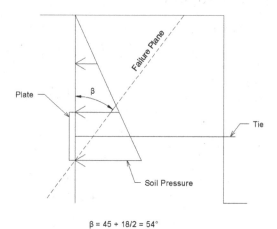

β = 45 + 18/2 = 54°

FIGURE 5.4   Passive pressure diagram against plate.

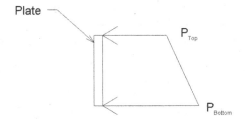

FIGURE 5.5 Trapezoidal pressure diagram on plate.

loading conditions. Determine the P_{max} at the top and bottom of the plate and the total capacity of the plate:

$$\gamma_{in\ situ} = 110\ pcf$$

$$c = 250\ psf$$

$$\Phi = 18°$$

The 8-ft-wide × 6-ft-high steel plate is buried 10 ft deep to the top of the plate.

Step 1: Calculate $K_p = 1 + \sin\ 18/1 - \sin\ 18$

$$K_p = 1.894$$

Step 2: Calculate the pressure at the top of the plate.
Top $P_p = 110\ pcf \times 10' \times 1.894 + 2 \times 250\ psf \times \sqrt{1.894} = 2772\ psf$

Step 3: Calculate the pressure at the bottom of the plate.
Bottom $P_p = 110\ pcf \times 16' \times 1.894 + 2 \times 250\ psf \times \sqrt{1.894} = $ 4022 psf

$$Plate\ capacity = \left[\left(\frac{P_{top} + P_{btm}}{2} \right) \times plate\ w \times plate\ h \right] / FOS$$

$$\frac{[((2772 + 4022)/2) \times 8' \times 6']}{1.5} = 108,672\ lb\ (54.3\ tons)$$

Example 5.4 Intersecting Slip Planes
In order to determine a safe distance behind an excavation for equipment so that the equipment surcharge load does not influence the shoring load, calculate the horizontal distances created by both passive and active slip planes. The goal would be to maintain a zero or positive distance between X_1 and X_2 (see Figure 5.6).

Given information should be the depth of excavation, the two angles β and α, and the depth to the bottom of the anchor plate. From this point, the calculation is strictly geometry and trigonometry:

$$D = 30\ ft\ to\ bottom\ of\ excavation$$

$$d = 20\ ft\ to\ bottom\ of\ plate$$

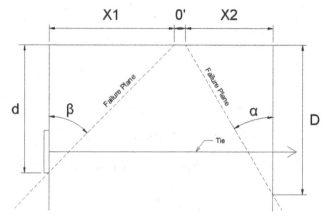

FIGURE 5.6   Failure planes for active and passive.

$\beta = 54°$

$\alpha = 45 - \dfrac{18}{2} = 36°$

$X_1 = \tan\ 54 \times 20\ \text{ft} = 27.53\ \text{ft}$

$X_2 = \tan\ 36 \times 30\ \text{ft} = 21.80\ \text{ft}$

$X_1 + X_2 = 14.53 + 41.29 = 49.33$ (safe distance for equipment behind shoring wall)

Active Slip Plane with Soil Anchors   When soil anchors are used to tie back a support of excavation, the length of the anchor is critical to its capacity. The wedge of earth between the failure plane and the vertical shoring face cannot be relied upon for support because this soil is already unstable. Therefore, once the soil anchor length is determined, the wedge width should be added to this minimum length in order to achieve the full drilled depth of the anchor (see Figure 5.7).

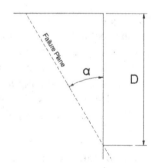

FIGURE 5.7   Active slip plane.

Example 5.5   Active Soil Pressure on Sheet Pile Wall with a Granular Soil
We are supporting an SOE for a 30-ft-deep excavation with a sheet pile wall. The
following describes the soil properties and loading conditions. Determine the P_{max} at
the bottom of the excavation. Refer to Figure 5.8 for the pressure diagram:

$$\gamma_{in\ situ} = 120\ pcf$$

$$c = 0\ psf$$

$$\Phi = 35°$$

$$D = 30\text{-ft-deep excavation}$$

Step 1: Calculate $K_a = 1 - \sin\ 35/1 + \sin\ 35.$

$$K_a = 0.271$$

Step 2: Calculate the pressure at the bottom of the excavation.
Bottom $P_a = 120\ pcf \times 30' \times 0.271 - 2 \times 0\ psf \times \sqrt{0.271} =$
975.6 psf

$$\alpha = 45 - \frac{35}{2} = 27.5°$$

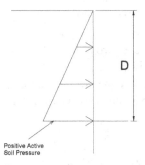

FIGURE 5.8   Active soil pressure diagram.

Example 5.6   Active Soil Pressure on Sheet Pile Wall of Sandy Clay
We are supporting an SOE for a 30-ft-deep excavation with a sheet pile wall. The
following describes the soil properties and loading conditions. Determine the P_{max} at
the bottom of the excavation. Refer to Figure 5.9 for pressure diagram.

$$\gamma_{in\ situ} = 130\ pcf$$

$$c = 500\ psf$$

$$\Phi = 10°$$

$$D = 30\text{-ft-deep excavation}$$

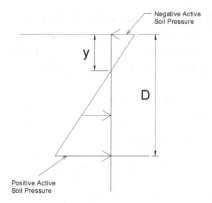

FIGURE 5.9   Active soil pressure diagram.

Step 1: Calculate $K_a = 1 - \sin\ 10/1 + \sin\ 10$.

$$K_a = 0.704$$

Step 2: Since there is cohesion, calculate the pressure at the top of the excavation.
Top $P_a = 130\ \text{pcf} \times 0' \times 0.704 - 2 \times 500\ \text{psf} \times \sqrt{0.704} = -839\ \text{psf}$

Step 3: Now calculate the pressure at the bottom of the excavation.
Bottom $P_a = 130\ \text{pcf} \times 30' \times 0.704 - 2 \times 500\ \text{psf} \times \sqrt{0.704} =$ 1907 psf

Step 4: Calculate the distance from the surface to where the pressure crosses 0 psf (the vertical line).

Step 5: Draw two similar triangles with the upper and lower portion of the pressure diagram (see Figure 5.10). Calculate the unknown vertical distance y.

$$\frac{y}{839} = \frac{30'}{2746} \qquad y = 9.2\ \text{ft}$$

From the top of the excavation to 9.2 ft below the surface, theoretically the soil is supporting itself.

$$\alpha = 45 - \frac{10}{2} = 40°$$

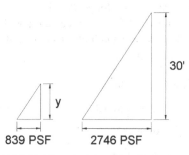

FIGURE 5.10   Similar triangles for y dimension calculations.

5.2.5 Soil Pressures Using State and Federal Department Standards

Curved Failure Planes We have discussed Rankine's theory in depth thus far. Before Rankine (1857), Coulomb had developed his own theories that are still recognized today. However, it was not until 1943 that Terzaghi brought a better understanding to how the original earth pressure theories work. Coulomb's and Rankine's theories assumed a wedge-shaped failure plane, whereas Terzaghi introduced a more logarithmic failure plane shape, shown in Figure 5.11. This is the same failure plane discussed under Rankine's theory described as a slip plane.

Field experiments have shown that Rankine and Coulomb theories are very accurate when they are applied to active soil pressure cases but only accurate in passive cases with clean dry sand.

As mentioned earlier, soil pressures can also be developed using other theories. The following theories have been developed by engineers and adopted by state and federal agencies. Their diagrams appear to have been simplified; however, field testing validates their accuracy. These theories are classified by soil type and support condition. The soil types are granular and fine grained. The support conditions are unrestrained (cantilevered) and restrained (single support and multiple supports). The unrestrained systems rely of passive soil resistance below the bottom of the excavation as they push back against the structure. In addition, the bending capacity of the structure itself is also a large component of the strength of the system. These systems are usually not effective with depths greater than 15–18 ft. The restrained systems typically use the same vertical shoring members, but instead of relying on passive soil support, they rely on walers (horizontal beams) and struts (compression members) or tiebacks (tension members) to complete the system. With restrained systems, contractors can support excavations over 60 ft in depth safely and economically.

Both types of support systems count on the vertical members toeing into the subgrade for some distance. However, this is definitely more necessary with the unrestrained system since this is where it gets its passive soil resistance.

Earlier in this chapter, the defining line between granular soil and fine-grained soils was discussed. In addition, the difference between sands and gravels was defined.

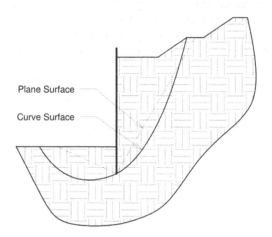

Plane Surface

Curve Surface

FIGURE 5.11 Curved failure plane.

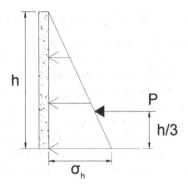

FIGURE 5.12   Triangular pressure diagram.

As far as these theories go, sands and gravels are combined into the same theory. Lateral pressures for both active and passive cases are determined by the soil properties and the type of shoring system.

The triangular pressure diagram (Figure 5.12) that varies with depth has a resultant one third of the depth from the bottom of excavation. The maximum pressure in this case is at the bottom of excavation. Horizontal stress is calculated by

$$\text{Horizontal stress } (\sigma h) = \gamma_h K$$

$$\text{Resultant} = \sigma_h h/2$$

Soil typically is classified as a liquid even though hydrostatic groundwater pressure is added to the lateral soil pressure. When soil pressure diagrams are added together, the normal triangular shape becomes a series of triangular and rectangular shapes as shown in Figure 5.13, which shows soil and water pressure, both separately and combined.

Standard Pressure Formulas Simplified   This text has categorized just a few shoring conditions that cover the majority of common situations while avoiding very complex situations. Shoring conditions that are more extreme than the examples used in this text should be analyzed and designed by a registered professional engineer. The shoring embedment below the bottom of excavation has been simplified to a

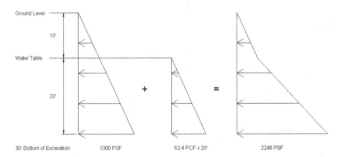

FIGURE 5.13   Combined soil and water pressure.

TABLE 5.6 Most Common Soil Pressure Formulas

Category No.	Description	Formula
1	Cantilevered and single brace, granular soil	$P_a = K_a \gamma H$
2	Cantilevered and single brace, cohesive	$P_a = K_a \gamma H - 2c$
3	Multiple brace, granular or cohesive soil	$P_a = 0.65 K_a \gamma H$

where P_a = maximum active soil pressure
0.65 = reducing multiplier
K_a = active soil coefficient
γ = soil unit weight
H = depth to bottom of excavation
C = shear strength of cohesive soil (psf)

ratio to avoid complicating the loading diagram. The categories selected are listed in Table 5.6.

Some assumptions are made when using these theories. The main assumptions are as follows:

The wall and the soil adhere to each other through cohesion and friction.
The ground elevation is level to the top of the shoring ($\beta = 0°$).
The failure wedge is a function of the soil's internal friction angle.
The lateral pressure acts perpendicular to the shoring.

The soil types that follow either fall into the granular category or the cohesive category. These differences, combined with the support type, make up the differences between the theories. It should be noted that the granular-type soils are represented well between the Rankine and the Coulomb theories. However, when cohesive soil is present, a modified Rankine theory is better suited. The credit for this modification, which was shown previously in this chapter, goes to Bell in 1952. The original Rankine formula, also shown above, uses this modification in the second part of the formula ($\pm 2cK_a$ or K_p). It should be repeated that this is how the top of the excavation in an active soil pressure case can be a negative number. This formula modification also does not take into account the effects of groundwater and hydrostatic pressure.

Lateral soil pressure acting on a wall with a height of h should never be less than 25% of the vertical effective stress, which is $\sigma = \gamma h$. Vertical stress is equal to unit weight of soil (γ) times the depth (h). The only way an engineer can use a lesser value would be if extensive laboratory testing can verify that cohesion is higher than specified.

Wall friction is the presence of bonding between the shoring and the soil. Coulomb's theories diminish in credibility because the theory does not consider the effects of wall friction. In fact, the theories in this text ignore wall friction calculations and use additional factors of safety to compensate for the fact that wall friction varies with soil type.

Figure 5.14 shows the combination between active and passive soil pressure when a curved failure plane is considered.

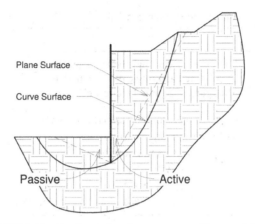

Plane Surface

Curve Surface

Passive   Active

FIGURE 5.14   Passive and active pressures with curved failure.

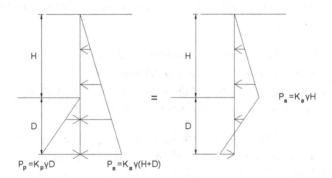

$$P_a = K_a \gamma H$$

$$P_p = K_p \gamma D \qquad P_a = K_a \gamma (H+D)$$

FIGURE 5.15   Cantilevered shoring system.

Cantilevered Shoring Systems   Shoring systems using either sheet piling or soldier beams can typically cantilever (without support) 15–18 ft in depth. Beyond these depths, the engineer should consider tiebacks or a strutting 1.5–2.0 times the length of vertical shore supporting the excavation depth. In other words, if an excavation is 16 ft deep, the vertical shoring component (sheet pile or soldier beam) could be 40 ft [(16 × 1.5) + 16] to 48 ft [(16 × 2.0) + 16] long, depending on the required embedment toe.

Embedment calculations can be complicated; therefore, many professional engineers use computer software to help calculate these values. This text will attempt to simplify this process. As mentioned previously, one way to simplify this distance is to use a multiplier from 1.5 to 2.0. This range can determine this dimension safely in most cases. This is especially common when an engineer is doing a projected cost estimate and is only trying to get close to the quantity of shoring required. The construction manager, in many cases, is trying to accomplish just that in order to get a rough, but conservative, estimate of materials so that an educated cost estimate can be achieved.

The example in Figure 5.15 requires a distance of D in order to calculate the active and passive pressures at the different depths. The pressure most important for calculating shoring component sizes is the maximum active soil pressure at the bottom of the excavation $P_a = K_a \gamma H$. This value can be used for pressure calculations in granular soils

1. Cantilevered and Single Brace System in Granular Soil Shoring in granular soil (both dense and loose) should be designed for a rectangular load where $P_a = K_a \gamma H$. Figure 5.16 illustrates this soil pressure graphically, where P is lateral pressure, K_a is active soil coefficient, γ is the soil unit weight, and H is the depth of excavation.

2. Cantilevered and Single Brace System in Cohesive Soil When a cantilevered system is supported in a cohesive material such as clay, the soil load can be represented by $P_a = K_a \gamma H - 2c$. Figure 5.17 shows this pressure diagram, and the maximum soil pressure is shown by the largest pressure area in the middle of the diagram. The upper portion of the pressure can be negative due to subtracting the cohesion $(-2c)$. This is usually just assumed to be zero pressure because there can't actually be zero pressure.

Supported Shoring Systems Once it is determined that a system is too deep to support itself without walers, tiebacks, and/or struts, the soil loading conditions

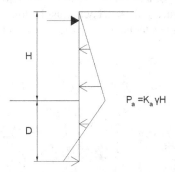

FIGURE 5.16 Cantilevered system in granular soil.

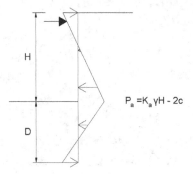

FIGURE 5.17 Cantilevered and single brace system in cohesive soil.

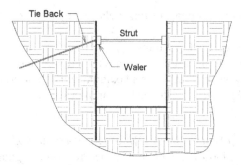

FIGURE 5.18 Supported shoring system components.

change and so does the support system. When more than one brace is required, another formula can assist the engineer in determining an accurate lateral soil pressure. Bracing consists of walers, struts, and/or tiebacks. These components will be briefly introduced here, then actually designed in Chapters 6 and 7. Figure 5.18 shows how these components correspond to a supported shoring system.

Walers: Horizontal support member that supports the vertical shoring (sheet piling or soldier beams) or transfers the shoring load to tiebacks or struts. Figure 5.19 is a photo of a double steel channel waler using tiebacks. The double waler is being supported by "lookout" brackets. These brackets support the waler until the soil pressure is applied.

Tiebacks: Tension rods or strands that support walers or soldier beams and transfers load into the supporting soil. A tieback must be long enough to extend through the failure wedge of soil and adequately into the competent earth. Tieback design will be mentioned later.

FIGURE 5.19 Double steel channel waler with tiebacks.

Struts: Compression pipes, beams or timbers that support walers and transfer loads to the other side of the excavation. Since struts act like columns, a high slenderness ratio will limit the struts buckling capacity. In this case, tiebacks should be considered or the struts may have to be supported vertically as well.

3. Multiple Brace, Granular or Cohesive Soil The third system described in this section is the system that is deep enough to require more than one level of support. The maximum pressure for this situation is reduced by 65% because of the number of supports (>1). This assumes water is not present. If water is present, $P_w = 62.4$ pcf $\times H$, should be added. Also, the soil unit weight should be the saturated or the submerged unit weight. The pressure for a multiple supported system with either granular or cohesive soil is calculated as $P_a = 0.65 K_a \gamma H$. Figure 5.20 illustrates this pressure diagram.

Figure 5.21 is a photo of a multiple supported excavation in San Jose, California. The system had three levels of supports using walers and struts. Each level from top to bottom was installed as the excavation proceeds downward.

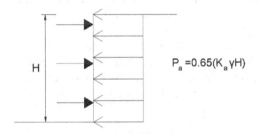

$$P_a = 0.65(K_a \gamma H)$$

FIGURE 5.20 Rectanglar pressure diagram.

FIGURE 5.21 Multilevel supported excavation.

Effects of the Water Table on Loading   Water table changes the unit weight
of the soil, which in some cases can lower the unit weight and in some cases increase
the unit weight. Water can also change the shoring system selection by forcing the
designer to use a water-tight system where, in some cases, another, less expensive,
system could have supported the load. Typically, an excavation has to be dry in order
to perform the work inside such as concrete placement. Pure hydrostatic water pressure must be added to the soil loads in a free draining case where granular soil is
dominant. In a cohesive soil case, the soil unit weight represents a saturated unit
weight, which is going to be much heavier than its dry weight, therefore, the γ value
takes this into account as far as pressure is concerned.

Surcharge Loads   Surcharge loads are typically from existing buildings/struc-
tures, construction equipment, traffic, or materials. This text will focus on construction equipment surcharge loads. Standard practice uses either a theoretical value or
an actual weight when a specific piece of equipment is specified.

   Theoretical surcharge loads are uniform loads predetermined by an engineering
counsel and covers standard equipment up to a certain weight. Equipment that is
heavier than the standard would have to fall into the actual weight category described
below. The minimum lateral construction surcharge load is 72 psf at 10 ft in depth
below the shoring system regardless if the system is subject to any surcharge loading
during construction. This would be the equivalent of soil that weights 109 pcf and
an active soil coefficient of 0.333 for a depth of 2 ft. For surcharge load to affect the
lateral loading the surcharge load should be within the active soil wedge. Figure 5.22
illustrates a surcharge load (Q) and how it relates graphically to a lateral load in the
H_s zone:

$$P \text{ horizontal} = K_a Q$$

where Q = equipment load per square foot within the limits of the active failure
                wedge
        K_a = active soil coefficient
$P_{\text{horizontal}}$ = lateral pressure on the shoring due to the surcharge load

   When equipment weight is being estimated, one can apply a series of uniform loads that change as the depths increase. This typical surcharge loading

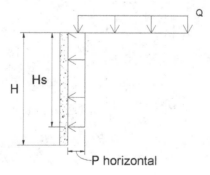

FIGURE 5.22   Surcharge loading.

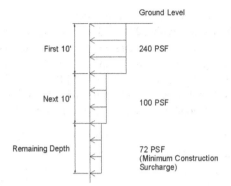

FIGURE 5.23 Standard surcharge loads by depth.

FIGURE 5.24 Equipment surcharge load on an excavation.

is recommended when standard, small equipment is planned for construction. Figure 5.23 shows the surcharge values decreasing as the depth of excavation increases. The values are higher closer to the surface and decrease as the excavation gets deeper. Figure 5.24 shows a photo of a long-reach excavator loading trucks next to a sheet pile support of excavation. This excavation is over 30 ft deep, so the excavator must position right up against the sheet piles.

Actual Equipment Weight Method There are many cases where a contractor designs a shoring system to support the lateral soil loads and an additional surcharge load for a particular crane or excavator, which is going to be the largest and heaviest used during the life of the shoring system in place. As an example, if a contractor plans to use a 150,000-lb crane with a $25' \times 18'$ base dimension, this would equate to $q_s = 333$ psf, where q_s is the actual surcharge load in psf. In practice, the surcharge causes a lateral pressure of $P = K_a \times q_s = 0.33 \times 333$ (33% of the load). Therefore, the designer would add a rectangular lateral load of 111 psf for the depth of the excavation. If this force is 72 psf or less, a minimum force of 72 psf should be used. In no case should a surcharge load be less than 72 psf. This will be explained further in the next section.

Equipment weight, 150,000 lb

Base dimensions, $25' \times 18'$

$q_s = 333$ psf

Surcharge load, $333 \times 0.33 = 111$ psf

The deeper the excavation is, the less effect the equipment weight has to the relative depth. There is theoretically a depth where the surcharge load has no effect. However, as discussed in the previous section, a minimum of 72 psf is still used.

Boussinesq Loads   Boussinesq developed three type of surcharge loads; the strip load, the line load, and the point load (Figure 5.25). These three types of surcharge loads were arranged for three particular cases. These cases are as follows:

Strip load—Highways, roads, or railroads parallel to the shoring

Line load—continuous wall or footing parallel to the shoring

Point load—outriggers from crane or pump

Table 5.7 shows three different surcharge loads and their totals when combined; 72 psf is the minimum surcharge load that can be used if equipment will be working within the limits of an excavation support.

OSHA Trench Classifications by Soil Type   This text focuses on designed shoring systems using sheet piling or soldier beams. In the case of trenching for utilities and other horizontal underground applications, state and federal OSHA regulations have

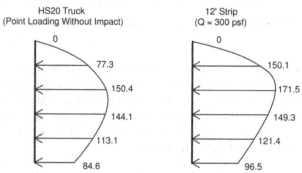

Conclusion: Strip loading of Q = 300 psf compares favorably to a point load evaluation for HS20 truck loadings.

FIGURE 5.25   Surcharge loads using Boussinesq.

TABLE 5.7 Surcharge Lateral Pressure from Equipment

Depth (ft)	$Q = 100$ (psf)	$Q = 200$ (psf)	$Q = 300$ (psf)	Sum (psf)
0.1	1.9	0.3	1.7	72 min
1	17.9	3.0	17.1	72 min
2	30.2	5.8	33.8	72 min
4	35.7	10.1	63.7	109.5
6	29.5	12.3	87.1	128.9
8	21.9	12.7	103.3	137.9
10	15.9	11.9	112.6	140.4
12	11.5	10.5	116.4	138.4
14	8.5	9.0	116.1	133.6
16	6.3	7.6	112.9	126.8

Source: Data from *Cal Trans Trenching and Shoring Manual* (2011).

TABLE 5.8 Soil Properties by Type

Soil Type	Unconfined Compressive Strength (TSF)[a]	Minimum Slope of Trench Wall
A	1.5	3/4:1
B	0.5	1:1
C	N/A	1 1/2:1 or greater

[a]Tons per square foot.

created guidelines. These guidelines classify soil types by type A, B, and C. Most jurisdictions require contractors to employ a "competent" person in soil identification. This person can then determine either sloping criteria or shoring types. These shoring types would be in the category of shielding, speed shoring, or a combination of both. Table 5.8 briefly describes the characteristics required for a soil to be a type A, B, or C.

The competent person on the project must go through a one-day (minimum) training course in order to be able to identify soil into one of the three categories listed in the table. This person is responsible for deciding on the type of shoring or sloping systems to be used and if engineering is required. The competent person's title should not be confused with a licensed, registered engineer. If engineering is required and the competent person does not meet the requirements of a licensed, registered engineer, that person should employ the proper person.

Other Project Examples The Figures 5.26–5.29 show some examples of work that involved support of excavation and soil type in addition to what was already discussed in this chapter. Figures 5.26 and 5.27 are examples of cemented soils in a cofferdam where sands and gravels were expected. Not only did the shoring installation suffer due to the unexpected soils conditions, Figure 5.29 shows how the excavation methods employed were very slow. A small track excavator was used with a 1000-lb breaker and the material was loaded into a skip box for removal from the cofferdam.

FIGURE 5.26   Soil conditions in a sheet pile cofferdam.

FIGURE 5.27   Methods of excavation for cemented soils.

Figure 5.28 shows a sheet pile cofferdam in a river with external walers. This system will be looked into further in Chapter 7. Dewatering pumps are shown in the background. It took two 4-in submersibles to keep up with the dewatering efforts due to split sheets and sheets that were dislodged from their interlocks.

FIGURE 5.28 Sheet pile cofferdam with external walers.

FIGURE 5.29 Slide rail system.

Figure 5.29 shows a slide rail system used in San Francisco to shore bridge pier footings. This type of system is preengineered and only works with its own components. Rental companies have systems like this available to rent and usually provide any engineering information required to submit to the owner/engineer for approval.

CHAPTER 6

SOLDIER BEAM, LAGGING, AND TIEBACKS

Now that soil properties and loading conditions have been covered, they will be incorporated with the study of two common support of excavation systems: soldier beam, lagging, and tiebacks in this chapter and sheet pile, walers (sometimes referred to as wales), and struts in the next chapter.

As with the soil loading, state and federal department trenching and shoring manuals are used in this text for shoring design. There are some cases where railroad and U.S. Navy standards may be more applicable. However, when not specifically stated, state and federal departments will be the standard. The examples in this chapter and the next have been fairly simplified so that the main concepts are not lost in the finer details.

6.1 SYSTEM DESCRIPTION AND UNITS OF MEASURE

6.1.1 Beams/Piles

Beams and piles are the foundation of the soldier beam and lagging system. They are the components that embed from original ground to below the subgrade of the excavation, providing anchor support for the system. Beams and piles are typically either drilled then set and grouted or driven. When there are sensitive environmental concerns, drilling seems to be the method of choice. However, when production is the only concern, driving seems to be most economical. Beams and piles are placed by the individual beam as a unit of each when estimating and tracking costs.

The beams and piles support the lagging, and their flanges make direct contact with the soil. However, for design purposes, the flanges of the beams support the lagging material, which is what supports the soil with which it makes direct contact.

6.1.2   Lagging

Lagging is the material in this system that makes direct contact with the soil. Lagging is typically wood planking or steel plates. Wood can be expensive to purchase initially but, if reusable, can be economical over several uses. Steel plates, per unit, are even more expensive than wood lagging but have a much higher probability of being reused multiple times. When estimating or tracking cost of this work, it is most common to quantify by the square foot (SF or ft^2). Wood prices can range from $0.60 to $1.20 per board foot, and steel plate prices can range from $0.45 to $0.75 per pound. Owning either of these materials, especially steel, is paramount to a company's success if this type of work is self-performed frequently.

Steel plates can be rented from shoring and trench plate rental companies but can be very expensive over the span of the rental period.

6.1.3   Tiebacks

Tiebacks, soil nails, and soil anchors are three different types of anchors used to support shoring systems. They range in size from small anchors (soil nails) to medium-size anchors (tiebacks) to larger ties (soil anchors). The larger, higher capacity the anchor, the more expensive the anchor and larger the shoring system components are. Alternatively, the smaller the anchor, the lighter the accompanying system is. The smaller the size of the system typically translates to a lower cost. However, if the system requires a large quantity of expensive ties, the cost increases. Therefore, ties are quantified by size for estimating or cost tracking purposes.

6.2   MATERIALS

6.2.1   Steel AISC

For temporary steel construction, the American Institute of Steel Construction (AISC) is the governing standards, except when some jurisdictions may require a higher factor of safety. The steel most commonly used for temporary structures is ASTM A36 steel, which has an ultimate strength of 58,000–66,000 psi and yield strength (F_y) of 36,000 psi. For rolled steel sections, A36 grade steel has the following allowable stress values:

$F_v = 0.4F_y = (0.4)36 \text{ ksi} = 14.4 \text{ ksi}$

$E = 29{,}000{,}000\text{–}30{,}000{,}000 \text{ psi (almost always 29,000,000 psi)}$

$F_b \text{ (noncompact sections)} = 0.6F_y, \ (0.6)36 \text{ ksi} = 21.6 \text{ ksi}$

$F_b \text{ (partially compact) between } 0.6F_y \text{ and } 0.66, F_y = (21.6 - 23.76 \text{ ksi})$

$F_b \text{ (compact sections) } 0.66F_y = (0.66)36 \text{ ksi} = 23.76 \text{ ksi}$

No matter which method for bending is used, the lowest value is 21.6 ksi. Therefore, when in doubt, using 21.6 ksi allowable bending stress is recommended. Different F_b values can be required when the compression flange of the beam is unsupported or not fully supported. Flange buckling stress will be covered later in this text.

6.2.2   Wood Species—National Design Specifications (NDS) for Wood Construction

Wood lagging is used quite often in soldier beam and lagging systems. In most cases, standard Douglas fir–Larch lumber is used. However, pressure-treated wood can be used when the system is installed for longer periods of time or even for permanent construction. When a species of lumber is not specified, the following values are the maximum allowable stress values this book will use in shoring applications:

$$F_{cll} = 480{,}000 \, (L/D)^2 \text{ psi; not to exceed 1600 psi}$$

$$F_b = 1800 \text{ psi} > 8'' \text{ nominal depth}$$

$$F_b = 1500 \text{ psi} \leq 8'' \text{ nominal depth}$$

$$F_t \text{ (direct tension)} = 1200 \text{ psi}$$

$$F_{c,\text{perp}} = 450 \text{ psi}$$

$$V_{\text{hor shear}} = 140 \text{ psi}$$

$$E = 1{,}600{,}000 \text{ psi}$$

Contractors commonly know what material will be used on a system before construction. Therefore, they should use values for that specific material as listed in the NDS.

Some agencies, such as railroads, may use slightly different values than those used for highway transportation. In these cases, the values are sometimes lower than the standard for the NDS. For example, railroad only allows 1710 psi for bending instead of 1800 psi. Deflection requirements are not as strict with shoring as in, say, concrete formwork, which usually leaves a permanent product (concrete wall) exposed to view, thus requiring tighter tolerances. The soldier beam and lagging shoring wall can deflect without being out of specification.

Short-Term Loading   Soldier beam and lagging systems (see Figure 6.1) are one of the more economical shoring systems that can be installed for $15–$25 per SF of shored face. The cost range is affected by whether steel or wood lagging is used, whether it is cantilevered or braced, and whether the beams are drilled or driven, at a minimum. Many other factors not mentioned here can affect cost and should be looked at case by case.

The length of time many shoring systems are in place ranges from a few days to several months to a couple years. Systems that are in use for 3 months or less benefit from a short-term allowance. For short-term loading on shoring, allowable stress can be increased by 33% except when:

Three months is exceeded.

Nearby construction equipment is causing vibration.

Excavations are adjacent to railroads.

Note: Strut design, which is covered later, is not offered this allowance.

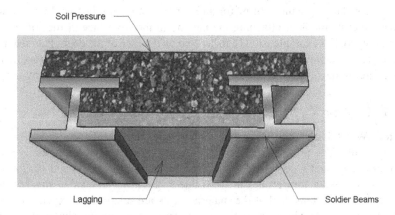

Soil Pressure

Lagging

Soldier Beams

FIGURE 6.1   Soldier beam and lagging components.

Soil Arching   Soil arching occurs when the supported soil is kept tight to the back of the soldier pile (see Figure 6.2). This condition helps the lagging in that the lagging does not experience the whole pressure of the soil. Engineers have concluded that the soil pressure is only 60% effective; therefore, the maximum moment or pressure can be multiplied by 0.60. Here is how the common maximum moment calculation is affected by this reduction:

$M = (0.6)wl^2/8$, a 0.6 multiplier is used with lagging to account for soil arching.

The lagging must be strong enough to support the soil load and transfer this load to each adjacent soldier beam. The two most common types of lagging are wood ($2 \times 10'$ to $6 \times 12'$) and steel plate ($1''$ to $1\,3/4''$).

Lagging calculations are somewhat simplified because the free-body diagram of a single lagging board or steel plate is a true case of a simply supported beam, as shown in Figure 6.3. The boards or plate rest behind the flange of the soldier beams. In rare

Soil Arching

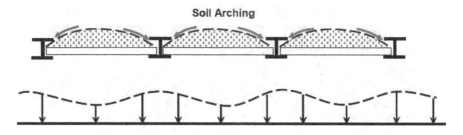

FIGURE 6.2   Soil arching between soldier beams.

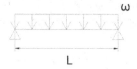

FIGURE 6.3   Typical loading diagram for lagging.

cases where lagging cannot fit or for another reason cannot be installed behind the flange, other methods of attaching the lagging to the front face of the soldier beam are employed. For example, studs can be used on the face of the beam flange to hold a steel plate that captures the lagging. This would only be for wood lagging applications and is rare. In most cases, the maximum bending moment can be found by $wl^2/8$.

6.2.3 Lagging

Wood Wood lagging is used for temporary shoring walls and for permanent shoring walls. The primary differences between permanent and temporary use are (1) pressure-treated wood is normally used for permanent walls and (2) the board thickness is greater for permanent walls due to the lower allowable stresses. Wood lagging is labor intensive, but the material costs are relatively inexpensive and the system is very versatile because the wood can be cut into any shape to fit around obstructions such as utilities and other shoring systems and stepped to go down slopes. Figure 6.4 also shows the versatility of wood for going down slopes.

Whenever wood is used for design, shear and bending are analyzed. However, because lagging boards are loaded in their weak direction (flat), shear is usually not a concern. The boards loaded in this manner will always fail in bending before failing in shear. The cross-sectional area (shear) is the same either way, but the section modulus (bending) is very different between the strong and weak direction of application.

Wood Lagging Design A few common formulas (also mentioned in the first two chapters) are necessary for wood lagging design:

$$M = wL^2/8 \quad \text{(max bending moment for a simply supported beam)}$$

FIGURE 6.4   Cantilevered soldier beam system.

where w = uniform load
        L = span from center of one soldier pile to another

$$F_b = \frac{M}{S} \quad \text{(max bending stress in a beam)}$$

where M = max moment
        S = beam section modulus

$F_v = 1.5\,V/A$ (max shear stress, which is not necessary for lagging because they are always loaded in the weak direction)

where V = max shear from the FBD
        A = cross-sectional area of beam

$S = bd^2/6$ (section modulus in the direction of loading shown as a 12″ strip in Figure 6.5)

where b = base dimension
        d = beam depth

Lagging boards always are assumed to be a 12″ strip. This way, the units in calculations shown above will always be in 1-ft increments. The same concept is used with steel plate. In other words, it does not matter whether the lagging is a 2 × 8, 2 × 12, and so forth, it is always assumed to be a 12″ strip. For this reason, the base (b) dimension in the section modulus formula is 12 in as shown in Figure 6.5. Therefore, the only unknown is the depth of the lagging. Figure 6.6 provides a good example of how a wood lagging system was used in a downtown location excavation. Let's look at some simple steps to determine the minimum depth of a wood lagging board.

Step 1: Determine the uniform load on one lagging board using $w = (P_{max})$ (1 ft). For this example assume the soil pressure on the shoring system is 600 psf

where P_{max} = 600 psf soil loading
        W = 600 psf × 1′ strip = 600 plf (since w in the moment
           formula is pounds per linear foot)
        L = 7 ft
        F_b = 1800 psi for construction-grade Douglas fir–Larch

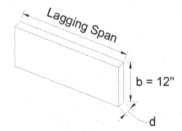

FIGURE 6.5 Typical lagging board loading.

FIGURE 6.6   Lagging system in place.

FIGURE 6.7   Lagging FBD.

Step 2: Draw an FBD (Figure 6.7) and calculate the maximum moment:

$$M = (600)(7')^2/8 = 3675 \text{ ft-lb}$$

Step 3: Set $f_b = M/S$ and solve for d. The depth of the lagging is the only unknown at this point.

Don't forget to convert the feet to inches in the formula because the allowable bending stress is in psi. The formula should look like this:

$$F_b = M/(12'' \times d^2/6) \quad \text{(solved for } d)$$
$$1800 \text{ psi} = (3675 \text{ ft-lb})(12''/\text{ft})/(12''d^2/6)$$
$$d = 3.5 \text{ in}$$

Conclusion: An S4S 4 × lagging board can be used because the actual depth is 3.5 in. Typically this would be a 4 × 12 so that the number of boards to be placed is minimized for labor economy. The less number of boards to be handled, cut, placed, and removed, the less expensive the system. As an example, if the shoring wall was 15 ft high, the difference between 4 × 8′s and 4 × 12′s would be (25−16 boards) 9 boards.

It should be noted that if the solution is not desirable, as with most temporary structure design, the variable information can be modified in order to change the results. For instance, in this problem we can change the type/specie of wood lagging, which changes F_b to a higher or lower value (though 1800 psi is about as high as we can go). We can also change the spacing of the soldier beams to lessen the bending moment. We also did not take advantage of soil arching. If we did, we could have reduced the 600 psf to 60% or to 360 psf. It should be determined if soil arching will occur by consulting a soils engineer.

Steel Plate   Steel plate lagging is more expensive to buy than wood lagging; however, contractors often prefer it because of the savings in labor cost to install. Like most decisions in temporary structures, the most economical system will be chosen when considering materials, labor, and equipment needed. Sometimes the more expensive material costs can be offset by lower labor rates. The down side to steel plate lagging is its lack of flexibility to conform around obstructions and other irregularities.

The same formulas as those used in wood lagging calculations are used. One big difference is that for steel, $F_v = V/d \times tw$. This will not change the problem because bending will always govern when the lagging is loaded in the weak direction (flat). The other more applicable difference is that the F_b allowable bending stress for steel is so much more than for wood. The same 12″ strip of steel is assumed as was the 12″ strip with wood. This makes calculations simple and the units remain per foot.

Steel Plate Lagging Design   Let's design a steel plate lagging system with the following parameters and using Figure 6.7 as a reference:

$$P_{max} = 800 \text{ psf}$$

$W = 800 \text{ psf} \times 1' \text{ strip} = 800 \text{ plf}$
     (since w in the moment formula is pounds per linear foot)

$L = 8 \text{ ft}$

$F_b = 21,600 \text{ psi for A36 steel (compact section } 0.60F_y)$

Step 1: Draw an FBD and calculate the maximum moment:

$$M = (800)(8')^2/8 = 6400 \text{ ft-lb}$$

Step 2: Set $F_b = M/S$ and solve for d. The depth of the lagging is the only unknown at this point.

Don't forget to convert the feet to inches in the formula because the allowable bending stress is in psi. The formula should look like this:

$$21,600 \text{ psi} = \frac{6400 \text{ ft-lb } 12''/\text{ft}}{12'' \times d^2/6}$$

$$d = 1.33 \text{ in}$$

Conclusion: If steel plate is rolled in increments of $\frac{1}{8}''$, then one could use $1\frac{3}{8}''$ plate. If they come in $\frac{1}{4}''$ increments, then we would need to use $1\frac{1}{2}''$ plate because 1.33″ is the minimum.

If soil arching was considered, we could have reduced the pressure to 480 psf (0.60w).

Steel plate lagging can be combined with wood lagging. Let's say we have a 15-ft-deep excavation and we have 12 × 8 steel plates available. The plates would be placed vertically, thus the top 3 ft would be unsupported. In this case, it makes sense to place three to four wood lagging boards at the top of the shoring to get the system to the original ground (OG) level. Figure 6.8 shows an example of this. The same calculations would have to be performed. However, one should take advantage of the fact that the wood lagging is at the top 3 ft of the excavation, which is where the soil pressure is at its lowest value. Keeping in check with our minimum values, a reduced pressure of say 400 psf may be acceptable at the top section.

Table 6.1 is from *Caltrans Shoring Manual* and shows recommended board thicknesses of Douglas fir with soil arching for different types of soils taken from the Unified Soils Classification System. These values assume there is no surcharge loading, only lateral soil loads. Surcharge loading is less common with these types of systems, especially when they are supporting a bank or slope for a road widening project where access to the top of the wall is impossible or impractical.

6.2.4   Soldier Beam Design

Once the lagging type and thickness is computed, the designer can consider the soldier beam sizes. Soldier beams can be any W (wide flange) or HP (hinge point) shapes, but HP shapes are usually considered first. HP shapes are available up to HP14 × 117,

FIGURE 6.8   Combined wood and steel plate lagging.

TABLE 6.1 Recommended Lagging Board Thickness (without Surcharge)

Recommended Thickness of Wood Lagging When Soil Arching Will Be Developed (for locations without surcharge loadings)

Soil Description Classification	Unified	Depth	Recommended Thickness of Lagging (rough cut) for Clear Spans of $5', 6', 7', 8', 9', 10'$
Competent Soils			
Silts or fine sand and silt above water table; sands and gravels (Medium dense to dense)	ML, SM – ML GW, GP, GM, GC, SW, SP, SM	0' to 25'	2″, 3″, 3″, 3″, 4″, 4″
Clays (stiff to very stiff); nonfissured clays, medium consistency and $\gamma H/C < 5$.	CL, CH CL, CH	25' to 60'	3″, 3″, 3″, 4″, 4″, 5″
Difficult Soils			
Sands and silty sands, (loose); clayey sands (medium dense to dense) below water table	SW, SP, SM		
Clays, heavily over-consolidated fissured	SC	0' to 25'	3″, 3″, 3″, 4″, 4″, 5″
Cohesionless silt or fine sand and silt below water table	CL, CH ML; SM – ML	25' to 60'	3″, 3″, 4″, 4″, 5″, 5″
Potentially Dangerous Soils (appropriateness of lagging is questionable)			
Soft clays $\gamma H/C > 5$; slightly plastic silts below water table; clayey sands (loose), below water table	CL, CH ML SC	0' to 15' 15' to 25' 25' to 35'	3″, 3″, 4″, 5″, 3″, 4″, 5″, 6″, 4″, 5″, 6″

Source: Adapted and revised from the April 1976 Federal Highway Administration Report No. FHWA-RD-130 and the *CalTrans Shoring Manual*.

which has a maximum section modulus in the strong direction of 172 in³. Beyond this section modulus, wide flange (W shapes) need to then be considered. The method of beam placement should also be considered. As mentioned earlier, beams can be "drilled and dropped" and backfilled with slurry or driven with a vibratory or impact hammer. The beams in Figure 6.9 were drilled and dropped. Technically, the placement method would change the soldier beam design slightly because the beam width is widened due to the slurry in the drilled hole; but in this text, the methods will result in the same design.

The temporary structure student must first understand how the uniform load is placed on the soldier beam. As we look at the load path that begins with the soil load,

FIGURE 6.9   Soldier beam and plate with dewatering system.

we notice that the soil loads the lagging and the lagging transfers that load to the soldier beams. This means that each soldier beam has to support the load halfway to each adjacent soldier beam. In other words, the soldier beam tributary width is the same as the soldier beam spacing. Figure 6.10 shows this tributary width as it applies to one soldier beam.

From this understanding, the designer must multiply the P_{max} pressure by the tributary width in order to continue with the soldier beam design. For example, if the maximum soil pressure is 800 psf and the soldier beam spacing is 6 ft on center, then the uniform load on one soldier beam would be:

$$w = 800 \text{ psf} \times 6 \text{ ft} = 4800 \text{ plf}$$

The loading diagram would look like Figure 6.11.

Now that the uniform load on a single soldier beam is understood, the size of the beam can be determined. In order to size a beam, the maximum bending moment needs to be calculated. Besides the uniform load, the designer would have to

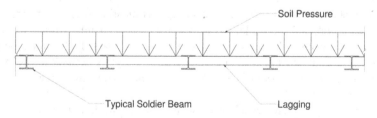

FIGURE 6.10   Tributary width on a single soldier beam.

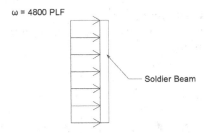

FIGURE 6.11 Vertical loading on a single soldier beam.

determine the length of the beam and where it is supported. In Chapter 5 we studied cantilevered beams and supported beams. Let's look at the following examples.

Cantilevered Soldier Beams Cantilevered soldier beams must rely on a fixed portion of the beam supported by the passive soil pressure coming from the embedded portion of the beam below the subgrade (dredge line) of the excavation. For this purpose, it will be assumed that the beam is embedded deep enough to develop full fixity of the beam. This could be the 1.5–2.0 times the excavation depth mentioned in Chapter 5 and would provide a large enough FOS. Also, the beam fixity begins at a point near the subgrade (SG), which can be relied on to offer a fixed connection. An acceptable practice is to assume a point 3 ft below the dredge line, called the hinge point (HP). The 3-ft point is assumed so we can rely on a substantial passive force. This first 3 ft of soil from the SG to the HP is assumed to fail along a 45° failure plane and cannot be relied upon for support. Figure 6.12 is a good example of a cantilevered soldier beam system. The maximum moment for a cantilevered soldier beam assuming proper fixity below subgrade is

$$M = \frac{wL^2}{2}$$

It should be pointed out here that L includes the supported wall height plus the additional 3 ft to the HP. In this example, if the excavation depth was 12 ft, then $L = 12' + 3' = 15$ ft.

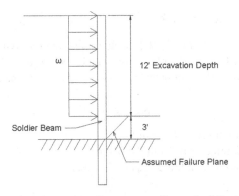

FIGURE 6.12 Soil loading on a cantilevered soldier beam.

Example 6.1   Soldier Beam Cantilevered
Step 1: Draw a proper loading diagram (see Figure 6.13) and determine the maximum bending moment in the beam.

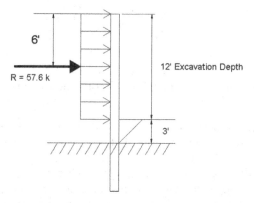

FIGURE 6.13   FBD of cantilevered soldier beam.

If the excavation depth is 12 ft from OG to the subgrade and the hinge point is 3 ft below subgrade, then the full cantilever distance L is equal to $12' + 3' = 15$ ft. Now, let's assume the soil pressure is the same as above with the 6-ft soldier beam spacing. This resulted in a uniform soldier beam load of 4800 plf.

The resultant $R = 4800$ pfl $\times$ 12 ft $= 57,600$ lb or 57.6 k

The resultant location of a rectangular load is at the midpoint of the diagram height, or 6 ft from the bottom of the excavation; but the 3 ft to the hinge point needs to be added in order to represent the proper moment arm, which is from the HP to the resultant.

Implementing two different cantilevered max moment formulas, the answers are within 4% of each other. In most cases, the designer would be prudent to use the higher, more conservative value of 540 ft-k. This would obviously produce the larger soldier beam:

$$M_{max} = P \times L \quad 57.6 \text{ k} \times 9 \text{ ft} = 518.4 \text{ ft-k}$$

$$M_{max} = wL^2/2 \quad 4.8 \text{ klf} \times (15)^2/2 = 540 \text{ ft-k}$$

Using a maximum moment of 540 ft-k, determine the minimum section modulus of the soldier beam. It is preferable to find an HP beam first. However, if there is not an HP beam with sufficient section modulus, then a W shape beam should be considered:

$$S_{min} = \frac{M}{F_b}$$

where S_{min} = minimum section modulus
M = maximum moment in the beam
F_b = allowable bending stress of the steel

If A36 steel is assumed and a compact section is desired, then $F_b = 0.6F_y$, or 21.6 ksi:

$$S_{min} = \frac{540 \text{ ft-k } (12''/\text{ft})}{21.6 \text{ ksi}} = 300 \text{ in}^3$$

The AISC manual contains most beams manufactured and their properties. Appendix 1 should be used for beam selection. It is evident that there are no HP shapes with a minimum section modulus of 300 in^3. Therefore, the W shapes should be considered. Since compact sections are desired for soldier beams, the designer should begin with a 12″ or 14″ beam. For the most economical beam, the weight per foot is a concern. The available W shapes are shown in Table 6.2.

TABLE 6.2   Available Beams Meeting Section Requirements

Nominal Depth (in)	Beam	Section Modulus (in^3)
12	W12 × 230	321
14	W14 × 193	310
16	W16 × ?	Not applicable
18	W18 × 158	310

Based on these results, the designer has the option of compactness vs. economy. Since soldier beams are continuously supported by the earth and lagging, flange buckling should not be an issue and compactness is not as important. As far as economy, at steel prices between $0.55 and $0.75 per pound (at the time of this book's publishing), the lighter beams would save the project a great deal of beam costs. Let's say 20 of these beams are required and they are 34 ft in length. The difference between the W14 and the W18 using a steel price of $0.65/lb is

$$20 \text{ ea} \times 34 \text{ ft} \times (193 - 158 \text{ plf}) \times \$0.65/\text{lb} = \$15,470$$

The same comparison can be calculated between the 12″ and 14″ beams:

$$20 \text{ ea} \times 34 \text{ ft} \times (230 - 193 \text{ plf}) \times \$0.65/\text{lb} = \$16,354$$

This is a significant cost savings.

Supported Soldier Beam System   Cantilevered soldier beams have their limitations and where the depths exceed these limitations, supports must be added to the system. Supports, in this case, are either struts or tiebacks.

Single Support/Tieback   Figure 6.14 provides an example of a single supported tieback system. Lets design another system using a single level tieback and the given information in Example 6.2 that follows.

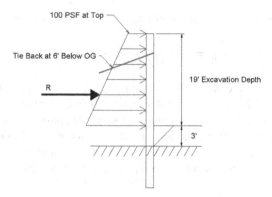

FIGURE 6.14 Single supported system.

Example 6.2 Supported Soldier Beam
Given information:

Soldier beams at 6 ft on center
$P_{max} = 1200$ psf and the pressure diagram is triangular
Surcharge load of 100 psf as a rectangular load from OG to dredge line
Depth = 19 ft
Upper support located 6 ft from OG
Lower support is at HP, 3 ft below dredge line
$P_{max} = 1200$ psf + 100 psf = 1300 psf
R_1 (soil load) is located $\frac{1}{3}$ up from bottom of dredge line
R_2 (surcharge) is located halfway up from the bottom of the dredge line

Step 1: Calculate resultant, which is the area of the triangular load plus the 100 psf rectangular surcharge load (see Figure 6.15).
R_1 = 6-ft spacing × (1200 psf × 22 ft/2) = 79.2 k located $22'/3$ from bottom = 7.3 ft.
R_2 = 6-ft spacing × (100 × 22 ft) = 13.2 k   located $22'/2$ from bottom = 11 ft.

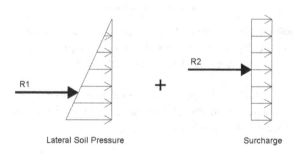

Lateral Soil Pressure   Surcharge

FIGURE 6.15 Total load from soil and surcharge.

Step 2: Total the moments about the subgrade location (fixed point) on the right. The drawing in Figure 6.16 has been turned 90° for simplicity. A shear and moment diagram would be time consuming and could produce errors.

Another method to determine the maximum moment in the beam that is very conservative is to convert the soil and surcharge load to a uniform load across the longest span of the beam and use the simply supported beam moment formula $(wl^2/8)$. In this case, the uniform load would be $1200 + 100 = 1300$ psf over a 16-ft span (L). The load diagram would look like Figure 6.17.

$$M = (6' \times 1300 \text{ plf})(16')^2/8 = 249{,}600 \text{ lbs}$$

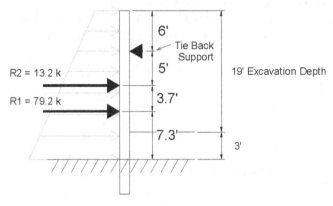

FIGURE 6.16 FBD of soldier beam.

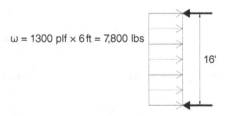

FIGURE 6.17 Alternate FBD (simplified).

The beam size for this maximum moment would be

$$S_{min} = \frac{(249.6 \text{ ft-k})(12''/\text{ft})}{21.6 \text{ ksi}} = 138.7 \text{ in}^3$$

Table 6.3 lists some available beams and their beam section modulus in cubic inches from Appendix 1.

TABLE 6.3 Available Beams

Nominal Depth (in)	Beam	Beam Section Modulus (in^3)
12	W12 × 106	145
14	W14 × 90	143
14	HP14 × 102	150
16	W16 × 89	155
18	W18 × 76	146

Beam Software

Beam software can be used in this case as well. Figure 6.18 illustrates a computer-generated diagram. Even though this is not an indeterminate structure, using software such as this saves a lot of hand calculations. The surcharge load has been input as a 100-psf rectangular load, and the soil pressure is represented by a triangular load ranging from 0 psf at the top to 1200 psf at the bottom. The upper hinged support is a tieback 6 ft down from OG, and the lower support is fixed and at subgrade.

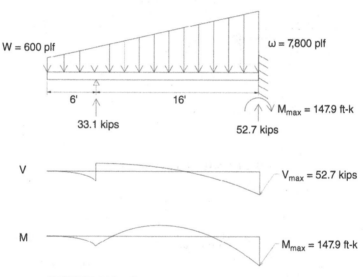

FIGURE 6.18 Computer-generated S and M diagram.

Max moment = 147.9 ft-k

Besides the program calculating the maximum moment in the beam, the program also calculates other useful information. For example, the maximum shear value in the beam is determined. In addition, the reactions are calculated, which in this case determines the tieback load per 6 ft of wall. The shear and moment diagram shown above indicates a 35.4-k tie load. If this was a strut, the load would be the strut force per 6 ft of wall. So, if the tiebacks/struts were placed at every other soldier beam, the load would be 2 × 35.4 k or 70.8 k.

There is an obvious difference between the two maximum moments. The conservative method produced a 249.6-ft-k moment and the beam program produced a 160-ft-k moment. These two methods led to two very different size beams. Economically speaking, this also results in a large beam cost difference.

Soldier beams, depending on their size and the soil type can cantilever between 12 and 15 ft. This method can be very economical if the higher limits are reached. For instance, a 15-ft cantilever with steel plates would be very inexpensive to install. If the materials were owned by the installing company, then the costs of this system could be very favorable.

Soldier Pile without Lagging

If the soil has enough cohesion, soldier piles can be spaced closer together and the wood or steel lagging eliminated. The author is not recommending this method due to the high cost of steel soldier pile; however, the goal here is to show different options in construction. If a company owned these steel soldier pile and they were to be removed at the conclusion of the project, then an argument (cost analysis) could be made to justify such a system. Figure 6.19 shows this unique case where soldier beam spacing was reduced to 3.5 ft on center to eliminate the need for lagging. Of course, the soil type has to be cohesive enough to allow this type of system.

FIGURE 6.19 Soldier beams without lagging.

6.2.5 Tiebacks and Soil Nails

When excavation depths exceed the option to cantilever (up to approximately 15-ft depth) or to strut to the opposite side of the excavation, the need for "tying-back" is usually one of the only options. When there is no obstruction underground such as utilities, a swimming pool, roadways, a building foundation, high groundwater, and the like, tiebacks are fairly versatile.

The sketch in Figure 6.20 shows the three main components of a tieback system. The soil or rock is what is drilled through to create the hole for the tieback. The grout

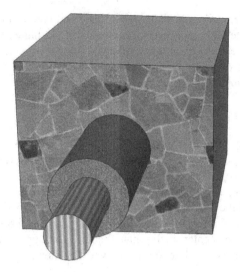

FIGURE 6.20 Tieback isometric.

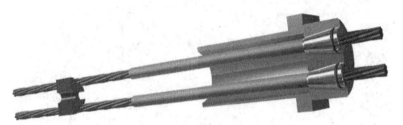

FIGURE 6.21 Two-pair strand tieback system.

joins the soil or rock to the steel tension members (strand or rod). Finally, the tension members, which are rods or strand, join the grout to the structural framework of the shoring, such as walers or soldier beams.

Figure 6.21 provides an example of a two-pair strand system. The materials that make up these systems are usually high-strength steel ranging in allowable stresses between 60 and 120 ksi.

Determining Tieback Loads from Soil Pressure Diagrams Pressure diagrams are essential for designing beam components, but they have other benefits as well. The reactions that are calculated on a pressure diagram typically represent the support of the beam in question. Since the next step of the design is to size the support system or apply the reaction load to the next member, these values give you the information needed for this step. For instance, the information from Example 6.2 can be used in the next step. If we look a little closer, we notice the reactions at the left and right are 35.4 and 57 k, respectively. The values could represent either the forces from this beam to its supporting beam or the forces required in a tieback or strut system if this is the final portion of the design. Shoring loads always terminate at a tieback or strut.

Once the system load requirements are known, the tieback can be designed. The factors that contribute to the strength of a tieback are the length (L), the shear strength of the soil or rock (psi or psf), the diameter of the drilled hole, and the strength of the steel tension members. In addition, a factor of safety should be included between 1.25 and 2.5 : 1.0. A good tieback design would closely match the strength of the tension member to the strength of the grout vs. soil/rock matrix. Also, it is assumed that the grout material matches or exceeds the strength of the soil/rock.

The strength of a tieback can be determined by combining the factors mentioned above into a formula. The formula, shown below, basically determines the contact surface area of the earth and the grout. First, the circumference of this contact zone multiplied by the length of the hole gives us the area of this contact surface in square feet. Next, the shear strength in psf is multiplied in order to determine the allowable load of the system. Finally, an FOS is divided into this load in order to obtain the *safe working load* (SWL) of the anchor.

$$P_{all} = L \times D \times t \times \pi \ / \ \mathrm{FOS}$$

where P_{all} = SWL of the anchor
$\quad\quad L$ = length of the anchor
$\quad\quad D$ = Diameter of the drilled hole
$\quad\quad\ t$ = shear strength of the soil or rock
$\quad\quad \pi$ = 3.1416

and FOS is between 1.25 and 2.5.

Most of the components of this formula are self-explanatory, except for shear strength of soil or rock. The values used for shear strength have been studied in great length. The shear strength of the soil is a main component of the tieback strength, which is shown in section in Figure 6.22. The failure plane is shown in this figure and where the bonding zone of the tieback is located.

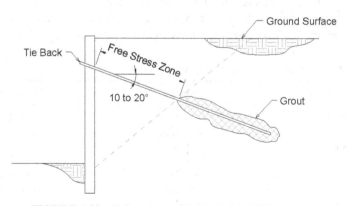

FIGURE 6.22   Soil anchor embedded behind failure plane.

FIGURE 6.23   Soldier beam shoring with tiebacks.

Shear Strength of Soil and Rock   According to theories from Mohr (1900) and Coulomb (1776), material fails from a combination of shear and normal stresses. Depending on the material type (cohesion and angle of internal friction values varying), the shear strength can fluctuate due to these stresses. Like everything else we have seen so far, when soil properties are different—whether granular, cohesive, or silty—all will produce different strength characteristics. Similarly, very soft to very hard soils will produce a range of different strengths the designer can rely on to estimate the strength of a soil/rock anchor or soil nail. Figure 6.23 shows two strand tiebacks used with a cantilevered soldier beam and lagging system for a road widening project.

For literally hundreds of years, engineers and scientists have been developing theories about soil strengths. The intention of this text is to minimize confusion and illustrate the basic concepts behind the subject. In this case, we need to understand what kind of shear strength one can expect from different soils and how one determines allowable loads on a system relying on these shear strengths. Table 6.4 shows

TABLE 6.4   Sample Material Strengths

Shear Strength	Approximate psf	Approximate tons/SF
Very hard rock	>5200	>2.6
Rock	3000–5200	1.50–2.6
Hard soil	3140	1.57
Stiff soil	1560–3140	0.78–1.57
Firm soil	840–1560	0.42–0.78
Soft soil	420–840	0.21–0.42
Very soft soil	<420	<0.21

a range of material strengths. This table is just a guide; actual strengths should be derived from field and laboratory testing.

Using the theory of soil nail design above, the following examples will work through the process of either determining soil anchor capacity or determining the required length of a soil anchor.

Example 6.3 Determine the allowable capacity of an anchor with the following characteristics:

P_{all} = ?

Shear strength of the soil is 1800 psf

Length is 30 ft

Diameter of hole is 4″

FOS = 1.5

P_{all} = (30 ft)(4″/12″/ft)π(1800 psf)/1.5

P_{all} = 37,700 lb (37.7 k)

Example 6.4 Determine the necessary length of a rock anchor requiring 10,000-lb capacity:

L = ?

Required P = 10,000 lb

Shear strength of the soil is 4000 psf

Diameter of hole is 2″

FOS = 1.25

10,000 lb = (L)(2″/12″/ft)π(4000 pfs)/1.25

Solving for L_{min}

L = (10,000 lb)(1.25)/(2″/12″/ft)π(4000 psf)

L_{min} = 5.97 ft long

In Chapter 5, the subject of slip planes was discussed, and it was mentioned that tieback lengths should be measured from behind the slip plane so that the unstable wedge of soil does not influence the failure of the tieback. It is also common practice in temporary and permanent design to sleeve the hole that passes through the failure wedge to avoid aiding this wedge into more failure. In the examples above, the distance L is the anchored length of the soil anchor. This concept is well illustrated in Figures 6.22 and 6.24. If the thickness of the failure wedge is added, then the drilled hole would be $L' = L+$ "wedge thickness." Figure 6.24 shows these various dimensional properties of the bonded and unbonded lengths.

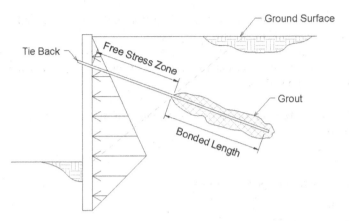

FIGURE 6.24   Bonded and unbonded zones of a tieback.

CASE STUDY: SOLDIER BEAM AND PLATE

While BART was being extended from Fremont to San Jose, California, there were many challenges to the excavations along some of the narrow right-of-ways. Several different support of excavations were employed. Along one 1000-linear foot long stretch, a soldier beam and plate system was put in place.

Equipment used is shown in Figure 6.CS1.

FIGURE 6.CS1   Soldier beam and plate system.

Plate installation is shown in Figure 6.CS2.

FIGURE 6.CS2 Plate installation.

FIGURE 6.25 Deep soil mixing equipment Presidio Project, San Francisco.

Other Support of Excavation Methods   This chapter has discussed one of the more common types of shoring systems when excavation depths do not exceed 30–40 ft. Figures 6.25–6.29 show other support systems for projects requiring excavation. These systems were not discussed in this chapter; however, they are frequently used in the industry. The next chapter will review sheet pile and walers in more detail.

FIGURE 6.26   Multilevel walers and struts supporting soldier beam and deep-soil mix wall.

FIGURE 6.27   Deep soil mixing with soldier beams, Owner, Valley Transit Authority (VTA).

FIGURE 6.28 Valley Transit Authority.

FIGURE 6.29 Soldier beam and lagging wall with tieback and no waler, Enloe Hospital, Chico, California.

CHAPTER 7

SHEET PILING AND STRUTTING

7.1 SHEET PILING BASICS

7.1.1 Materials

In this chapter, the materials discussed will follow the same allowable stress values as Chapter 6. Sheet pile systems are predominantly steel systems. The sheet pile are either hot or cold rolled steel, walers are WF beams or double steel channels, and struts are steel pipe, wide flange (W), or HP beams. As mentioned in Chapter 2, there are different types of steel for temporary and permanent construction. The grade of steel can range from 36,000 psi to over 80,000 psi. As in the previous chapters, A36 steel will be the most widely used steel type.

7.1.2 System Description and Unit of Measure

When describing sheet pile systems, the load path can help explain the components in the order in which the load travels. The soil pressure is applied to the sheet pile itself. In this case, as in most cases, the sheet pile acts as the sheeting material. It is the first contact between the soil and the system itself. Typically, the sheet piles are installed before the excavation begins. Therefore, the sheeting does not encounter the pressure until the excavation is deep enough to develop soil pressure at a particular depth. At this point in time, the sheet piling either has to be designed to support the soil pressure or the excavation stops in order to install a waler/strut system. Figure 7.1 illustrates the sequence of a typical excavation with one level of bracing.

Hot rolled sheet piling, shown in Figure 7.2A, is the more desired sheet pile for its strength and interlock tightness. As already mentioned, the cost for hot rolled sheets is slightly higher but that comes with better quality.

Drive Sheets   Begin Excavating   Finish Excavating
                      and install Bracing

FIGURE 7.1   Excavation sequence.

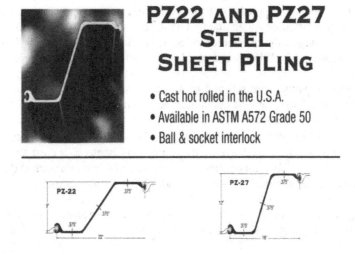

PZ22 AND PZ27 STEEL SHEET PILING

- Cast hot rolled in the U.S.A.
- Available in ASTM A572 Grade 50
- Ball & socket interlock

FIGURE 7.2A   Hot rolled sheet pile.

Sheet piling comes in two basic forms: hot rolled and cold rolled. Hot rolled sheets are formed when the steel is molten metal (liquid). The interlocks are shaped while the steel is formable. This makes the interlock tight, which means less water or soil can penetrate the interlock joint. The cold rolled sheets are bent from flat plate when the steel is cold. The interlocks are bent as shown in Figure 7.2B with machinery to create

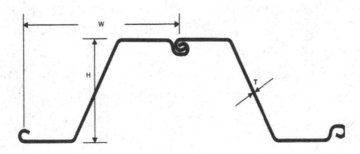

FIGURE 7.2B   Cold rolled sheet pile.

the interlock. These sheets are less expensive; however, they are not as watertight as the hot rolled sheets.

Sheet piles are quantified by the number of pairs to be driven and by the total pounds of steel to be purchased or rented. The contractor has to install each pair and remove each pair by the method chosen and described in the next section of this chapter. In addition, sheet pile are either purchased for $0.60–$0.90 per pound or rented for a monthly rate per pound. When estimating this type of work, the sheet pile liquidation cost should be added to the price of the sheet piling. Liquidation is the amount of sheet pile that gets damaged and cut off during installation and removal procedures. For example, a contractor rents 40-ft-long sheet pile. Then after the project is completed and the sheets are extracted, there is typically about a 2-ft section of sheet that is damaged due to the installation and extraction process. A damaged sheet pile is often unusable due to a damaged interlock. The 2 ft that becomes scrap metal is charged the liquidation price. It is seldom economical or feasible to repair the damaged 2 ft of the sheet and it is difficult to justify the labor cost.

Walers Walers (sometimes referred to as wales) are horizontal steel beams that support the sheet piling when the depth becomes too much for a cantilevered system. Walers can be any steel sections such as wide flange, HP shapes, or double channels at a minimum. If a strutting system is going to be used, then wide flange and HP shapes are common. If a tieback system (mentioned in Chapter 6) is used, then double channels will allow a space for the tieback rod or strand to pass through and load the waler concentrically. Figure 7.3 shows a double channel waler sytem.

Strutting Strutting accepts the load from the waler and transfers this load to the waler on the other side of the excavation or cofferdam. Struts are loaded axially,

FIGURE 7.3   Double channel waler system before excavation.

FIGURE 7.4 Sheet pile, waler, and strut system.

so they need to have substantial buckling resistance. Pipes are very popular as struts because they have the same radius of gyration 360° around. Square tubing offers the same symmetry with an increased cost. In comparison, a beam, which has a strong r_x and a weak r_y, always has to be checked for the weak axis or braced along the weak axis.

Figure 7.4 shows a three-level waler and strut system. The waler is a W24 wide-flange beam at 12-ft centers and the struts are 18-in-diameter pipe at 16-ft centers. These levels were placed as the excavation proceeded downward. Struts always pose an excavation production hindrance because the excavator has to work between them and take care not to make contact with them, which typically slows down production. Tiebacks (also discussed in Chapter 6) can be used in lieu of strutting. However, tiebacks can be much more expensive, and they disrupt the production of the excavation more because they take longer to install than struts. The upside to using tiebacks is that they leave the contractor more room to work within the excavation.

7.1.3 Driving Equipment

Although this is a design textbook, it is sometimes important to briefly discuss installation methods. Three methods of sheet pile driving are briefly covered here: vibratory, hydraulic impact, and diesel hammer impact.

Vibratory Z-shaped sheet piles are generally driven as pairs. This means that two Z sections are threaded together and driven as a pair (at the same time). The hammer secures and impacts the center of the pair, just above the interlock. The traditional vibratory hammer is the most common method of driving sheet piling, as long as the soil type is favorable. With fairly soft to medium compacted soils, sheet piles can

FIGURE 7.5 Pile driving operation using vibratory hammer.

be installed with good success. Figure 7.5 shows a pile driving operation using an Ape 200 vibratory hammer. There are also fewer chances of obstructions with soil of this type.

Hydraulic Hammer When the soil type is harder than a vibratory hammer can penetrate, the use of impact hammers becomes necessary. This method is only used when a vibratory hammer is not practical. With this method, more time is spent in the lining, driving, and cutting or "fresh heading" the sheets when damage occurs. Fresh heading means the top of the pile is damaged; therefore, the driving has to stop, the hammer is removed, and the pile driving crew cuts the bent part of the sheet pile in order to start with a flat, uncrimped top. This process is costly as it interrupts production of driving the sheets. However, if the operation continued to drive the damaged pile, even more time would be lost in production because the energy between the hammer and the pile would be lost.

Hydraulic hammers can be single and double action. This means the hydraulic ram either pushes one direction (down) or both directions (up and down). The impact the hydraulic hammer provides is not as violent as the diesel hammer mentioned next. The force that this type of hammer provides can also be regulated in order to minimize the damage to the top of the sheet pile or slow the hammer down in sensitive conditions. The double-acting hydraulic hammer has more strokes per minute, so even though it does not impact with as much force, part of this can be made up with more blows to the sheet pile per minute. The lower impact also lowers the chance of crimping the top of the sheet pile. The hydraulic hammer does require a hydraulic power pack to operate the system's hydraulic fluids. This becomes an extra piece of equipment to the operation.

Diesel and Steam Hammer The diesel and steam hammers are the last method that a pile driving crew would prefer to use. The diesel hammer can also

FIGURE 7.6   Sheet pile driving with an impact hammer.

be double acting. The time to loft the pile, place the hammer on the pile, and drive the pile takes much longer than the vibratory hammer and a bit more time than the hydraulic hammer. Figure 7.6 shows a sheet pile driving operation using a Delmag D36 diesel hammer. With the diesel hammer, fresh heading is very common and is probably necessary one to three times per sheet pair, depending on the soil type and experience of the crew. The impact force that the hammer supplies has a great deal of energy (up to 90,000 ft-lb); and if the hammer is not perfectly plumb, the top of the sheet pile will crimp, thus require fresh heading. Diesel and steam hammers do not require a hydraulic power pack to operate the hydraulics.

Templates   The quality of the installation of sheet piling ranges from not very important to very critical. If the excavation is designed to be much larger than the room needed to work on the permanent structure, then the location and vertical alignment of the sheeting is not that critical. However, if the sheeting is designed as a permanent structure, or there is little room between the permanent structure and the sheeting, then quality control can be very critical. When the latter is the case, a sheet pile template can be used, as shown in Figure 7.7. A template can be made many different ways. They can be multiple tiers tall or a single beam lying on the ground surface. External walers (which will be covered later) can be used as a single-tier template because they rest on the ground level and can be used as a template as well as function as a waler. This is similar to the method shown previously in Figure 7.5.

    The idea behind the template is that it is surveyed in and plumbed using precise instrumentation and then used to guide the sheets into their final location. The template needs to be designed strong enough that if the sheet pile bumped it, it would not lose line or plumb. This does not mean that it has to be designed by a certified engineer, but is should be checked by an experienced superintendent, project manager, or project engineer.

FIGURE 7.7   Two-tier driving template.

Driving Sheets in Hard Soils   Hard soils can be very difficult to drive sheet pile into unless the contractor takes one or some of the following measures:

Incorporate a heavier sheet pile section.

Jet the tip of the sheet pile with high pressure.

Penetrate the sheet pile line with a spud beam.

Use an impact hammer to drive the sheets.

Predrill the sheet pile line with a full flight auger, as depicted in Figure 7.8.

Preexcavate the inside of the cofferdam to reduce skin friction.

All of the above-mentioned methods are costly. The decision to use sheet piles in the first place was derived from other factors, and lowest cost is usually not one of the reasons. In many cases, sheet piling is the only option, so the goal is to make the system as inexpensive as possible.

Two types of designs will be illustrated in this chapter: the cantilevered sheet pile system and the braced sheet pile system. The soil loading has been previously discussed so the soil loading values used for the following examples will be given by the author, and it will be assumed that they were calculated using the methods in Chapter 5.

Cantilevered Sheet Piles   The cantilevered method for sheet piling simply means that there is no need for a waler/strut brace, and the sheets can support the pressure on the cantilever without failing in bending. A rule of thumb for excavation depths using the cantilevered method is between 10 and 18 ft, depending on the soil type and sheet pile size. A cantilevered system relies heavily on the amount of sheet embedded below the dredge line (bottom of excavation) and the soil type

FIGURE 7.8   Predrilling a sheet pile line in the Feather River.

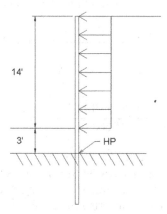

FIGURE 7.9   FBD of a cantilevered sheet pile.

as the passive pressure on the toe of the sheet pile. Because of this reliance, the passive shear strength of the soil is crucial as it pushes against the back of the sheet. Once embedded properly, the sheet pile is "fixed" at the hinge point, which, in this book, is 3 ft below the dredge line. From this point, the soil load is rectangular or triangular, depending on the soil type, and applies this uniform pressure to the back of the sheets similar to Figure 7.9.

Sheet Pile Design with a Cantilever   Sheet piles are like other sheeting materials in temporary structures because they are analyzed by a 12-in strip. The sheet pile tables that will be used for sheet pile design are in Appendix 5. These tables consist of lists of AZ and PZ sheet piling. The AZ and PZ designations are specific to the manufacturers, and these are just two of several manufacturers. The tables provided by these manufacturers basically have the same information categories, except the values are going to be different per sheet type. The following is the most common information needed about sheet piles:

The section name (PZ 22, AZ 17, etc.)

The sheet dimensions (width, depth, and thickness)

The weight per square foot or linear pile (psf, plf)

The section modulus (in^3)

Sometimes the section name matches another property of the sheet pile. For instance, the PZ sheet sections are PZ 22, PZ 27, and so forth. The 22 and the 27 represent the sheet's respective weight per square foot. This makes calculating weight very quick for the designer and estimator.

The first sheet pile design example will be a cantilevered system.

Example 7.1   Design a sheet pile with a uniform load of 1150 psf and cantilevered as shown in Figure 7.9. Add the excavation depth to the 3-ft hinge point to achieve the maximum cantilever.

Maximum moment at fixed end of sheet pile that can't rotate is $M = wL^2/3$:

$$M = \frac{(1150 \text{ ft})(14' + 3')^2}{3} = 110,783 \text{ ft-lb}$$

$$S_{min} = \frac{M_{max}}{F_b}$$

$$S_{min} = \frac{(110,783 \text{ ft-lb})(12''/\text{ft})}{21,600 \text{ psi}} = 61.55 \text{ in}^3$$

According to Appendix 5, the AZ 36-700(N) has a section modulus of 66.8 in^3.

This sheet would be acceptable, but at 34.61 psf, the steel costs may be higher than desired. On the other hand, if the alternative is adding a waler and strut system, an estimate would have to be done comparing the larger sheet costs to the lower sheet costs with the added waler and strut system.

Braced Sheet Piles   A braced system relies on the sheet strength but also is dependent on most of the forces transferring to the walers. In this type of system, the toe or embedment of the sheet pile is not as deep. This depth is from the subgrade/dredge line to the tip of the sheet. Figure 7.10 shows a sheet pile operation prior to the excavation being complete. The waler has been installed and the tiebacks are in progress. Once the tiebacks are stressed, the excavation can continue.

Braced Sheet Pile Design   The maximum moment of the sheet pile with two or more spans can be estimated using $M = wL^2/10$. This is the basic bending moment for beams with multiple supports. If there are only two supports, then $M = wL^2/8$.

FIGURE 7.10 Double channel waler system with tiebacks.

Example 7.2 Design a sheet pile with a uniform load of 1500 psf and supported as shown in Figure 7.11. Add the excavation depth to the 3-ft hinge point to achieve the maximum space between walers. Use this distance for L if it is the longest unsupported length. Since 12 ft + 3 ft is greater than any of the other spans of sheet piling, then $L = 15$ ft.

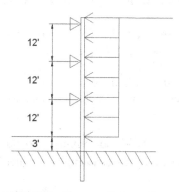

FIGURE 7.11 Supported sheet pile system.

The first step is to determine the maximum moment using the more-than-two support moment formula:

$$M = \frac{(1500 \text{ psf})(12' + 3')^2}{10} = 33{,}750 \text{ ft-lb}$$

$$S_{\min} = \frac{M_{\max}}{F_b}$$

$$F_b = 0.6F_y \qquad F_b = 21,600 \text{ psi}$$

$$S_{min} = \frac{(33,750 \text{ ft-lb})(12''/\text{ft})}{21,600 \text{ psi}} = 18.75 \text{ in}^3$$

According to Appendix 5, the AZ 12 has a section modulus of 22.3 in³.

The PZ 27 has a section modulus of 30.2 in³ (the PZ 22 is slightly under the 22.3 in³ with 18.1 in³).

It becomes apparent that supported systems can use lighter sheet piling even when the loading is greater. This is mostly due to the reduced bending moment. This same example with only a top waler and the lower hinge point would produce a larger bending moment; but without doing a calculation, it is difficult to know by how much. If the excavation was the same depth, this would pose a large challenge because L would become $36' + 3' = 39$ ft. However, the bending moment would then be $M = wL^2/8$.

Rework the previous example with the following new information:

$$M = \frac{(1500 \text{ plf})(39 \text{ ft})^2}{8} = 285,188 \text{ ft-lb}$$

$$S_{min} = \frac{(285,188 \text{ ft-lb})(12''/\text{ft})}{21,600 \text{ psi}} = 158.44 \text{ in}^3$$

Appendix 5 does not have a sheet pile with such a large section modulus. There are several choices at this point:

Redesign with a smaller sheet pile combined with a steel pipe pile also know as a combi-wall.

Lower the upper waler to reduce the bending moment on the sheet.

Add at least one row of struts and walers.

It should be noted that the methods of determining the maximum moment in the examples above are somewhat conservative. If a more accurate moment is required, then the designer should:

Employ engineering methods for indeterminate structures.

Employ a beam software program as mentioned earlier in this text.

Walers   Walers are the structural members that transfer the loading from the sheet pile to the struts or tiebacks. The waler design is dependent on the uniform load, which is determined by its spacing and the P_{max} load from the soil. The further the walers are from one another, the larger the linear load. The greater the soil pressure, the larger the linear load as well. Figure 7.12 shows the amount of load each waler is supporting. The linear load on the L_2 waler is determined by the soil load times the uniform width of the L_2 waler. The design on the waler is also dependent on the spacing of the struts because the struts are the supports of the walers and dictate L. A waler can be a simply supported beam if it only has two supports, one at each end.

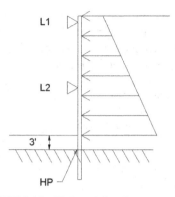

FIGURE 7.12 FBD multilevel support system.

A waler can also be a continuous beam with multiple supports (struts). Whichever is the case, the linear load will have to be determined.

Since walers make continuous contact with the sheet pile, lateral buckling is usually not a concern. The two are usually welded in some fashion to temporarily hang the waler and keep it from falling into the cofferdam before the soil pressure compresses it with the struts.

In the sketch shown in Figure 7.12, assume the distance between the L_1 waler and the L_2 waler is 12 ft and the distance between the L_2 waler and the HP at the bottom is 16 ft.

The tributary width on L_2 is half the distance to L_1 plus half the distance to the HP:

$$L = \tfrac{1}{2}(12') + \tfrac{1}{2}(16') = 14 \text{ ft}$$

If the average uniform load from the soil pressure is 1200 psf, the linear load can be determined by

$$W = 1200 \text{ psf} \times 14 \text{ ft} = 16,800 \text{ plf}$$

Once the linear load on a waler is computed, the first part of the waler design is complete. The second part is determining the support condition of the waler. As mentioned above, it can be simply supported if the shoring wall is not very long, or it can be a continuous beam with multiple supports, which in this case would be struts or ties (Figure 7.13). As we have learned, with the same uniform load (W) and the same unsupported lengths (L) the maximum moment of [7.3(a)] a simply supported

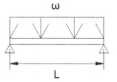

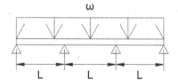

A) Simply Supported   B) Continuous

FIGURE 7.13 Support conditions for a waler.

beam and [7.13(b)] a continuous beam with multiple supports is different, and the continuous beam will always be less if the conditions described are true.

Using the linear load calculated above and $L = 18$ ft, determine the bending moment of walers in Figure 7.13(a) and 7.13(b).

[7.13(a)]
$$M = \frac{wL^2}{8}$$

$$M = \frac{16,800(18)^2}{8} = 680,400 \text{ ft-lb}$$

[7.13(b)]
$$M = \frac{wL^2}{10}$$

$$M = \frac{16,800(18)^2}{10} = 544,320 \text{ ft-lb}$$

Example 7.3   Using the case in Figure 7.13(b), determine the minimum size L_2 waler (16,800 plf) that is required for bending. Assume the waler is attached to the sheet pile continuously and lateral buckling is not an issue.

$$S_{min} = \frac{M_{max}}{F_b}$$

$$F_b = 0.6F_y \qquad F_b = 21,600 \text{ psi}$$

$$S_{min} = \frac{(544,320 \text{ ft-lb})(12''/\text{ft})}{21,600 \text{ psi}} = 302.4 \text{ in}^3$$

Refer to Appendix 1 and select three beams that have a section modulus equal to or greater than 302.4 in^3. Some beams that may work here are W18 $\times$ 158, W21 $\times$ 147, W24 $\times$ 131 and W30 $\times$ 129.

Internal Walers   Internal walers are the most common type of waler in cofferdams, whether they are on dry land or in a marine environment. The sheet piles are loaded from the outside, inward. So it makes sense to allow the structural components to push against each other. The size of the waler determines the spacing of the struts that support the waler. Another way to look at this is to say the strut spacing determines the waler size. Like most decisions in temporary structures, the spacing and sizes of materials makes up a large portion of the total cost of the system. The larger the members, the less number of pieces required. On the other hand, the smaller members have to be spaced closer together, thus requiring more pieces to complete the system. The final decision can't be made until a complete quantity takeoff is performed and a cost analysis done for each option including materials, labor, and equipment. Then, at this time, an educated decision can be made.

Internal walers can be used as driving templates and then converted to a waler once the sheets are in place. Figure 7.14 shows a river cofferdam. The additional pile (pin pile) inside the cofferdam are used to temporarily support the template (waler) during the sheet pile driving process. Once the sheets were driven, the template is attached to the sheet pile and the pin piles are removed. The template now becomes the waler and the excavation can begin.

FIGURE 7.14 Internal waler in a river cofferdam.

The downside to internal walers is that the waler takes up more room inside the cofferdam. The cofferdam usually needs to be larger because the waler must clear the permanent construction. A larger cofferdam means there will be more excavation and backfill quantity and also means at least four more pairs of sheet pile (one for each side = 3–4 ft) have to be driven and removed.

External Walers External walers are used when there is only support needed at the top of the sheet pile and the sheet pile can handle the bending stress from the top (OG) to the hinge point, 3 ft below the dredge line (subgrade). Figures 7.15 and 7.16 are two examples of external waler systems.

The clamps that attached the top of the sheet to the waler can be a bridge style C clamps shown in Figure 7.15, a section of steel plate cut into a C shape (pac man) as shown in Figure 7.16, or even a coil rod installed into a burned or drilled hole through the sheets and walers. The first two options cause less damage to the steel beams and sheet pile. Whenever possible in temporary construction, damage to the steel should be minimized or eliminated all together. When steel is cut unnecessarily, one or more of the following occur:

The steel has to be repaired if possible.

The member can't be used in the future to its maximum design potential.

The member is shorter than its original length.

The contractor has to pay liquidation costs.

For the same reason an internal waler causes a larger cofferdam, the external waler system allows a smaller cofferdam when there is only one top level waler. The smaller cofferdam reduces the earthwork and sheet pile quantities, thus making the excavation, as a whole, more economical. Also, the excavation crew does not have to work

FIGURE 7.15 External waler system using bridge style C clamps.

FIGURE 7.16 External waler using plate clamps.

around the internal waler and risk damaging or compromising its strength. In addition, while the permanent structure is being built, the trades have clear access with equipment and ladders. Finally, if there is a concrete foundation to be placed inside the excavation, the concrete can be poured against the sheet pile, thus eliminating additional internal support and the cost of formwork and back filling within a small space (between the footing and sheet pile). If needed, a strut can still be installed with

FIGURE 7.17 Strut supporting walers.

external walers as long as the struts make contact with the walers and not just the sheet pile. Figures 7.15 and 7.16 show how difficult it would be to add an internal strut.

Struts So far in this text we have looked at some axial loads on columns and struts. Figure 7.17 shows a very short strut used in a sheet pile cofferdam system. We know that the design of a strut is determined by:

The axial load
The unsupported length of the strut
The weakest radius of gyration
The end connections

With all of the information above, a strut can be designed for any system. There are cases where practicality tells the designer that the strut is extremely long and either can't carry the axial load without buckling or needs to be supported for its own weight and weak direction axis. These problems can be overcome with proper design considerations.

In Chapter 2, column and strut design was introduced. In this section, the application learned earlier will be applied to a practical case. The first challenge when designing struts is to determine the axial compressive load on the strut. This can be accomplished with one of three methods:

Using the linear waler load per foot
Calculating the pressure within the tributary area of one strut (see Figure 7.18)
Adding the reactions of the free-body diagram from each side of the strut

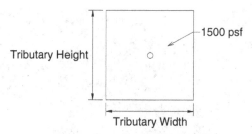

FIGURE 7.18   Tributary area of a single strut.

Example 7.4   Using the load in the previous waler example, determine the load on a strut if they are spaced at 18 ft on center.

No matter which method is used, the same result should occur.

Method 1:   Using the linear waler load from the previous waler example, multiply this linear load times the strut spacing:

$$P = 16,800 \text{ plf} \times 18 \text{ ft on center (OC)} = 302,400 \text{ lb}$$

Method 2:   Using the design load of 1200 psf, multiply this times the tributary area on one strut. $P = P_{max} \times$ Trib. width (strut spacing) $\times$ Trib. height (waler spacing):

$$P = 1200 \text{ psf} \times 14 \text{ ft} \times 18 \text{ ft} = 302,400 \text{ lb}$$

Method 3:   Add the left reaction to the right reaction from the free-body diagram:

$$151,200 \text{ lb} + 151,200 \text{ lb} = 302,400 \text{ lb}$$

These three methods should match; if they do not, a mistake has been made. Now that the axial load has been determined, the unsupported length of the strut is needed. This is usually achieved by calculating the cofferdam size and subtracting the waler depth on each side. In other words, if the cofferdam is 30 ft across, sheet pile to sheet pile, and a W24 is used for the waler system, then the strut length would be approximately $30' - 2' - 2' = 26$ ft long. This length will be used for the rest of this example. Figure 7.19 shows struts used in a floodplain cofferdam.

To determine the size strut required, the trial-and-error method can be used. This means that the designer will choose a steel member, either a beam or a pipe and check to see if it works. As in Chapter 2, Appendix 4 will be used. This table has precalculated F_b (ksi) values using a factor of safety and A36 steel with $E = 29,000$ ksi. The first trial and error will be for an HP beam and the second will be with steel pipe. These values can be found in Appendixes 2 and 3. However, if the pipe used is larger than those shown in Appendix 2, the value will have to be calculated using common formulas for r and A. The following steps will be used:

1. Assume $r_y = 3.0''$, weak direction radius of gyration. This is an average of a range of beams that are commonly used as struts or columns.

FIGURE 7.19 Struts used in a floodplain cofferdam.

2. Using $r_y = 3''$ and $k = 1.0$, calculate the slenderness ratio (SR) of the strut (SR = kl/r).

3. Using Appendix 4 for allowable buckling stress, determine the maximum buckling stress of the H pile beam.

4. Using $f = P/A$, calculate the required (A)rea of the cross section.

5. Find an H pile section that is close to this required (A)rea in Appendix 3.

6. Recalculate the new SR with the actual r_y.

7. Go back to Appendix 4 for the new, actual F_b (ksi).

8. Using $f = P/A$, look up the area of the new strut in Appendix 3, and calculate P allowable.

9. Compare P_{all} to P_{actual}.

If $P_{all} < P_{act}$, repeat the last five steps (5–9). When $P_{all} > P_{act}$, the strut works.

This method will always find a strut that is adequate for the load, but it is not guaranteed to find the best or most economical strut. In step 5, when the designer is looking for a strut to match the minimum area required, there are several beams from which to choose. Since the area is used just to get the designer close, there are cases when a slightly lower area will produce an adequate strut and be more economical because a lower area will always be lighter per linear foot. Since steel is purchased by the pound, economy means less steel weight.

Example 7.5 Using the method described above, find an HP and a pipe strut that work to support 303 k with $L_e = 26'$.

H Pile Strut

Assume $r_y = 3''$, $k = 1.0$

SR $= [(1.0)(26' \times 12''/\text{ft})]/3.0''$ SR $= 104$

Appendix 4, $F_b = 12.47$ ksi

Using $f = P_{\text{act}}/A$ and $P_{\text{act}} = 302$ k, solve for A; $A = 303$ k$/12.47$ ksi $= 24.3$ in^2.

Select an HP shape in Appendix 3 with an area close to 24.3 in^2, say HP 13 $\times$ 87 (25.5 in^2)

HP 13 $\times$ 87 has an $r_y = 3.13''$ New SR $= [(1.0)(26' \times 12''/\text{ft})]/3.13''$ SR $= 100$ F_b for a 26-ft-long HP 13 $\times$ 87 is 12.98 ksi

Using an HP 13 $\times$ 87, $A = 25.5$ in^2, $P_{\text{all}} = 12.98$ ksi $\times$ 25.5 in$^2 = 331$ k > 303 k (OK)

Steel Pipe Strut

When the designer is using steel pipe, the pipe properties have to be available. Sometimes this is not the case, so the properties have to be achieved using statics. The following is a review of the area (A) and radius of gyration (r) of a cylinder, where D_o (OD) is the outside diameter and D_i (ID) is the inside diameter.

One can determine the area and radius of gyration of a cylinder by using the following formulas:

$$A = \tfrac{1}{4} \pi (D_o^2 - D_i^2)$$

$$r_x = r_y = \tfrac{1}{4}\sqrt{D_o^2 + D_i^2}$$

Let's look at an $18'' \times \frac{1}{4}''$ wall pipe strut as shown in Figure 7.20 to support the long sides of this cofferdam. If a large pipe is designated as being $18''$ in diameter, the $18''$ dimension is the outside diameter. Therefore, the inside diameter is $18'' - \left(2 \times \frac{1}{4}''\right) = 17.5''$. Using this information, the area and radius of gyration can be determined:

$$A = \tfrac{1}{4}\pi \,[(18)^2 - (17.5)^2] = 13.94 \text{ in}^2$$

$$r_x = r_y = \tfrac{1}{4}\sqrt{(18)^2 + (17.5)^2} = 6.28 \text{ in}$$

Trial and error is more difficult if the designer does not have a list of the pipe properties. For example, with pipe, a radius of gyration assumption of 3.0 would be very low. As you can see, the pipe chosen above has a radius of gyration of $6.28''$. This is more than double the assumption in the previous example. Instead of the trial-and-error method used above, let's determine the capacity of this pipe and adjust from there.

Determine the capacity on an $18'' \times \frac{1}{4}''$ wall pipe.

SR $= [1.0 \times 26' \times 12''/\text{ft}]/6.28 = 50$

Using Appendix 4, $f_b = 18.35$ ksi

FIGURE 7.20 Aerial view of a strutted cofferdam.

If $f = P/A$, then $P_{all} = 18.35$ ksi $\times 13.94$ in^2

$P_{all} = 256$ k

If the design called for a minimum capacity strut of 303 k as previously used, this A36, $18'' \times \frac{1}{4}''$ wall pipe would not be acceptable. The radius of gyration of this pipe is substantial, but the cross-sectional area is small. Compare the normal stress between the actual strut load and the actual capacity of the pipe:

$$f_c = \frac{303 \text{ k}}{13.94 \text{ in}^2} = 21.74 \text{ ksi}$$

$$f_c = \frac{256 \text{ k}}{13.94 \text{ in}^2} = 18.36 \text{ ksi}$$

This normal stress with either case is at or below the bending limit of 21.6–24 ksi and below the compression limit of 28.8–32.4 ksi.

Determining the capacity of an axial loaded column or strut is another way to use material that is available to the contractor. However, with this method, the available strut will determine the spacing and the waler size. Therefore, this is not the preferred method because most of the money is in the sheet pile and waler design. This method is desirable when a designer just wants to know what a particular column or strut capacity is. To determine the maximum axial capacity of a standard 12'' diameter, 26-ft-long pipe, use Appendix 2 and follow these steps:

1. Find $r = 4.38''$.
2. Find $A = 14.6$ in^2.
3. Select the strut member 12'' standard from Appendix 2.

4. $SR = [1.0 \times 26' \times 12''/\text{ft}]/4.38 = 71$.
5. Go to Appendix 4, $f_b = 16.33$ ksi.
6. If $f = P/A$, then $P_{\text{all}} = 16.33$ ksi $\times 14.6$ in^2.
7. $P_{\text{all}} = 238$ k

The conclusion is that a 12″ standard pipe with $l_e = 26$ ft can support an axial load of 238 k. If this pipe was supported with a brace, say in the center of its length restricting movement 360°, the capacity would change. The designer would have to repeat steps 1 through 4 above and change the 26 ft in step 4 to 13 ft.

────────

Web Yielding   In some of this chapter's photos, one might notice plates welded against the beam web between the top and bottom flanges. These are stiffener plates that are added when the strut (or any support member) force concentrates enough load on the beam or waler that the web of that same beam yields (collapses) as shown in Figure 7.21.

Allowable web yielding stress is a percentage of the steel's yield strength, usually 75–80%. Using 75%, the allowable yield stress for A36 and grade 50 steel is as follows:

$$F_{\text{wy}} = 0.75\,F_y$$

$$\text{A36: } F_y = 0.75\,(36\text{ ksi}) = 27\text{ ksi}$$

$$\text{A572, Gr. 50: } F_y = 0.75\,(50\text{ ksi}) = 33\text{ ksi}$$

Web yielding stress is calculated by projecting the concentrated load into the top flange of the supporting beam at a 2.5 : 1 angle each direction. Figure 7.22 shows this

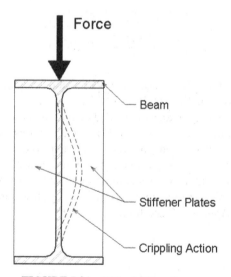

FIGURE 7.21   Web yielding diagram.

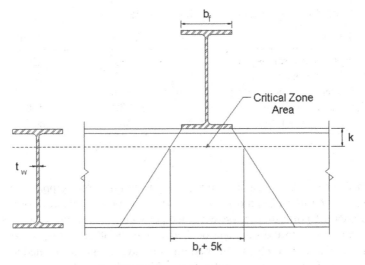

FIGURE 7.22   Web yielding zone.

2.5 : 1 projection from the bottom flange of the top beam downward in both directions to the bottom beam. These two projected lines cross the imaginary horizontal line k distance from the top of the bottom beam. The k value represents the distance from the top of the beam to where the web thickness begins to change (at the tangent) and can be found in the AISC steel manual under beam properties or Appendix 1. Because the projected 2.5 : 1 line slopes each direction, the horizontal distance is 5 times k.

The zone of concern is along the k horizontal line times the web thickness, which is the area in square inches. In order to find the length of this area, the bottom flange of the top beam should be known. This is also found in Appendix 1 as the b_f dimension. The web thickness of the lower beam is t_w.

Therefore,

$$f_{wy} = \frac{P}{(t_w)[b_f + (5k)]}$$

where P = concentrated load from the beam above
   t_w = thickness of the lower beam's web
   b_f = flange width of the upper beam
   k = the lower beam

Assuming a load (P) of 133 k from a top W21 × 147 onto a lower HP 14 × 89, determine the web yielding stress of the HP 14 × 89:

$$F_{wc} = \frac{133 \text{ k}}{(0.615'')[12.51 + (2 \times 1.31'')]} = 14.3 \text{ ksi}$$

14.3 ksi < $0.75 F_y$

14.3 ksi < 27 ksi   (OK)

TABLE 7.1 Shoring Costs for Different Systems

System Type	Low $/ft^2	High $/ft^2
Shotcrete and soil nailing	$15	$20
Soldier beam and plate	$15	$25
Soldier beam and wood lagging	$18	$28
Sheet pile, walers & struts	$25	$50
CDSM (Cement Deep Soil Mixing)	$20	$50

System Costs When costs are compared between different supports of excavation systems, the unit of comparison is cost per "exposed" square foot (SF). The exposed area (SF) of a shoring system is the perimeter in feet times the depth from the original ground to the subgrade or dredge line. In other words, the amount of sheet pile or soldier beam below the bottom of excavation is not counted toward the square footage. For example, if a cofferdam is 22 ft wide by 46 ft long and 19 ft deep, the square foot of exposed surface is perimeter length times the excavation depth or $[(22 \times 2)' + (46 \times 2)'] \times 19' = 2584$ ft^2. Table 7.1 gives the reader a rough idea of shoring costs for different systems. Actual costs should be calculated using actual project parameters. Even though some costs are much higher, these systems may be chosen because the less expensive system is not practical. This could come from the depth, water tightness requirements, or existing conditions.

CASE STUDY: SHEET PILE COFFERDAM

In San Jose, California, the WWTP was being expanded in 2006. Because of the tight job site conditions, much of the excavation required shoring so that real estate was not lost from open-cut excavations. The project even used a tower crane on rails to eliminate having to have enough room for crawler and rough terrain type cranes.

One particular excavation was over 38 ft deep. The excavation required a long-reach excavator, which is very slow when it comes to production. Also, because even the long-reach excavator could not reach the whole excavation, a smaller excavator had to be hoisted into the excavation to complete the subgrade and help place the aggregate base rock at the bottom. Figure 7.CS1 shows this excavator being hoisted into the excavation.

The excavation was eventually a three-tier waler and pipe strut supported excavation. Figure 7.CS2 shows this same excavator working at the bottom of the excavation, feeding the long-reach excavator material to remove.

FIGURE 7.CS1   Sheet pile cofferdam (courtesy of Kiewit Infrastructure West).

FIGURE 7.CS2   Sheet pile cofferdam (courtesy of Kiewit Infrastructure West).

(continued)

CASE STUDY: SHEET PILE COFFERDAM (*Continued*)

When the excavation was complete, the walers were also used to support the dewatering pumps and discharge lines (Figure 7.CS3).

FIGURE 7.CS3   Sheet pile cofferdam (courtesy of Kiewit Infrastructure West).

CHAPTER 8

PRESSURE AND FORCES ON FORMWORK AND FALSEWORK

The committee that oversees formwork design is the American Concrete Institute Committee 347 (ACI 347) and is made up of 30–40 members. Its findings and decisions are recorded and published in *Formwork for Concrete* by M. K. Hurd. This text is in its ninth edition in order to keep up with the advancements in technology and materials. This same textbook has been used as the main reference for these writings.

For these next three chapters, the term formwork will refer to (1) vertical forms supporting the hydrostatic pressures of the unhardened concrete and (2) horizontal forms and components supporting the gravity dead loads from the concrete and live loads from personnel, equipment, and tools.

8.1 PROPERTIES OF MATERIALS

8.1.1 Unit Weights

The unit weight of concrete will be 150 pcf for hydrostatic conditions. For falsework, the unit weight of concrete will be increased to 160 pcf to include additional weight of the reinforcing steel and formwork, not including steel beams.

The additional 10 pcf from reinforcing steel, form weight, comes from an average amount of reinforcing used in structural concrete designs. Here are some sample calculations when determining this additional weight, which begins with an estimate of how many pounds of rebar are in one cubic yard of concrete. The additional 10 pcf includes reinforcing. Table 8.1 shows sample weights to provide a guide in case the reinforcing steel is excessive and the engineer needs to add more weight for unusual circumstances.

From Table 8.1, one can adjust the unit weight of concrete on falsework due to the amount of steel per cubic yard if 10 pcf is exceeded. This should be checked with a

TABLE 8.1 Rebar Weight per Cubic Yard $[CY(yd^3)]$ of Concrete

Amount of Rebar	Pounds of Rebar (lb/CY)	Pounds per Cubic Foot (pcf)
Mildly reinforced	150	5.6
Light	175	6.7
Medium	200	7.4
Heavy	300	11.1
Very heavy	>500	>18.5

registered engineer, but in no case would the 160 pcf be reduced. A simple quantity takeoff can be performed to determine how much reinforcing steel is actually in the elevated slab or supported concrete element. A complete takeoff is not necessary. Simply select a typical section $(20' \times 20')$ and calculate the weight of the steel in that 20×20 section, and then divide that number by the volume of concrete in the same section.

Example 8.1 Determine the amount of steel in a $20' \times 20'$ section of elevated slab with a thickness of 12 in and No. 6 bars top and bottom at 12" on center, each direction. The No. 6 bars weigh 1.502 lb/lf per Table 8.2.

TABLE 8.2 Reinforcing Steel Weight

Bar Size Designation (No.)	Nominal Bar Dimension (lb/ft)	Nominal Weight (lb/ft)
3	0.375	0.376
4	0.50	0.668
5	0.625	1.043
6	0.75	1.502
7	0.875	2.044
8	1.00	2.670
9	1.128	3.400
10	1.270	4.303
11	1.410	5.313
14	1.693	7.650
18	2.257	13.60

Source: Erico International Corporation (Lentun®), 34600 Solun Road, Solun OH 44139.

Reinforcing Steel:

Top 20 ft long × 21 bars × 2 directions × 1.502 plf = 1262 lb

Bottom 20 ft long × 21 bars × 2 directions × 1.502 plf = 1262 lb

Total = 2524 lb

Volume of Concrete:

$$20' \times 20' \times (12''/12''/\text{ft})/27 \text{ ft}^3/\text{yd}^3 = 14.81 \text{ yd}^3$$

$$\text{Amount of steel} = 2524 \text{ lb}/14.81 \text{ yd}^3 = 170.4 \text{ lb}/\text{yd}^3$$

Based on the figures above, the unit weight for this slab would be $150 + 6 = 156$ pcf. Since the value is less than 160 pcf, the value used for weight would remain 160 pcf.

8.1.2 Forces from Concrete Placement

In this chapter, pressure formulas will be used and referred to as formula 1 and formula 2. These two formulas will be described in detail at that time.

Hydrostatic Pressure on Formwork The pressure that is imposed on vertical forms can never be overestimated. Since concrete weighs approximately 150 pcf (varies with mix design), the forces that are generated while it is in its liquid state can be extremely high. Concrete remains in a liquid state for a short period of time, but during this period, due to its unit weight, it is very heavy and can produce very high lateral pressures on formwork. Concrete is also unique to soil and other construction loads as it is liquid for a relatively short period of time before it begins to take an initial set and eventually (after a few hours or days) is able to support itself. The initial set of concrete occurs when the concrete is no longer imposing lateral pressure on forms. In terms of falsework loads, the initial set time does not lessen the forces of gravity on a falsework system. In this case, the concrete must support its own dead load before forms can be removed. Other solid materials, such as wood and steel, maintain the same forces throughout its useful life. Even soil, other than when the water table affects its stability and unit weight, is a relatively consistent force.

When the hydrostatic pressure of concrete is in question, several factors have to be considered. Among these is the ambient air temperature and concrete temperature at the time of the pour, the rate at which the concrete is placed, the mix design of the concrete including its slump, the way in which it is placed and consolidated, and the height of the placement. Let's look at each one of these factors individually.

Concrete Temperature and Ambient Air Temperature For formwork design, temperature has a large effect on the lateral pressure on formwork. The warmer the temperature, the faster the concrete will take its initial set. To the contrary, the colder the temperature, the longer it takes the concrete to set. It is common for form failures to take place in the winter months. This is because the form designer may have designed the forms in the warmer months and underestimated the temperature at the time of anticipated construction. If the form designer designed the forms in August and the ambient temperature at that time was 90°F, he may have used 60–80°F as the controlling temperature in formula 1 or 2. However, if the work is performed in January and the ambient temperature is recorded below 30–40° this could bring the concrete temperature down below 50°F, which would put more pressure on the formwork than was originally planned.

When temperatures are expected to be either very high or very low, batch plants take measures to reverse the effects of either one. For instance, ice or liquid nitrogen can be added to the mix to cool down the concrete temperature. If the concrete is being placed in cold weather conditions, the batch plant can add hot water to the mix to attempt to raise the temperature of the concrete. In either case, the form designer should consider these attempts at temperature manipulation and how they will affect the lateral pressure of the concrete. These effects are also increased when adverse temperatures are combined with the use of flyash or retardant admixtures. Flyash is a by-product of the coal firing process. The finely divided residue is used in ready-mixed concrete and can replace a certain percentage of the specified cement content by weight. Many mix designs will accept up to 40% of the cement content as flyash. Admixtures are chemicals added to a mix for various reasons. Two of the reasons are better workability and slower set time.

When ACI 347 discusses temperatures for formula 1 or 2, it references the "concrete temperature in Fahrenheit degrees in the forms." This is to mean that the ambient air temperature could be 60°F, but the concrete in the forms is over 90°F. This poses two problems: (1) the temperature in the formula would lower the maximum pressure possibly more than ACI 347's intentions, and (2) most concrete mix design specifications do not allow concrete to get warmer than 80–90°F. If the concrete temperature is going to reach temperatures above the specifications, the concrete supplier or contractor must take measures to lower the temperature. In this case, the form designer will consider this and make appropriate adjustments.

CASE STUDY: THE BENICIA–MARTINEZ BRIDGE—CONCRETE TEMPERATURE

The Benicia–Martinez Bridge project (2000–2007) required mix designs with up to 11 sacks of cement per cubic yard of concrete and a design strength of 11,000 psi. This amount of cement was unprecedented. The concrete temperature was exceeding the specification limitations and peaked at approximately 190°F (water boils at 212°F at normal atmospheric pressures). Excessive temperatures during the concrete curing process are known to shorten the long-term strength and durability of the concrete. Excessive variances in temperatures within a single mass could also cause significant thermal cracking and reduction in quality. The project required that cooling measures be installed to prevent the concrete from reaching a maximum temperature threshold, as well as a maximum temperature variance. Measures included injecting liquid nitrogen at the batch plant to precool, and cooling tubes were installed within the structure to continue even cooling during curing. Cooled water was pumped through the cooling tubes to keep the temperature as low and consistent as possible. Even with those measures in place, the concrete temperature was still recorded to peak above 140°F. The following Figures 8.CS1 and 8.CS2 illustrate this project under construction and completed.

The State of California and the contractor eventually had to accept this "best attempt" at lowering the temperature and maintaining the temperature

FIGURE 8.CS1   Benicia–Martinez Bridge under construction looking from Benicia, California (courtesy of Kiewit Infrastructure West).

FIGURE 8.CS2   Benicia–Martinez Bridge completed looking from Martinez, California (courtesy of Kiewit Infrastructure West).

after placement so the differential between the outside temperature and the concrete was not excessive. To monitor how well these attempts worked, the contractor employed real-time measurement of the concrete temperature from the moment of batching, throughout the initial concrete curing phase,

(continued)

when temperatures reach their peak. Once concrete was placed in the truck, handheld thermal-sensing devices were used to check the temperature during transport. Within the concrete structure, thermocouple wiring was installed within the rebar framework and connected to recording devices. These devices monitored and recorded the temperatures at multiple locations within the poured structure. Workers adjusted the water flow and temperature as necessary to maintain proper cooling on a 24/7 basis. If any recorded temperatures or variances exceeded the maximum recommendation, the contractor and the state performed an engineering analysis to decide whether that portion of concrete could remain in place.

Concrete cooling work at the Benicia Bridge required over 1.2 million feet of PVC pipe and over 700,000 ft of thermocouple wiring.

Now if the form designer on the Benicia–Martinez Bridge had used 140° in its maximum pressure calculations (either formula 1 or 2), P_{max} could be less than 600 psf, the forms would be well underdesigned, and there would be a high potential for disaster. The design pressure should be discussed and agreed upon with a registered engineer on what temperature to use in the pressure calculations.

Placement Rate   Placement rate is how fast the concrete is placed and vibrated in the forms. The value is measured in vertical feet per hour of placement and designated by R. Placement rate is typically decided on and controlled by the contractor. This decision is communicated to the form designer for accurate pressure calculations. It would not make sense for the designer to design the forms for a pour rate of 4 ft per hour and the contractor to place the concrete at 8 ft per hour because the forms would be overloaded and bending, shear, and/or deflection on the forms would be jeopardized. On the other hand, it would not make sense for the form designer to design at a rate of 6 ft per hour when the contractor can only place, at best, 4 ft per hour. In this case, the forms would be overbuilt and not as economical ($/SF).

Placement rate is controlled by many factors. A few of these factors will be discussed here. Access to the forms or to the pump is crucial. The faster the mixer trucks can get to the pump, the faster the pump can provide concrete to the forms. If the contractor can continuously keep two trucks at the pump for the duration of the pour, more concrete can be supplied to the forms at a faster rate. To do this, there has to be plenty of space for the two trucks and room for turning around. Sometimes small earthen ramps at the pump's hopper can raise the mixer trucks discharge just enough to speed up the flow of the dispersed concrete. Crane and bucketing concrete, when pumps are not economically or geometrically feasible, is an option and controls the rate at which concrete is placed. Buckets range in size from 1 to 4 yd^3 and larger. A cubic yard of concrete weighs approximately 4050 lb. Therefore, crane capacities for this method should be substantial since a 4-yd^3 bucket full of concrete could weigh over 17,000 lb (8.5 tons).

The thickness of the elements being placed will have a very large effect on how fast the placement rate is. For instance, an 8-in-thick wall will pour much quicker in vertical feet per hour than a 24-in-thick wall. Let's compare the two walls if they were both 12 ft tall and 40 ft long:

$$\text{Wall 1: } 8'' \times 12' \times 40'$$

$$\text{Volume} = 11.9 \text{ CY}$$

The amount of vertical feet placed by a 9.5-CY load of concrete $9.5/[(0.67' \times 40')/27 \text{ CF/CY}] = 9.5$ ft
This would result in less than half a truckload to place at 4-ft per hour rate.

$$\text{Wall 2: } 24'' \times 12' \times 40'$$

$$\text{Volume} = 35.7 \text{ CY}$$

The amount of vertical feet placed by a 9.5-CY load of concrete $9.5/[(2.0' \times 40')/27 \text{ CF/CY}] = 3.2$ ft
This would result in a little over one truckload to place at 4-ft per hour rate.

As you can see from this example, the thickness of the element is crucial to the estimated placement rate. This puts a huge responsibility on the placement crew and the engineer ordering the concrete. The concrete delivery speed must meet the placement rate so there are no mistakes in this rate, and forms are not overloaded per the design.

Mix Design   Specifications will go into great lengths regarding the components and their quantities in a cubic yard of concrete. Cement types, aggregate sizes, and admixtures are among a few specifications that can change the unit weight of concrete and the lateral pressure on forms. ACI 347 makes reference to these items and their effects on form design.

The design formulas used in this chapter assume type I cement is used, the slump is less than or equal to 4 in, internal vibrators are used, and no pozzolan or other retarding admixtures are used. Pozzolan can possess cementitious qualities when mixed with water and calcium hydroxide. The higher slump, the more liquid the concrete is, thus increasing the pressure and the time before initial set occurs. As mentioned previously, set retarders keep the mix more workable for longer periods of time while still achieving the same strength requirements intended for the design of the structure by the engineer. Workability of the concrete is essential for the labor force placing it and for proper consolidation of the concrete ingredients, particularly in cases where reinforcing steel is very congested.

Placement and Consolidation Methods   ACI 347 recognizes that the most common way to consolidate concrete in forms is with the use of internal vibrators. These vibrators produce the frequency and amplitude necessary to consolidate the concrete into a solid mass without increasing the lateral pressure above an acceptable amount. Spading of concrete actually produces a slightly smaller pressure, while external air or electric vibrators induce an extreme stress to the forms by shaking the form components so much that they can be stressed beyond their initial intent, which is resisting the lateral pressure of the concrete.

Placement Height In recent editions of ACI 347, the committee updated their pressure formula to include a limit on the height of an individual wall. They determined that a form higher than 14 ft should utilize a more conservative pressure calculation. Standards now state that if the placement height is greater than 14 ft, the form designer must use formula 2. The pressure calculations later in this chapter take into account this change.

Placement heights (form heights) can also affect the way a form is constructed. The spacing of the form components is determined by design calculations on individual beam members. When a form height is over 18–20 ft, additional considerations need to be addressed to make sure the form can support itself vertically without unnecessary deflections or undulations.

Adjustments to Lateral Pressure Adjustments to the lateral pressure from concrete itself and how much pressure is applied to the forming system are commonly used in formwork design. Two factors are most important: unit weight coefficient adjustment (C_w) and chemical coefficient adjustment (C_c). The unit weight adjustment is used when the unit weight of concrete is lower or higher than the 150 pcf assumed. Typically, this factor will lower the unit weight of concrete by 1–10 pcf instead of raising the unit weight. The unit weight of concrete rarely, if ever, rises above 150 pcf; however, when a lightweight aggregate is used, the unit weight can be lowered.

The chemical adjustment is for different types of cements used in the mix design in conjunction with any types of retarders used. Different cement types will change the way the concrete reacts, changes the cohesion between ingredients, and affects the pressure on the forms. Retarders include any admixtures that delay the setting of the concrete. This includes, but is not necessarily limited to, water reducers, mid-range water reducers, high-range water reducers, and superplasticizers. Tables 8.3 and 8.4 show some example adjustments that would be used for both unit weight and chemical coefficients.

With the addition of these adjustments, the lateral pressure formula used will be

$$P_{max} = (C_w)(C_c)(\text{lateral pressure})$$

TABLE 8.3 Unit Weight Coefficient (C_w)

Unit Weight of Concrete, w (pcf)	Formula Where w = unit wt	Adjustment (C_w)
Less than 140	$0.5[1 + (w/145 \text{ lb/ft}^3)]$, not less than 0.80	0.80–0.99
140–150	None	1.0
More than 150	$w/145 \text{ lb/ft}^3$	>1.0

Source: American Concrete Institute.

TABLE 8.4 Chemical Coefficient (C_c)

Cement Type	Adjustment (C_c)
Types I, II, and III without retarders	1.0
Types I, II, and III with a retarder	1.2
Other types containing less than 70% slag or 40% flyash without retarders	
Other blends containing less than 70% slag or 40% flyash with a retarder	1.4
Blends containing more than 70% slag or 40% flyash	

Source: American Concrete Institute.

CASE STUDY: THE FOLSOM DAM—MIX DESIGN

The Folsom Dam Phases I–IV involved lowering the main gates of the original structure, adding a new auxiliary spillway chute, and improving other aspects of the dam in order to allow the lake level to be lowered and ultimately take on more water from its watershed area.

The contractors involved with all the phases of construction had the challenge of designing formwork that would meet the challenging schedule requirements while supporting concrete mix designs that required mix set retarders and high flyash weights. The mix design was not that flexible by specification and was designed for workability and consolidation and not for formwork design economy.

By ACI requirements, concrete lateral pressures should be increased by up to 1.40 times if there is 40% or more flyash and retarders used.

The wall heights exceed 14 ft and in some cases were up to 60 ft tall and could be poured as such, as long as the wall thickness did not exceed 5 ft (Figure 8.CS3).

FIGURE 8.CS3 Folsom Dam (courtesy of Kiewit Infrastructure West).

(continued)

CASE STUDY: THE FOLSOM DAM—MIX DESIGN (*Continued*)

The form designers on site, calculated a 6-ft pour rate at 60°F concrete temperature for a wall greater than 14 ft in height.

Before the chemical coefficient (C_c) was included, the lateral pressure was

$$P = 150 + (43{,}400/65°F) + [2800(6\text{ ft})/65°F] = 1076\text{ psf}$$

Since the mix design had over 40% flyash and included retarders, the full 1.40 chemical coefficient (C_c) had to be used. After the C_c was included, the lateral pressure was increased to

$$P_{max} = P(C_c) = 1076 \times 1.40 = 1507\text{ psf}$$

Forty percent more load added 431 psf to the form design.

As a result, the contractor's plywood thickness could not be less than $\frac{3}{4}'' - 1''$ and the stud spacing, depending on the plywood selected, could not exceed 9–10 in on center. Most important, depending on the tie size selected, ties were reduced to 4-ft spacing or tie diameter had to be increased to $1\frac{1}{4}''$ in diameter.

The costs associated with the mix design choice were very high when formwork was concerned. This does not mean that the overall decision was wrong. It could be argued that the workability of the concrete and the increased resistance to water velocity was well worth the cost incurred by the formwork.

Hydrostatic Pressure on Concrete Formwork Pressure on formwork acts perpendicular to the face of the sheeting, which in most cases is plywood or thin steel plate. The form designer must know the temperature, T, of the concrete at the time of placement, the average rate, R, of pour and how tall (h) the wall or element will be for this individual placement. ACI 347 has developed two formulas to represent two distinct conditions.

Formula 1   For formwork less than 14 ft (h) in height *and* a pour rate (R) of less than 7 ft per hour uses the following formula:

$$P_{max} = C_w C_c [150 + 9000(R)/T].$$
Not to exceed 2000 psf.
Not to exceed $P = \gamma h$ (concrete unit weight × height of placement).
Minimum pressure to be used in any case is 600 psf.

Formula 2   For formwork greater than or equal to 14 ft (h) in height *OR* a pour rate (R) of greater than 7 ft per hour uses the following formula:

$$P_{max} = C_w C_c [(150) + (43{,}400/T) + (2800R/T)].$$
Not to exceed 2000 psf.

Not to exceed $P = \gamma h$ (concrete unit weight × height of placement).

Minimum pressure to be used in any case is 600 psf.

Column Forms For columns less than or equal to 15 ft in height, use the following formula:

Use formula 1:

$P_{max} = C_w C_c [150 + 9000\ (R)/T]$.

Not to exceed 3000 psf.

Not to exceed $P = \gamma h$ (concrete unit weight × height of placement).

where P = pressure on the formwork in psf

R = average placement rate in vertical feet per hour

T = average concrete temperature at time of placement

C_w = unit weight coefficient

C_c = chemical coefficient

Minimum pressure to be used in any case is 600 psf.

The derivations of these formulas are not included as they are not published by ACI 347. It should be noted that there is a minimum and a maximum lateral pressure in all cases. For columns greater than 15 ft, use formula 2.

Pressure Diagrams Pressure calculations and diagrams are extremely useful in determining the shape and dimensions of the pressures on the forms and at which point in the form's height the pressure begins to taper down to zero pressure, which always occurs at the top of the placement (top of form). To draw a proper pressure diagram, follow these steps:

1. Calculate P_{max}, the lateral pressure using formula 1 or 2 (considering T, R, and mix designs).
2. Draw the form height as a vertical line and pressure as a horizontal line (it is preferred to draw to the same scale).
3. Add a diagonal line from the top of the form to the maximum pressure, which indicates $P = 150\ h$, full liquid pressure.
4. Draw a vertical line from the maximum lateral pressure (P_{max}) derived in step 1 to the diagonal line represented by $P = 150\ h$. Make sure you use the lesser of the three values described earlier in this chapter.
5. Calculate the distance from the top of the form to the point at which the max pressure begins its decline to zero pressure. This distance is equal to $P_{max}/150$.
6. Label the form height, the $P = 150\ h$, the max design pressure (P), and the distance from the top of the form to the point where the max pressure begins to reduce.

Example 8.2 Draw a pressure diagram for a 15′ high form (h), concrete placed at a rate of 5′ per hour (R), and at a temperature of 70°F (T).

Since $h > 14$ ft, use formula 2

$$P = 150 + 43{,}400/70°\text{F} + 2800(5 \text{ ft})/70°\text{F} = 970 \text{ psf}$$

$$(150 + 620 + 200 = 970 \text{ psf})$$

This value is less than 2000 psf and $P_{\text{max}} = 150(15') = 2250 \text{ psf}$; therefore, 970 psf is the designed pressure.

Figure 8.1 shows how this pressure diagram would be drawn.

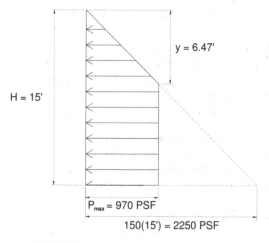

FIGURE 8.1   Example pressure diagram.

To calculate the dimension from the top of the placement to the maximum pressure value (top of the vertical), divide the maximum pressure by 150 pcf.

$$y = \frac{P_{\text{max}}}{150}$$

$$y = \frac{970}{150} = 6.47 \text{ ft}$$

Example 8.3   Draw a pressure diagram for a $12'$ high form (h), concrete placed at a rate of $6'$ per hour (R) and at a temperature of $80°\text{F}$ (T).

Since $h < 14$ ft and $R < 7$, use formula 1.

$$P = 150 + 9000 \ (6 \text{ ft})/80° \ \text{F}; \ P = 825 \text{ psf}$$

$$(150 + 675 = 825 \text{ psf})$$

This value is less than 2000 psf and $P = 150(12) = 1800 \text{ psf}$; therefore, 825 psf is the designed pressure.

Figure 8.2 shows how this pressure diagram would be drawn.

$$y = \frac{P_{max}}{150}$$

$$y = \frac{825}{150} = 5.5 \text{ ft}$$

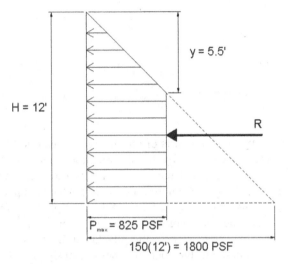

FIGURE 8.2 Example pressure diagram.

Full Liquid Pressure There are several conditions where lateral pressure should be calculated using the full liquid pressure formula, $P = wh$, where P is the pressure in psf, w is the unit weight of the concrete, and h is the form height. Some conditions that require full liquid pressure are as follows:

Use of external vibrators (internal vibration) is considered normal.

Use of admixtures that slow the initial set of the concrete such as superplasticizers, mix retarders, and newly introduced admixtures (when little field experience is present).

Concrete that is pumped from the bottom using the pressure of the pumping device. In some cases, 25% should be added due to the pressure from the pump piston. This includes tunnel forms.

Slumps higher than 4 in.

When revibration of the freshly placed concrete is required.

Figure 8.3 shows what a pressure diagram would look like with full liquid pressure. The diagram remains triangular.

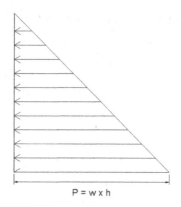

P = w x h

FIGURE 8.3   Full liquid pressure diagram.

Cantilevered Formwork   When lifts of concrete are placed on top of previously placed lifts of concrete, such as during dam construction when monoliths are placed on monoliths, concrete loads and the type of forms used can introduce design challenges. Figure 8.4 shows a monolith form ready for the next placement lift and anchored to the previous lift of concrete with a Williams pigtail anchor. It is common in these cases to attach the bottom of the form to an anchor placed in the top of the previous pour. The anchor used to make this attachment has to be strong enough to resist the force at the bottom of the pour, and the previously placed concrete has to

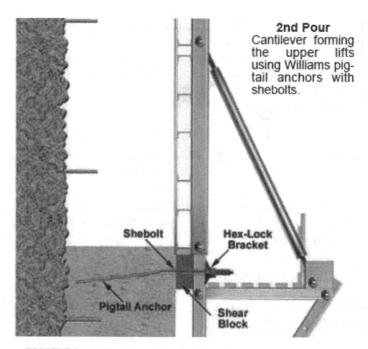

FIGURE 8.4   Cantilevered formwork anchored to previous lift.

FIGURE 8.5   Monolith dam construction cantilevered form.

have achieved adequate strength prior to the placement of the next lift. Usually this would require the lower concrete to have at least 3000–5000 psi in strength.

The pressure diagram in a case like this could either be a full liquid head, triangular diagram, or a trapezoidal diagram from formula 1 or 2. This would depend on the same parameters mentioned earlier. If the form designer thinks the lift can be poured very quickly (within one hour) or one or more of the full liquid head criteria apply, then he would create a triangular diagram. If the form designer recognized that, based on the amount of concrete needed for the lift, a particular placement rate will not be exceeded; he can then develop a diagram based on formula 1 or 2.

The statics involved in this type of problem are relatively simple.

Sample Cantilevered Form Pressure   A 6-ft concrete lift will be placed on a previous lift (like a dam monolith). A basic form will be used and a steel vertical waler will support the form at 5′ 0″ on center horizontally, as shown in Figure 8.5.

Example 8.4   Refer to Figure 8.6 and determine the force (T) in the anchor and the force (C) in the compression point at the bottom of the waler. Assume full liquid pressure from the 6-ft lift. This could be because the concrete could be placed in one hour or because of one of the five reasons stated above when full liquid head should be considered.

Determine the load diagram. In this case, it is assumed full liquid head. Therefore, the pressure diagram shown in Figure 8.6 will be used. This is a triangular shape from zero pressure at the top to a P_{max} value at the bottom. Since the lift is 6 ft,

$$P_{max} = (1.0)(1.0)(150 \times 6) = 900 \text{ psf}$$

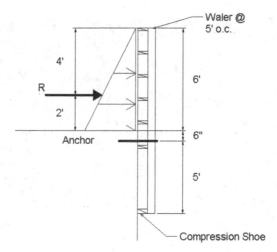

FIGURE 8.6   Form drawing of cantilevered formwork.

Note that this value is for 1 ft of horizontal form only.

Calculate the resultant (R). For a triangular pressure diagram, the resultant is located one third of the height from the bottom ($\frac{1}{3}$ of 6 ft = 2 ft from the bottom). The resultant of a triangular pressure diagram is the area of the triangle ($b \times h/2$).

$$R = 900 \text{ psf} \times 6 \text{ ft}/2 = 2700 \text{ plf of horizontal form}$$

The resultant value can now be multiplied by the waler spacing of 5 ft. It can also be done later in the problem, but it is recommended that it be done now.

$$2700 \text{ lb} \times 5 \text{ ft on center} = 13,500 \text{ plf of wall}$$

This is the resultant value on one waler spaced at 5 ft on center.

Now we have three reference points: Resultant (R), compression shoe (C), and tension anchor (T). Since the resultant has just been calculated, we have one known value and two unknowns, which means we can total the moments about one of the unknown points and solve for the second unknown point. Let's add the moments about (T). In order to do this, the distances need to be known.

$$(13,500 \times 2.5 \text{ ft}) - (C \times 5 \text{ ft}) = 0; \text{ therefore } C = 6750 \text{ lb}$$

Now add the forces in the horizontal direction.

$$13,500 + 6750 - (T) = 0; \text{ therefore, } (T) = 20,250 \text{ lb}$$

Draw a shear and moment diagram similar to Figure 8.7.

The FBD shows what the required load on the embedded anchor needs to be. A factor of safety of at least 2.0 would need to be used. This would require an anchor with a safe working load of 42 k. This is a rather heavy force for which to design. If there is not an anchor that has this capacity, the designer can adjust the criteria. The first

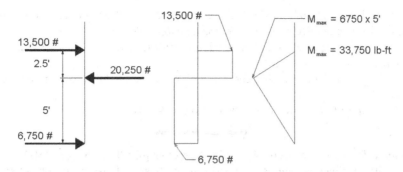

FIGURE 8.7   Free-body, shear, and moment diagram of Example 8.4.

might be to close the spacing of the vertical walers in order to reduce the amount of resultant load on one waler. The second option would be to add an anchor at the top of the form. This would change the whole problem and the statics would then have to be recalculated. This would definitely reduce the load on the bottom anchor while the upper anchor would have to assume some of the load. The challenge would be to find an anchor location for an upper tie. The last option might be to change the pour rate, which would give us a different pressure diagram with a lower maximum pressure. This is not always a good option as it increases the labor cost; and if there are many placements with this form, changing the pour rate could be very costly over the life of the project.

From the shear and moment diagram in Figure 8.7, the designer can determine the size of the vertical waler. From the maximum moment of 33,750 ft-lb, the minimum required section modulus can be determined, and a beam can be selected. The following two examples illustrate double waler selection. A double waler is used so that the anchors can pass through the gap between the beams and concentrically load the two beams as illustrated above in Figure 8.5.

Example 8.5 Double steel waler using American channels. Determine the beam size based on bending only (shear will not be checked due to use of steel).

Determine the type of steel and the factor of safety (FOS). Assume A36 steel with a 1.5 : 1.0 FOS.

$$F_b = \frac{F_y}{1.5}$$

therefore

$$F_b = \frac{36,000 \text{ psi}}{1.5} = 24,000 \text{ psi}$$

Calculate the minimum section modulus required for a single channel.

$$S = \frac{M}{F_b}$$

therefore,

$$S = \frac{33,750 \text{ ft-lb} \times 12''/\text{ft}}{24,000 \text{ psi}} = 16.875 \text{ in}^3$$

Determine the size of a single steel American channel. The minimum section modulus needs to be divided by two since it is a *double* channel.

$S/2 =$ single channel minimum section modulus, therefore, $S = 16.875 \text{ in}^3/2 = 8.44 \text{ in}^3$.

Look up this channel in Appendix 1 with a section modulus equal to or greater than 8.44 in³.

Conclusion: Use an American channel size C 9 × 13.4.

Example 8.6   Double, wood 4 × waler using Douglas fir with $F_b = 1800$ psi and $F_v = 150$ psi. Since we are using wood walers, shear will also need to be checked. First select the material size based on bending requirements and then check the shear strength of that same size beam. The beam selection has been reduced to 4 × material only from experience.

The material has been determined and the values already include FOS.

Calculate the minimum section modulus of the double waler beam.

$$S = \frac{M}{F_b}$$

therefore,

$$S = \frac{33{,}750 \text{ ft-lb} \times 12''/\text{ft}}{1800 \text{ psi}} = 225 \text{ in}^3$$

Determine the size of a single Douglas fir waler. The minimum section modulus needs to be divided by two since it is a *double* wood waler.

$S/2 =$ single channel minimum section modulus, therefore, $S = 225 \text{ in}^3/2 = 112.5 \text{ in}^3$.

Assuming a base dimension of 3.5 (4 × S4S dimension) inches on one of the beams, determine the minimum depth of the Douglas fir single beam.

$$S = \frac{bd^2}{6}$$

therefore,

$$112.5 \text{ in}^3 = (3.5'')(d)_2/6$$

Solve for d, $d = 13.88$ in (shear has not been checked in this example).

Assuming shear is not the controlling factor, next decide if this size beam is feasible. Based on these calculations, this waler would have to be a Douglas fir 4 × 16. Because this beam size is very large, the designer would most likely decide to change some of the controlling parameters in order to use a smaller beam. This could include reducing the vertical spacing or increasing the minimum dimension to a 6×. Another option would be to go with steel double channels as used in Example 8.5.

One-Sided Forms   One-sided forms are used when a concrete wall is required against a shoring system, sloped earth, or vertical rock faces. In other words, one-sided forms are used when the concrete only has to be formed on one side because the other side is up against a stable surface. Figure 8.8 shows a one-sided

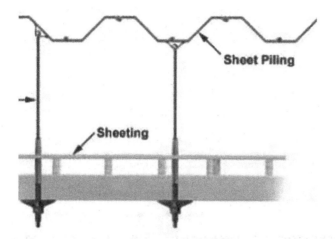

FIGURE 8.8 One-sided form.

form attached to a sheet pile wall. One-sided forms are typically designed the same as a two-sided form except the designer usually attempts to minimize the number of ties. Ties for a system like this are typically more expensive than standard ties because of the added cost in labor and materials to attach the tie to the shoring wall or rock face. As in Figure 8.8, each tie must be attached to a welded bracket against the sheet pile. The pressure on a one-sided form is calculated the same way as a two-sided form. It is based on form height, pour rate, and temperature.

The formwork placement and removal costs for a one-sided wall are higher than a two-sided wall form by approximately 40–60%, mostly due to the ties/anchors attaching the form to the other side. In order to reduce the cost of the one-sided form, the designer typically increases the size of the anchors/ties in order to reduce their quantity. When the number of rows are reduced, not only does the ties size increase but the double waler size would also increase due to the spacing between ties becoming larger and increasing the maximum moment on that vertical or horizontal waler. This will be discussed more in Chapter 9.

Concrete Pressure Underwater Concrete placed under water has different effects on formwork because concrete is lighter under water than on land. For the most part, the difference in unit weight between concrete under and above water is the difference between the unit weight of concrete and the unit weight of the water (150 − 62.4 pcf = 87.6 pcf). The pressure of the water at 62.4 pcf counters the concrete pressure.

The placement heights are always different, but underwater placement can be simplified if the engineer assumes full liquid pressure at 87.6 pcf times the height of placement. Water depth (atmosphere) can affect this value, but it is negligible.

Example 8.7 Calculate the placement pressure under water on a form with concrete 6 ft high. Then draw a pressure diagram (Figure 8.9).

$$P = 6 \text{ ft} \times 87.6 \text{ pcf} = 525.6 \text{ psf}$$

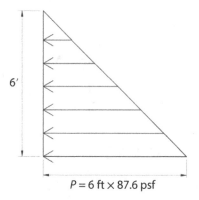

$P = 6 \text{ ft} \times 87.6 \text{ psf}$

FIGURE 8.9 Underwater pressure diagram.

CASE STUDY: CHRYSTAL SPRINGS PROJECT

A project in northern California called Chrystal Springs required concrete placement under water up to depths over 100 ft. This project was upgrading the infrastructure of the water supply system to the San Francisco Bay Area. Included in this work was the demolition and reconstruction of several underwater structures that were built in the late 1800s and early 1900s. Most of the underwater structures that existed were built by brick and mortar construction. (Figure 8.CS4).

FIGURE 8.CS4 Barge arrangement for underwater work. (courtesy of Kiewit Infrastructure West)

The project staff had to design forms to support the hydrostatic forces caused by the underwater placement.

The formwork for underwater work was selected to be steel forms so that buoyancy of lighter materials was not an issue. The forms selected were rented from EFCO Corporation and came with all the bolts, plates, and other accessories for attaching to one another.

Falsework Loads Elevated concrete loads are a combination of the dead load (DL) of the concrete, the live Load (LL) from the tools and workers, and additional weight of the falsework beams:

$$max = DL + LL + steel\ beams$$

where DL = slab thickness (feet) × 160 pcf (160 = concrete unit weight including
           rebar and formwork weight, not including steel)
        LL = depends on the type of construction (ranges from 25 to 75 psf)
  Light duty = 25 psf (painters)
   Med duty = 50 psf (carpentry)
Heavy duty = 75 psf (masonry)
      Special = designed by a registered engineer

In this chapter, a flat slab concrete deck (Figure 8.10) and a T section of deck (Figure 8.11) will be used in the examples to demonstrate basic falsework loads. In Chapter 10, bridge falsework will be introduced as a more sophisticated problem.

Live Load = 50 PSF

Dead Load = 400 PSF

FIGURE 8.10   Matrix diagram of falsework loading.

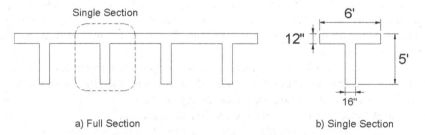

a) Full Section   b) Single Section

FIGURE 8.11   Elevated concrete in T sections.

Example 8.8   Calculate the design load of a 30″ slab with medium live load.

$$DL = 30''/12''/\text{ft} \times 160 \text{ pcf} = 400 \text{ psf}$$

$$LL = 50 \text{ psf}$$

$$Total = 450 \text{ psf}$$

Example 8.9   Determine the weight of a 1-ft-long T section shown in Figure 8.11 if the webs are equally spaced at 6 ft on center. The other properties of the structure are as follows.

Overall height = 5 ft
Top slab thickness = 12″ thick
Web = 16″ thick
Top flange (6 ft × 12″/12″/ft)(1 ft)(160 pcf) = 960 plf of T
Web (5 ft − 1 ft)(16″/12″/ft)(1 ft)(160 pcf) = 853.3 plf of T
Total weight plf of T = 1,813.3 lb

When the falsework is designed for this elevated section, most likely one stringer beam will be designated per T section. This stringer will have to support 1813.3 plf plus the live load and its own weight, depending on the size of the beam. This will be covered in detail in Chapter 10.

Formwork and Falsework Removal   Project specifications typically control the time frame between concrete placement and form removal. In addition, there are at least two distinct conditions for form removal criteria. The first is how long a vertical form must remain in place before its removal. This is determined by the amount of time the concrete needs before it can support itself without the aid of side forms, which controlled the initial lateral pressure. The second case covered in the specifications is the amount of time before falsework can be removed. As discussed before, falsework supports the dead load of the concrete, the live loads from personnel and equipment/tools, and falsework materials. The concrete is usually reinforced either with rebar or a combination of rebar and posttensioning rods or strands. Therefore, the time that elapses before the concrete, reinforcing, and/or posttensioning can support the elevated slab or bridge can be between 14 and 28 days. It is more common for the engineer to require 28 days or when the concrete reaches its design strength, which is typically between 4000 and 6000 psi. The engineer may also require that both of these be met.

Posttensioning creates some situations that can increase removal time and apply additional loads on the falsework that possibly were not planned for during the design phase. In bridge construction, cure time of approximately 10–15 days must elapse after the last placement before the posttensioning strand can be stressed to the design loads. After tensioning takes place, then the falsework can be removed. The posttensioning typically follows a curved pattern, similar to Figure 8.12, where the strands

FIGURE 8.12   Post-tensioning pattern of a typical two-span bridge.

are high at the supports and low in the mid-spans. This allows the strand to lift up the spans of the bridge, which in turn applies a downward force at the piers and abutments. This increased downward force adds load to the falsework support members located on each side of the piers and adjacent to the abutments, which could make removal more difficult. At this time the bridge is self-supporting, so the effects on the design of the falsework can be ignored.

The next two chapters will implement the concepts learned in this chapter as they apply to formwork and falsework, respectively. The student will have several opportunities to apply loads and pressures to design situations used in the industry today.

CHAPTER 9

CONCRETE FORMWORK DESIGN

In Chapter 8, we covered the pressures and loads that are generated by liquid concrete. From that information, the reader should understand that these forces can be extremely large; and if not planned and designed for adequately, they can be very costly as well as dangerous. This chapter will use the load information from Chapter 8 and begin to design the most economical system for the forming condition at hand. Since this chapter relates to formwork only, hydrostatic pressure will be the main focus. Loads will be given to enable students to spend more time on the design process and decision making. Like most temporary structure decisions, there are multiple solutions to formwork design. The challenge is to find the most economical.

This chapter will include several wood structure and composite design examples that go beyond basic formwork design. In addition, there is a detailed discussion on formwork ratios and how they relate to form costs.

9.1   GENERAL REQUIREMENTS

9.1.1   Concrete Specifications

Specifications for concrete work are typically separated between formwork and cast-in-place concrete mix designs. The mix design of the concrete can occupy many specification subsections, depending on the engineer. Formwork specifications give the contractor guidelines while not attempting to require specific means and methods. Owners and engineers are usually reluctant to force a contractor into a particular way to build unless the risk to the design is so great that they must intervene through the use of the specifications.

Formwork specifications tend to be slightly different from engineer to engineer; however, specifications covering the most critical aspects of concrete formwork tend to be consistent. The following sections will illustrate some of the similarities from one concrete specification to another. These items, at a minimum, should give the student a basic idea of what to look for in concrete specifications.

Submittals   Submittals have always been a requirement for falsework design. CalTrans and other agencies may define submittal requirements based on these elements: height, maximum span length, whether spanning water or existing roadways, and the like. Formwork design, on the other hand, is not always a required submittal. However, of late, more engineers are seeing value in formwork design reviews. Construction litigation and risk has driven the owners and engineers to take a more proactive approach to contract management.

Materials   Engineers may only allow particular form sheeting or not allow another. By doing this, they may be trying to assure that their standard of concrete quality is maintained. Formwork ties are usually an item that is seen after completion, and the type and spacing used could be a concern to the design engineer.

Loading   It is fairly common that an engineer would suggest a minimum load on falsework or formwork. The engineer may insist on a minimum live load (say 50 psf) and possibly a minimum dead load (say 100 psf). Anything below these two values would be designed at a minimum of 150 psf, combined DL + LL.

Execution of the Work   There is usually a quality discussion so that the contractor understands the expectations of the owner/engineer. This can easily be done by referring to ACI 347 standards for quality. The manner in which the forms are built, placed, and removed may be specified. The length of time a form must remain in place after concrete placement is typically specified as a minimum duration in days or hours. The engineer would also reiterate the expectation of form cleanliness prior to concrete placement so that debris is not incorporated into the final product.

The location and distance between construction joints is critical to the water tightness of water-retaining structures, and these parameters would be limited by the owner/engineer. This could simply be to reduce the chances of cracking.

The number of times a form can be used may be specified, or at least a discussion would take place concerning when a form is no longer acceptable for reuse.

California state or federal OSHA is usually referenced to the contractor so it is clear that safety standards will be maintained in accordance with the proper authorities.

Specifications will vary, but the previously mentioned items are some common items that are found in most general concrete provisions.

9.1.2   Types and Costs of Forms in Construction

Table 9.1 lists formwork types used in construction along with a range of the cost ($/SF) for the formwork material, not including labor, equipment, and so forth. Not all these elements would be design considerations in this chapter, but all are integral to concrete construction in one form or another. Typically, an element becomes a concern to form designers when it is elevated, very tall or thick in dimension, or unusual.

TABLE 9.1   Cost of Form Elements

Element	Type of Form	Estimated Material Cost per SF
Footing	Wood, braced to soil	$1.00–$1.50
Slab on grade	Wood, braced to soil	$0.75–$1.25
Pier foundation	Steel or wood forms	$2.00–$4.00
Small walls	Hand set forms	$2.50–$8.00
Large walls	Gang forms	$8.00–$30.00
Columns/piers/pilasters	Steel, wood, and fiberglass	$10.00–$15.00
Elevated slabs	Shoring frames, wood post	$2.00–$5.00
Bridge superstructure	Falsework and soffit	$3.00–$6.00
Bridge deck	Carrier beam (horse)	$2.00–$4.00
Construction joints	Wood, throw away, stay-form	$1.50–$3.00
Edge of deck	Wood, braced to deck	$1.00–$1.50

9.2   FORMWORK DESIGN

9.2.1   Bending, Shear, and Deflection

We have learned previously that beam design has to pass three different tests: bending, shear, and/or deflection. Unlike earth support, formwork has more restrictions because the finished product is permanent in nature, thus exposed to the public eye. In earth support systems, shear is not an issue for the steel components and the lagging are used flat, so shear is not the weak link. As for deflection, earth shoring can deflect slightly without being a problem unless it causes excessive settlement. This leaves bending as the most critical element. However, in formwork design, all three of these tests are critical to the success of the form design. Shear becomes important because wood is a common forming material. Deflection becomes important for aesthetics.

Since formwork is simply a series of beams, each one has to comply with all three checks to assure design adequacy. The three checks will be discussed in more detail. Formwork members are also typically continuous over multiple supported members. For instance, plywood spans multiple studs; studs span multiple walers, and so forth. Therefore, the formulas used are those that consider a continuous beam over more than two or three supports. We call these indeterminate structures because there are more than two unknowns. In some formwork design problems, when the loading conditions are very complicated, beam analysis software is used to speed up the engineering process. When special loading conditions are encountered, standard static practices should be used.

It should be noted that what we learn here with bending, shear, and deflection will also be used in Chapter 10, where applicable.

In formwork design, tables are used for material property information. The two types of tables most commonly used for formwork are (1) specie information with allowable stress values and (2) section property information, which is based on the beam's dimensions in cross section or plywood thickness.

Specie information would give the engineer the allowable stress values for bending, shear, compression perpendicular and parallel to grain, and modulus of elasticity at the least. These values may or may not be adjusted for the various adjustments mentioned in Chapter 2 such as duration, wet use, flat use, and temperature.

Section property information would give the engineer information about the dimensions of the sections, which would lead to the section modulus, the moment of inertia, and the cross-sectional area. For plywood, the cross-sectional area is replaced by rolling shear constant because plywood is a laminated product. The horizontal shear strength is largely dependent on the glued laminates and a material testing laboratory is generally used to determine these values from testing.

Bending Stress Bending stress is an important component of formwork. Since bending stress is a derivative of bending moment and section modulus, our basic formulas are used to solve for various unknowns. The following static and strengths of material concepts derive the bending formulas used in formwork design. In most cases we can solve for any variable. However, it is most common to solve for the span of the supports knowing the material sizes and specie type.

Maximum bending moment (M), simply supported (two to three supports), one to two spans, $M = wL^2/8$

Multiple supports (more than two spans), $M = wL^2/10$

Bending Stress, $f_b = M_{max}/S$

where w = uniform load

L = span from the inside face of one support to the inside face of the adjacent support

Section modulus (S)

Dimensional lumber, $S = bd^2/6$ from Appendix 6

Plywood (laminated), $S =$ from Table 9.3

where b = base dimension

d = beam depth

Since most formwork is a series of beams over multiple supports, the second bending moment formula shown above ($wL^2/10$) is more commonly used. In a case where the designer knows there will only be two supports, the first formula for bending would apply. With these three formulas, one can solve for any of the different variables as long as the information is provided that leaves only one unknown. For formwork, it is common to know what size material is desirable and have to solve for the spacing of the elements. In this case, if we know all the variables except L (span length), we can solve for the support spacing. Let's see how this would work.

Assume we are given the following:

Type of materials and their maximum bending stress values (F_b)

Uniform load on the beam in question (w)

Section modulus of selected materials based on size or thickness (S)

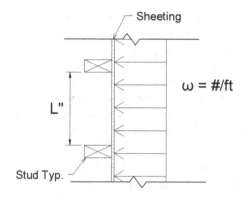

FIGURE 9.1 Span between studs.

We could then solve for L span by using the following formula. This formula is the same for plywood and dimensional lumber:

$$L = 10.95\sqrt{F_b S/w}$$

Once the span (L) is determined, the designer knows the maximum spacing of the materials supporting the beam section that was just analyzed. Figure 9.1 shows the span dimension, which is the inside face of one stud to the inside face of the other.

The variables in the span formula for bending come from the statics and strength formulas.

Shear Stress The same process can be used for shear stress, except the basic statics and strengths of materials concepts of shear will be used. This is what we know about shear:

Maximum shear (V)

Simply supported (two supports), $V = wl/2$ or $0.5wl$

Multiple supports (more than 2), $V = 0.6wl$

where w = uniform load
$\quad\quad L$ = span from the inside face of one support to the inside face of the
$\quad\quad\quad$ adjacent support

Shear stress, $f_v = 1.5V/A$

Dimensional lumber, $A = bd$

Rolling shear for plywood (laminated), Ib/Q

where b = base dimension
$\quad\quad d$ = beam depth

Assume we are given the following:

Type of materials and its maximum shear values (F_v)

Uniform load on the beam in question (w)

Cross-sectional area or rolling shear of selected materials based on size or thickness (Ib/Q) or b and d

Before we solve for L, ACI allows us to adjust the allowable shear stress by up to a factor of 2.0 for dimensional lumber. This basically means we can double the values given to us by the National Design Specification for Wood Construction. We call the new shear value $F_v' = 2.0 \times F_v$. For plywood, the value given by the plywood association is used and does not get an allowance. This value is typically between 57 and 75 psi.

Solve for L.

Plywood formula:

$$L = \frac{F_v(lb/Q)}{0.6w}$$

Dimensional lumber formula:

$$L = \frac{(F_v'bd)}{0.9w} + \left(\frac{2d}{12}\right)$$

Deflection Finally, the same process is used for deflection, with different variables coming from the deflection parameters as follows:

Maximum allowable deflection for forms using $L/360$

Maximum deflection $= 5wL^4/384EI$ (Simple span, two supports)

Maximum deflection $= wL^4/1740EI$ (Multiple supports)

where $w =$ uniform load

$L =$ span from the inside face of one support to the inside face of the adjacent support

$E = 1,300,000$ psi to $1,700,000$ psi depending on material

$I =$ moment of inertia

Assume we are given the following:

Type of materials and its maximum modulus of elasticity (E) in psi

Uniform load on the beam in question (w) in PLF

Moment of Inertia (I) in in.^{4}

Solve for L. The same formula is used for plywood and dimensional lumber.

$$L = 1.69\sqrt[3]{EI/w}$$

Load Path Load path for concrete formwork is not much different from other temporary structures. The path of the concrete pressure is as follows:

The concrete pressure is applied to the sheeting material (typically plywood, steel plate, or sometimes flat, dimensional lumber.

The sheeting loads the studs. Based on the stud spacing, the linear load on the stud is determined.

The stud loads the waler system. Based on the spacing of the walers, the linear load on the waler is determined.

The waler loads the ties or struts. Based on their spacing, the total load of the tie or strut is determined.

Uniform Loads on Form Members   The most important aspect when following the load path is determining the uniform load on the next beam member to which the load is transferred. For instance, if it is determined that studs are to be spaced at 14 in on center to support a plywood thickness, then the uniform load on the stud must take into account the 14-in spacing. The uniform load on the stud becomes $P_{max} \times 14''/12''/\text{ft}$. If P_{max} is 800 psf, then the uniform load on the stud is 933.3 plf. The unit is now pounds per linear foot since the spacing in feet is canceled.

Adjustment Factors

Adjustments to Material Stresses   The wood industry has developed a number of adjustment factors for permanent construction using lumber. ACI 347 has adopted a few of these, but not all of them, for temporary structures and formwork. These factors are multipliers to either raise or lower the allowable stresses. For example, if a particular factor is 1.20, then the allowable stress will be raised by 20%. Conversely, if the factor is more conservative, the multiplier would be less than 1.0, say 0.80; and in this case, the allowable stress would be reduced to 80% of its original value. More than one factor can apply to a single stress value. For instance, the short-term duration factor (for loads less than a few months) may increase the stress by 25%, but the wet service factor (for material with moisture content more than 19%) may lower allowable stress to 90%. Therefore, an allowable stress of $1200 \text{ psi} \times 1.25 \times 0.90 = 1350 \text{ psi}$.

Also, some adjustment factors only apply to some stresses. For example, one adjustment factor may apply to bending stress and modulus of elasticity but not shear stress, whereas another factor may apply to shear stress only. Table 9.2 lists the most common adjustment factors to permanent wood structures and temporary structures.

For the purpose of this textbook, three main adjustment factors will be used:

Load duration	1.25
Shear adjustment	2.0
Wet service factor	0.67–0.95

The stress values used in this chapter's examples include the adjustment factors already.

TABLE 9.2   Adjustment Factors

Factor	Symbol	Applies to
Load duration	C_d	F_b, F_c, F_v
Wet service	C_m	F_b, F_c, F_v, E
Temperature	C_t	F_b, F_c, F_v, E
Repetitive member	C_r	F_b
Shear adjustment	C'_v	F_v
Flat use	C_{fu}	F_b
Size factor	C_f	F_b, F_c
Beam stability	C_l	F_b
Column stability	C_p	F_c

Form Design Using Statics and Strengths of Materials The first type of form design discussed in this text will be what is considered form design through the use of statics and material standard principles. This means we will use all the basic statics and strengths concepts mentioned earlier in this chapter and apply it to an *all-wood* form design. Later in this chapter, steel and aluminum components will be introduced for larger type form systems. Once the designer knows what type and what size of material is available or preferred by the contractor, the three formulas for shear, bending, and deflection will be applied to determine the most effective design.

In order to begin this process, select the material type for the following components:

Form Plywood—Standard BB-OES (both sides have type B finish, and the ply wood has been oiled, edged, and sealed) Form Ply

Rolling shear strength (Ib/Q)

Bending stress (F_b)

Modulus of elasticity (E)

Dimensional Lumber—Stud and Waler Size

Shear stress (F_v')

Bending stress (F_b)

Modulus of elasticity (E)

Tie Material and Type

Tensile Strength

Innumerable species of wood are available throughout the world. However, for temporary structures, the designer considers the use and selects a particular species of wood that is more affordable in the project's geographical area. Common wood species for temporary structures are Douglas fir, Hem fir, and Southern Pine. Table 9.3 lists estimated stress values for the more common lumber used for formwork. These values were derived from the NDS literature.

TABLE 9.3 Common Wood Specie Allowable Stresses for Formwork (psi)

Specie	Bending, F_b	Compression Perp-grain, $F_{c\perp}$	Compression Parallel-grain, $F_{c\parallel}$	Horizontal Shear, F_v	Modulus of Elasticity, E
Douglas fir–larch	1000	625	1600	95	1,500,000
Douglas fir–south	1250	585	1600	90	1,600,000
Southern pine	1250	565	1600	90	1,200,000
Spruce–pine fir	975	405	1500	70	1,400,000
Hem-fir	975	405	1500	75	1,300,000
BB-OES plywood	1600	210 on face		60	1,500,000

Source: National Design Specification (NDS)

Table 9.3 includes important stress values that are used at one time or another during form design. The following allowable stresses used for basic form design:

Shear stress, F_v

Bending stress, F_b

Modulus of elasticity, E

Some additional stresses used when analyzing members in compression are parallel to grain ($F_{c\parallel}$) and perpendicular to grain ($F_{c\perp}$).

Plywood Design   When deciding on plywood thickness, the designer can follow one of three methods: (1) Select a particular thickness and type of plywood (often based on experience), (2) select material owned by the contractor, and (3) select material that will meet specifications. (For example, the concrete specifications may also lean the designer toward a smoother finish or a strict deflection requirement.) Figure 9.2 shows how excessive deflection in plywood and studs can be out of specification tolerance.

For the less experienced form designer, not all thicknesses of plywood are used. Plywood thicknesses range from $1/4''$ to over $1''$ in thickness. For form design though, most likely $\frac{5}{8}''$ to $\frac{3}{4}''$ will be the form ply thickness of choice. This text will use one of these two sizes. Plywood also has alternating laminates, usually three, five, or seven ply. The more plies, the better the quality of plywood. The number of plies is an odd number because each face of the plywood always has the grain of that ply running in the direction on the length of the sheet. When using plywood as a form sheeting material, it is best, whenever possible, to place the plywood in its strongest direction. This occurs when the outside face grain of the plywood is running parallel to the span of the plywood. Therefore, the weak direction of the plywood is when the face grain is perpendicular to the span of the plywood. A common mistake is to interpret these two options as parallel or perpendicular to the "stud direction." This is a costly mistake and should be avoided whenever possible. Most plywood charts give figures

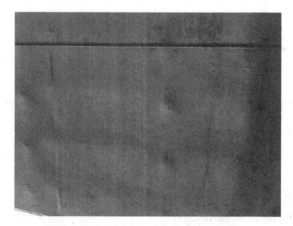

FIGURE 9.2   Plywood and stud deflection.

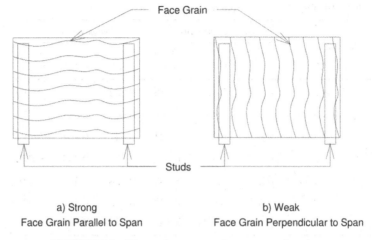

a) Strong
Face Grain Parallel to Span

b) Weak
Face Grain Perpendicular to Span

FIGURE 9.3   Plywood spans, weak vs. strong direction.

for both cases and allows the designer to select the one that fits his design. Figure 9.3 illustrates this concept of plywood direction.

Most Commonly Used Plywood Types in Concrete Formwork

BB-OES
Both sides type B sanded face
Oiled, edged, and sealed
One side medium-density overlay (MDO)
One side high-density overlay (HDO)

SPAN CALCULATIONS ON PLYWOOD   Bending stress for plywood:

$$L = 10.95\sqrt{F_b S/w}$$

Shear stress for plywood:

$$L = \frac{F_v(Ib/Q)}{0.6w}$$

Deflection for plywood:

$$L = 1.69\sqrt[3]{EI/w}$$

In order to use these span formulas, the form designer must have material allowable stress values and property values as mentioned earlier. Tables 9.3 and 9.4 include estimated values for plywood and dimensional lumber. These values have been estimated from NDS and ACI.

Table 9.4 contains properties for moment of inertia, section modulus, and rolling shear (Ib/Q). Respectively, these values are used for deflection, bending, and shear calculations. The table has a strong section and a weak section. As mentioned earlier,

TABLE 9.4   BB-OES Plywood Properties

Thickness (in)	Minimum Ply	Strong Direction			Weak Direction			Weight (psf)
		I (in^4)	S (in^3)	Ib/Q (in^2)	I (in^4)	S (in^3)	Ib/Q (in^2)	
$\frac{3}{8}$	3	0.027	0.125	3.088	0.002	0.023	3.510	1.1
$\frac{1}{2}$	3	0.077	0.236	4.466	0.009	0.087	2.752	1.5
$\frac{5}{8}$	5	0.129	0.339	5.824	0.027	0.164	3.119	1.8
$\frac{3}{4}$	5	0.197	0.412	6.762	0.063	0.285	4.079	2.2
$\frac{7}{8}$	7	0.278	0.515	8.050	0.104	0.394	5.078	2.6
1	7	0.423	0.664	8.882	0.185	0.591	7.031	3.0
$1\frac{1}{8}$	7	0.548	0.820	9.883	0.271	0.744	8.428	3.3

Source: American Concrete Institute (ACI).

the strong section is when the plywood exterior face grain runs parallel to the plywood span (more common), and the weak section is when the plywood exterior face grain runs perpendicular to the plywood span (not as common).

Stud and Waler Design   Stud and waler sizes have many more options, so the rules of material selection have more ranges. For instance, stud selection usually ranges from 2 × 4 to 2 × 6 and 4 × 4. Double walers would include 2 × 4, 2 × 6, 2 × 8, 4 × 6, and 4 × 8. Obviously, this still leaves a lot of decision making for the designer, but considering the number of sizes available, this narrows down the field considerably. This text provides the necessary properties of most available dimensional lumber sizes for design purposes. Like plywood, rectangular dimensional has its weak direction and its strong direction. This is usually very obvious because beams are stronger when their depth is greater than their base. The tables will give both of these values and the designer must be careful to use the correct size. As an example, take a 2 × 4 and compare its section modulus about the x axis (strong) and the y axis (weak).

$$S = bd^2/6$$

Strong: $S = \dfrac{(1.5'')(3.5'')^2}{6} = 3.063 \text{ in}^3$

Weak: $S = \dfrac{(3.5'')(1.5'')^2}{6} = 1.313 \text{ in}^3$

The strong direction in this case is 2.33 times stronger than the weak direction. The only time the weak direction would be preferred is when the board is acting as the sheeting as well as the stud, such as when footing and slabs are being formed.

Appendix 6 contains dimensional lumber properties for design purposes. Like Table 9.4, this appendix has area, section modulus, moment of inertia, unit weight, and the like. These values will be used for all dimensional lumber calculations.

This appendix also has a weak and strong section. The strong is using the material in the vertical loading position (*x-x* axis) and the weak is loading the material in the flat, horizontal position (*y-y* axis).

Wood Form Design   The process of form design will follow these steps:

1. Calculate the lateral pressure on the form.
2. Determine the plywood type and thickness.
3. Calculate the maximum stud clear spacing through shear, bending, and deflection. Add the stud thickness to determine the on-center spacing.
4. Determine the stud size and calculate the uniform load on one stud using P_{max}, the span, and the on-center spacing in step 3.
5. Calculate the waler spacing through shear, bending, and deflection.
6. Determine the waler size and calculate the uniform load on one double waler using P_{max}, the span, and the double waler thickness.
7. Determine the maximum load on the heaviest loaded tie.

Figure 9.4 shows typical wall forms using 2×4 and plywood construction.

The following formulas will be used when checking bending, shear, and deflection. The span between components is the unknown variable in this case. In order to know the other variables for each check, the designer must know the material sizes and properties. Figure 9.5 illustrates how multiple spans of plywood decrease bending moment and deflection.

FIGURE 9.4   Typical 2×4 and plywood formwork.

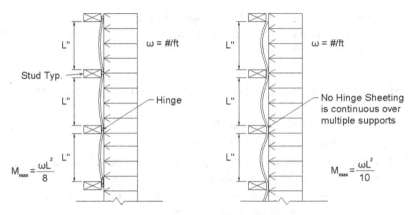

FIGURE 9.5   Plywood bending and deflection.

Bending

$$l = 10.95\sqrt{F_b S/w}$$

where l = span (in)
  F_b = allowable bending stress of material (psi)
  S = section modulus (in^3)
  W = uniform load on span (plf)
  10.95 = constant

Shear—Plywood

$$L = \frac{F_v(Ib/Q)}{0.6w}$$

where L = span (ft) multiply by 12 to compare in inches
  V = maximum shear (lb or k)
  F_v = allowable shear stress
  Ib/Q = rolling shear constant (in^2)
  W = uniform load (plf)
  0.6 = constant

Shear—Dimensional Lumber

$$L = \frac{F_v' b_d}{0.9w} + \frac{2d}{12}$$

where L = span (ft) multiply by 12 to compare in inches
  F_v' = allowable shear stress with shear factor of 2.0
  b = base dimension of wood section (in)
  d = depth dimension of wood section (in)
  w = uniform load (plf)
  0.9 = constant

Deflection

$$L = 1.69\sqrt[3]{EI/w}$$

where $L =$ span (in)

$I =$ moment of inertia (in^4)

$W =$ uniform load on span (plf)

$E =$ modulus of elasticity

$1.69 =$ constant

9.2.2 Form Design Examples Using All-Wood Materials with Snap Ties or Coil Ties

Example 9.1 Cases 1 and 2 will use a maximum lateral concrete pressure of 900 psf. The calculations for form pressure were discussed at length in Chapter 8. It is assumed that this value comes from a thorough pressure analysis, taking into account rate of placement, temperature, mix design, wall height, and coefficient adjustments.

Case 1 will use the plywood in the weak direction, small studs, small walers, and snap ties. This will represent a very lightweight system. The plywood direction is intended to show the student how important it is to run the plywood in the strong direction whenever possible. If not, it is either a mistake and the finished product will indicate high deflection and bending failures; or, if it is intentional, the form design will be less than economical as far as placement production and stud spacing.

Case 2 will use plywood in the strong direction, maximizing the plywood strength and stud spacing, larger studs and walers, and higher capacity ties.

The example will alternate between case 1 and 2 so that we can see the differences between the two at each step. Also, all spacing calculations will be rounded to the nearest tenth. In reality, these numbers will be rounded down to the nearest whole inch, which makes sense with constructability practices and simplifying dimensions for layout purposes. The fabrication of these forms must be done in an efficient manner. Therefore, the easier (using dimensions that are quicker to read on a tape measure) the spacing of the materials is, the lower the fabrication costs.

The material assumptions for cases 1 and 2 are recorded in Table 9.5.

TABLE 9.5 Case 1 versus Case 2 Assumptions

Case 1	Case 2
$\frac{5}{8}''$ BB-OES plywood used in the weak direction	$\frac{3}{4}''$ BB-OES plywood used in the strong direction
2 × 4 Douglas fir–larch studs	4 × 4 Douglas fir–larch studs
Double 2 × 4 Douglas fir–larch walers	Double 3 × 8 Douglas fir–larch walers
5000-lb capacity snap ties	She bolts

Step 1: Stud Spacing

Case 1: Plywood $\frac{5}{8}''$ thick BB-OES has been selected to be used in the weak direction. The important aspect here is that the values used for this plywood must be the lower values from Table 9.4, indicating that the plywood is not as strong when used in this manner. Use the column labeled "weak direction." For stress values use Table 9.3.

Case 1, Step 1: Check the maximum span between studs to comply with bending, shear, and deflection requirements.

Bending

$$l = 10.95\sqrt{(1600)(0.164)/900} = 5.91\,\text{in}$$

where $\quad l$ = span (in)

$\qquad F_b$ = allowable bending stress BB-OES plywood in psi from Table 9.3 (1600 psi)

$\qquad S$ = section modulus in in^3 from Table 9.4 (0.164 in^3)

$\qquad W$ = 900 plf

$\qquad 10.95$ = constant

The maximum span of the studs due to bending is 5.9 in when rounded to nearest tenth.

Shear—Plywood

$$L = \frac{(60)(3.119)}{0.6(900)} \times 12''/\text{ft} = 4.16''$$

where $\quad L$ = span (ft) divide by 12 to compare in inches

$\qquad F_v$ = maximum allowable shear stess from Table 4.2 (60 psi)

$\qquad Ib/Q$ = rolling shear constant (in^2) from Table 9.3 (3.119 in^2)

$\qquad W$ = 900 plf

$\qquad 0.6$ = constant

The maximum span of the studs due to shear is 4.2 in when rounded to the nearest tenth.

Deflection

$$l = 1.69\sqrt[3]{(1,500,000)(0.027)/900} = 6.01''$$

where $\quad l$ = span (in)

$\qquad I$ = moment of inertia (in^4) from Table 9.4 (0.027 in^4)

$\qquad W$ = 900 plf

$\qquad E$ = modulus of elasticity from Table 9.3 (1,500,000 psi)

$\qquad 1.69$ = constant

The maximum span of the studs due to deflection is 6.0 in when rounded to the nearest tenth.

Step 1 is complete for case 1 and the maximum span between studs is the *lowest* value as determined by the three checks above. The lowest of the three checks is 4.2″, which is governed by *shear*. From the clear span, the center-to-center spacing can be determined by adding the stud thickness (1.5″ for an S4S 2 × 4 on edge) to the span (*L*) dimension.

$$\text{Center to center} = L + b = 4.2'' + 1.5''\ (2 \times 4) = 5.7''\ \text{OC (on center)}$$

Case 2: Plywood—3/4″ thick BB-OES is being used in the strong direction; so the larger values are used from the tables, "strong direction."

Case 2, Step 1: Check the maximum span between studs to comply with bending, shear, and deflection requirements.

Bending

$$l = 10.95\sqrt{(1600)(0.412)/900} = 9.37''$$

where L = span (in)

         F_b = Allowable bending stress BB-OES plywood in psi from Table 9.3 (1600 psi)

         S = section modulus in inches^3 from Table 9.4 (0.412 in^3)

         W = 900 plf

      10.95 = constant

The maximum span of the studs due to bending is 9.2 in when rounded to nearest tenth.

Shear—Plywood

$$L = \frac{(60)(6.762)}{0.6(900)} \times 12''/\text{ft} = 9.02''$$

where L = span (ft) divide by 12 to compare in inches

         F_v = maximum allowable shear stress from Table 9.3 (60 psi)

    Ib/Q = rolling shear constant (in^2) from Table 9.4 (6.762 in^2)

         W = 900 plf

       0.6 = constant

The maximum span of the studs due to shear is 9.0 in when rounded to the nearest tenth.

Deflection

$$l = 1.69\sqrt[3]{(1,500,000)(0.197)/900} = 11.66''$$

where l = span (in)

        I = moment of inertia (in^4) from Table 9.4 (0.129 in^4)

       W = 900 plf

       E = modulus of elasticity from Table 9.3 (1,500,000 psi)

    1.69 = constant

The maximum span of the studs due to deflection is 11.7 in when rounded to the nearest tenth.

Step 1 is completed for case 2 and the maximum span between studs is the *lowest* value as determined by the three checks above. The lowest of the three checks is 9.0″, which is governed by *shear*. From the clear span, the center-to-center spacing can be determined by adding the stud thickness (3.5″ for an S4S 4 × 4) to the span (L) dimension.

Center to center = $L + b$, 9.0″ + 3.5″ (4 × 4) = 12.5″ OC (on center)

Case 1 Stud: Determine the minimum spacing between walers using a 2×4 Douglas fir–Larch stud. The spacing of the walers is dependent on the stud loading and size. From this information, the span between walers is determined. Since the variables have been defined in detail in step 1, the remaining steps will not show the formula in variable form (root formula).

Step 2: Waler Spacing

Case 1, step 2, span of walers. Use the same material in Table 9.3 as step 1. However, use Appendix 6, which has been placed in Table 9.6, for the properties of the dimensional lumber. The stud for case 1 is a 2×4.

TABLE 9.6   Properties of a 2×4 and 4×4 from Appendix 6

Material	2×4	4×4
Section modulus (in^3) S4S	3.06	7.15
Base dimension (in)	1.5	3.5
Depth dimension (in)	3.5	3.5
Moment of inertia (in^4) S4S	5.36	12.50

For Douglas fir–Larch construction grade:

$$F_b = 1000 \text{ psi}$$

$$F_v = 95 \text{ psi} \times 2$$

for short-term use factor (shear is allowed twice the normal allowable shear stress by ACI-347).

$$E = 1,500,000 \text{ psi}$$

The uniform load (w) for each stud should be determined based on the spacing of the stud and the P_{max} concrete pressure. The new w for the case 1 2×4 stud is

$$W = 900 \text{ psf} \times (5.7''/12''/\text{ft}) = 427.5 \text{ plf}$$

The uniform load is lower than the original P_{max} because the spacing of the 2×4 studs is less than 1 ft.

Bending

$$l = 10.95\sqrt{(1000)(3.06)/427.5} = 29.3''$$

The maximum span of the waler due to bending is 29.8'' when rounded to the nearest tenth.

Shear—Dimensional Lumber

There are two big differences in the shear calculation between plywood and dimensional lumber. The first is the allowable shear being doubled for F_v' when considering a shear factor multiplier. The second is the addition of twice the depth ($2d/12$) to the span calculation. If these two allowances were not used, the span solution would be extremely underestimated:

$$L = \frac{(F'_v b d)}{0.9w} + \frac{2d}{12}$$

where L = span (ft) multiply by 12 to compare in inches
   F'_v = 95 psi × 2.0 = 190 psi
   b = base dimension of wood section (1.50 in)
   d = depth dimension of wood section (3.50 in)
   w = uniform load (427.5)
   0.9 = constant

$$L = \frac{190(1.5)(3.5)}{0.9(427.5)} + \left(\frac{2\,(3.5)}{12}\right) = 3.173 \times 12''/\text{ft} = 38.08 \text{ in}$$

The maximum span of the waler due to shear is 38.1″ when rounded to the nearest tenth.

Deflection

$$l = 1.69\sqrt[3]{(1{,}500{,}000)(5.36)/427.5} = 44.94''$$

The maximum span of the waler due to deflection is 44.9″ when rounded to the nearest tenth.

Step 2 is completed for case 1 and the maximum span between walers is the *lowest* value as determined by the three checks above. The lowest of the three checks is 29.3″, which is governed by *bending*.

Case 2, Step 2: Determine the waler spacing for case 2 using a 4 × 4 stud.

The uniform load (W) for each stud should be determined based on the spacing of the stud and the P_{max} maximum concrete pressure. The new W for the 4 × 4 stud is

$$W = 900 \text{ psf} \times (12.5''/12''/\text{ft}) = 938 \text{ plf}$$

The uniform load is greater than the original P_{max} because the spacing of the 4 × 4 studs is slightly greater than 1 ft on center.

Bending

$$l = 10.95\sqrt{(1000)(7.15)/938} = 30.23 \text{ in}$$

The maximum span of the waler due to bending is 30.2″ when rounded to the nearest tenth.

Shear—Dimensional Lumber

$$L = \frac{190(3.5)(3.5)}{0.9(938)} + \left[\frac{2\,(3.5)}{12}\right] = 3.34' \times 12'' = 40.08''$$

The maximum span of the waler due to shear is 41.2″ when rounded to the nearest tenth.

Deflection

$$l = 1.69\sqrt[3]{(1,500,000)(12.50)/938} = 45.87''$$

The maximum span of the waler due to deflection is 45.9″ when rounded to the nearest tenth.

Step 2 is completed for case 2 and the maximum span between walers is the *lowest* value as determined by the three checks above. The lowest of the three checks is 30.2″, which is governed by *bending*.

Step 3: Tie Spacing

The difference between steps 2 and 3 is the fact that the waler is a double member so all the section properties (S, I, d, and b) are doubled. The waler is a double beam; therefore, it is twice as strong. Table 9.7 shows values for the waler material from Appendix 6.

TABLE 9.7   Properties of a 2 × 4 and 3 × 8 from Appendix 6

Double Waler Material	Double 2 × 4	Double 3 × 8
Section modulus (in³) S4S	3.06 × 2 ea	21.9 × 2 ea
Base dimension (in)	1.5 × 2 ea	2.5 × 2 ea
Depth dimension (in)	3.5 × 2 ea	7.25 × 2 ea
Moment of Inertia (in⁴) S4S	5.36 × 2 ea	79.39 × 2 ea

Before the spans can be calculated, the uniform load on the double waler must be determined. The tie spacing is dependent on the waler size, waler type, and the uniform load on each waler.

Case 1 Waler—Determine the minimum spacing between ties using a double 2 × 4 Douglas fir–Larch waler. The spacing of the ties is dependent on the waler loading and size. From this information, the span between ties is determined. Once again, since the variables have been defined in detail in step 1, the remaining steps will not show the root formulas.

Case 1, Step 3, Span of Walers (tie spacing)—Use the same material (Table 9.2) as step 1. However, use Appendix 6 for the properties of the dimensional lumber. The waler for case 1 is a 2 × 4.

The grade of lumber has not changed from the studs to the double walers. For Douglas fir–Larch construction grade:

$$F_b = 1000 \text{ psi}$$

$$F_v = 95 \text{ psi} \times 2 = 190 \text{ psi}$$

for short term use factor (shear is allowed twice the normal allowable shear stress by ACI-347).

$$E = 1,500,000 \text{ psi}$$

The uniform load (W) for each waler should be determined based on the spacing of the waler and the P_{max} concrete pressure. The new W for the case 1, 2 × 4 waler is

$$W = 900 \text{ psf} \times (29.39''/12''/\text{ft}) = 2205 \text{ plf}$$

The uniform load is greater than the original P_{max} because the spacing of the 2 × 4 double waler is more than 1 ft.

Bending

$$l = 10.95\sqrt{(1000)(3.06)(2)/2205} = 18.2 \text{ in}$$

The maximum span of the waler due to bending is 18.2 in when rounded to the nearest tenth.

Shear—Dimensional Lumber

$$L = \frac{190(1.5)(3.5)(2)}{0.9(2205)} + \left[\frac{2\,(3.5)}{12}\right] = 1.593' \times 12'' = 19.1 \text{ in}$$

The maximum span of the waler due to shear is 19.1 in when rounded to the nearest tenth.

Deflection

$$l = 1.69\sqrt[3]{(1,500,000)(5.36)(2)/2205} = 32.8''$$

The maximum span of the waler due to deflection is 32.8 in when rounded to the nearest tenth.

Step 3 is completed for case 1 and the maximum span between walers is the *lowest* value as determined by the three checks above. The lowest of the three checks is 18.2'', which is governed by *bending*.

Case 1 formwork is represented by Figure 9.6 where the forms are made from 2 × 4's and plywood except the plywood was used in the strong direction in this photo.

Case 2, Step 3: Determine the tie spacing for case 2 using a 3 × 8 double waler.

The uniform load (W) for each waler should be determined based on the spacing of the waler and the P_{max} concrete pressure. The new W for the case 2, double 3 × 8 waler is

$$W = 900 \text{ psf} \times (30.2''/12''/\text{ft}) = 2265 \text{ plf}$$

Bending

$$l = 10.95\sqrt{(1000)(21.9)(2)/2265} = 48.15''$$

The maximum span of the waler due to bending is 48.2 in when rounded to the nearest tenth.

Shear—Dimensional Lumber

$$L = \frac{190(2.5)(7.25)(2)}{0.9(2265)} + \left[\frac{2\,(7.25)}{12}\right] = 4.59' \times 12'' = 55.05''$$

FIGURE 9.6   Junction box formed with all-wood forms.

The maximum span of the waler due to shear is 54.4 in when rounded to the nearest tenth.

Deflection

$$l = 1.69 \sqrt[3]{(1,500,000)(79.39)(2)/2265} = 79.76''$$

The maximum span of the waler due to deflection is 79.8 in when rounded to the nearest tenth.

Step 3 is completed for case 2 and the maximum span between walers is the *lowest* value as determined by the three checks above. The lowest of the three checks is 48.2″, which is governed by *bending*.

Step 4: Tie Capacity

The tie capacity and size needs to be selected to complete this wood form design. The load on an individual ties is

$$P_{tie} = [\text{tie spacing (ft)}] \times [\text{waler spacing (ft)}] \times P_{max}$$

Case 1: $P_{tie} = [18.2/12''/\text{ft}] \times [29.3/12''/\text{ft}] \times 900 \text{ psf} = 3333 \text{ lb in tension}$

A 3500-lb tie will work for this case.

Case 2: $P_{tie} = [48.2\,/12''/\text{ft}] \times [30.2/12''/\text{ft}] \times 900 \text{ psf} = 9098 \text{ lb}$

A $\frac{1}{2}'' \times \frac{3}{4}''$ Shebolt will work for this case, capacity (safe working load includes 1.5 : 1.0 FOS) = 12,000 lb.

Solutions to wood form design example 1, cases 1 and 2, are shown in Table 9.8.

TABLE 9.8 Case 1 vs. Case 2 Solutions[a]

Material	Case 1 Center-to-Center Spacing (in)	Case 2 Center-to-Center Spacing (in)
Plywood	$\frac{5}{8}$ W	$\frac{3}{4}$ S
Stud spacing	5.7	12.5
Waler spacing	29.3	30.2
Tie spacing	18.1	48.2
Tie type and load	Snap tie (3333 lb)	She bolt (9098 lb)

[a]W = weak and S = strong.

9.2.3 Formwork Charts

Before the next example is illustrated, the use of wood formwork charts will need to be mastered. Such charts are very convenient because they eliminate much of the tedious calculation performed in the earlier examples, thus avoiding the risk of errors. These charts, based on NDS 2001, can be found in Appendix 7 and are provided with permission from Williams Form Engineering Corporation. The following items are needed to use the charts for wood formwork design:

Allowable bending stress of wood specie or plywood

Allowable shear or rolling shear stress of wood specie or plywood

Modulus of elasticity of wood specie or plywood

Uniform load (plf) on supporting beam

Number of spans for the supporting beam (single, 2, 3, or more)

Plywood and beam sizes

For typical all-wood form design, these three charts are necessary:

Plywood Chart—This includes two main sections, stronger plywood (higher allowable stress) and weaker plywood (lower allowable stress), each with two columns: one for plywood used in the strong direction and one in the weak.

Single-Beam Chart—This is used for studs, joists, and other single-beam components. The number of spans is an option because the chart will use either $wl^2/8$ or $wl^2/10$.

Double Walers—This is used for any two-beam component (double walers), for which all the applicable beam properties have been doubled to take into account the two beams working together.

The fine print on these charts indicates that the dimensions shown are center-to-center dimensions and not clear span, as long as the beam has a "relatively narrow" width (b) dimension. This author will loosely interpret this to mean a 2× beam up to roughly a 4× beam.

The form design process using these charts follows:

Determine maximum lateral pressure from the concrete (see Chapter 8).
Use the applicable plywood chart and determine the stud spacing.
Calculate the uniform load on one stud based on its spacing.
Use the applicable single beam chart and determine the double waler spacing.
Calculate the uniform load on a double waler based on its spacing.
Use the applicable double waler chart and determine the tie spacing.
Calculate the load on the heaviest loaded tie. Typically, these are the ties in the second row from the bottom of the form.

Example 9.2   Formwork Design Using Appendix 7 Charts (courtesy of Williams Form Engineering Corporation).
Design a wall form to meet the following conditions:

Wall form 12 ft high × 16 ft wide
$R = 5$ ft/h
$T = 65°F$

Materials:
Plywood $3/4''$:

Allowable Stress Values:
$F_b = 1930$ psi
Rolling shear $= 72$ psi
$E = 1,500,000$ psi
Deflection $\frac{1}{360}$, but not to exceed $\frac{1}{16}''$

Dimensional Lumber:
2 × 4 studs
Double 2 × 6 walers
Snap or taper ties

Allowable Stress Values:
$F_b = 975–1500$ psi
$F_v' = 180$ psi; $F_v' = F_v × 2$
$E = 1,600,000$ psi
Deflection $\frac{1}{360}$, but not to exceed $\frac{1}{4}''$

Step 1: Calculate P_{max}.
Use formula 1 from Chapter 8 since $h < 14$ ft and $R < 7$ ft/h

$$P_{max} = 150 + \frac{(9000 × 5)}{65} = 842 \text{ psf}$$

See Figure 9.7 for a diagram of the resulting pressure.

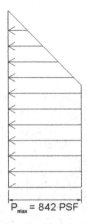

P_{max} = 842 PSF

FIGURE 9.7 Pressure diagram.

Step 2: Determine stud spacing using $^3/_4''$ plywood.
From Appendix 7, 842 psf, F_b = 1930 psi, shear = 72 psi
Space studs at 12″ OC

Step 3: Determine waler spacing with a 2 × 4 stud at 12″ OC.

$$W = 842 \text{ psf} \times 12''/12''/\text{ft} = 842 \text{ plf}$$

Stud and joist beam table in Appendix 7, 842 plf, f_b = 1500 psi, H = 180 psi
Space double walers at 28″ OC. Use 28″ OC.

Step 4: Determine tie spacing with 2 × 6 double walers at 28″ OC.

$$W = 842 \text{ psf} \times 28''/12''/\text{ft} = 1965 \text{ plf}$$

Double waler table in Appendix 7, 1965 plf, f_b = 1250 psi, H = 180 psi
Space ties at 37″ OC. Use 36″ OC.

The tables in Appendix 7 give you "not-to-exceed" numbers, center to center, so they should be adjusted to work with the panel size and to make the layout for the carpenters straightforward. Therefore, use the values that have been rounded down.

$$\text{Tie load} = 842 \text{ psf} \times 28''/12''/\text{ft} \times 36''/12''/\text{ft} = 5894 \text{ lb}$$

Final design:

$^3/_4''$ plywood (used in the strong direction)
2 × 4 studs at 12″ OC
2 × 6 double walers at 28″ OC
Ties at 36″ OC with minimum SWL ≥ 5894-lb capacity

FIGURE 9.8 Composite form using aluma beams and steel channels.

Formwork Design Using Lumber, Aluminum Beams, and Double Steel C-Channels

All wood forms are versatile and inexpensive. However, when a contractor desires a larger form, with materials that are more reusable, they might consider using a combination of plywood, dimensional lumber, or aluminum beams for studs and aluminum or steel double walers. This next section will use these materials for form design Examples 9.3 and 9.4. Figure 9.8 is an example of a larger type, composite form design. Most of the materials on this form are reusable from project to project.

Example 9.3 Composite Form Design

Using a lateral concrete pressure of 981 psf (which is assumed to be precalculated using the methods from Chapter 8), design a two-sided wall form using $3/4''$ BB-OES plywood in the strong direction, 4×4's for horizontal studs, double steel American C-Channels for vertical walers and she bolts. Do not exceed a $\frac{1}{2}''$ diameter inner rod she bolt. The following data should be used for the design criteria. *Ignore deflection in all calculations.*

Material Properties:

Plywood:

$F_b = 1600$ psi
$F_v = 60$ psi

Dimensional Lumber—Hem fir construction grade lumber

$F_b = 975$ psi

$F_v = 75$ psi, $F'_v = 150$ psi

$E = 1,300,000$ psi

A36 steel with a $F_b = 0.6$; $F_y = 21,600$ psi

Step 1: Determine the safe spacing of the 4×4 horizontal studs by analyzing a $12''$ strip of plywood. Also, check bending and shear, and ignore deflection. Do not use Appendix 7.

$$W = 981 \text{ psf} \times 1' = 981 \text{ plf}$$

Bending—Plywood

$$l = 10.95\sqrt{(1600)(0.412)/981} = 8.8 \text{ in}$$

Shear—Plywood

$$L = \frac{(60)(6.762)}{(0.6)(981)} \times 12''/\text{ft} = 7.9 \text{ in (governs)}$$

Space the 4×4's at $7.9'' + 3.5''$ (4×4 width) $= 11.4''$ use $11''$ OC

Step 2: Determine the safe spacing of the double steel channels. In other words, how far apart can the double steel channels be spaced so the 4×4's do not fail in bending or shear.

$$W = 981 \text{ psf} \times 11''/12''/\text{ft} = 899 \text{ psf}$$

Bending—Studs

$$l = 10.95\sqrt{(975)(7.15)/899} = 30.5 \text{ in (governs)}$$

Shear—Studs

$$L = \left[\frac{150\,(3.5)\,(3.5)}{0.9(899)}\right] + \left[\frac{2\,(3.5)}{12}\right] = 2.85' \times 12''/\text{ft} = 34.2 \text{ in}$$

Space the double steel channel's at $30.5''$ OC.

Step 3: Determine the double waler size and determine the tie capacity. At this point in the design, experience can come in handy because tie spacing can be estimated and the double channel can be sized appropriately, or the channel size can be known and the maximum tie spacing can be calculated. For this example, the ties will be spaced to stay within the $1/2''$ diameter requirement stated above and then the double channel will be sized.

A $\frac{1}{2}''$ inner rod tie for a she bolt can safely support $12,000$ lb ($1.5 : 1.0$ FOS) according to Williams Form Engineering Corporation. If the double walers are spaced at $30''$ OC, then the tie spacing can be estimated.

$P_{\text{capacity}} = $ Area $\times 981$ psf; area $=$ tie spacing $\times$ double waler spacing, therefore:

$$12,000 \text{ lb} = (30''/12''/\text{ft}) \times (\text{tie spacing}) \times 981 \text{ psf}$$

$$\text{Tie spacing} = 4.89 \text{ ft} \qquad \text{say } 4.75 \text{ ft}$$

The uniform load on the double waler is $W = 981 \text{ psf} \times 34/12''/\text{ft} = 2776 \text{ plf}$.

$$W = 2776 \text{ plf}$$

Figure 9.9 is an example of the FDB for this double waler.

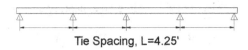

Tie Spacing, L=4.25'

FIGURE 9.9   FBD of double waler.

Waler size is based on the maximum moment generated by the FBD diagram in Figure 9.9. A computer program could be used, or the maximum moment can be calculated with $M = wL^2/10$.

M for a continuous beam over multiple supports (ties)
$M = 2776 \text{ plf } (4.25')^2/10 = 5014 \text{ ft-lb}$
$S_{min} = M/F_b$, $S = (5014 \text{ ft-lb})(12''/\text{ft})/21{,}600 \text{ psi} = 2.79 \text{ in}^3$

Since a double steel channel waler is a "double" beam, the minimum section modulus can be divided by 2 before searching for the best channel:

$$S_{min} \text{ for a single channel} = 2.79 \text{ in}^3/2 = 1.395 \text{ in}^3$$

Some channels that are in the range of this requirement are shown in Table 9.9. The *best* is shown in bold.

TABLE 9.9   Steel Channel Properties

Channel Size	Section Modulus (S_x)	Weight per foot	Conclusion
C 4 × 5.4	**1.93**	**5.4**	**BEST**
C 3 × 6	1.38	6	Section too small
C 4 × 7.5	2.29	7.5	OK, but heavy

The conclusion for this channel selection should be that the C4 × 5.4 is best for strength and economy. The C3 × 6 is very close; however, it is heavier than the channel chosen and it is also weaker.

Final design:

$3/4''$ plywood (strong)
4×4 studs horizontal at $11''$ OC
Double C4 × 5.4 channel walers at $30''$ OC
$1/2''$ she bolts at 4.75 ft OC, tie load not to exceed 12,000 lb

The tie lengths would have to be calculated so as to order the correct equipment. Figure 9.10 illustrates the hardware required for this design.

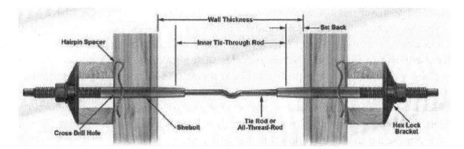

FIGURE 9.10 She-bolt cross section.

Here is how that would be accomplished for a she-bolt system:

Assume the wall thickness is 16″ reinforced concrete.
Available lengths for $^1/_2'' \times {}^3/_4''$ she bolts are 20″, 24″, and 28″.

TABLE 9.10 She-Bolt Length Calculation

Item	Length (in)
Plywood × 1	$\frac{3}{4}$
4 × 4 studs × 1	3.5
Double C-Channel × 1	4
Plate washer × 1	1.5
Wing nut × 1	0.75
Excess threads × 1	2
Required she bolt L	= 12.5
Add 2″ of taper inside wall	2
Net system L	14.5

The minimum thread length for this size tie is 9″. Therefore, there is plenty of thread at the end on the tie so that the wing nut will not "bottom out" (run out of thread).

Use the $^1/_2 \times {}^3/_4$ she bolt × 20″ in length.

Finally, the coil rod would be cut the wall thickness less 1-in cover from each face of wall, $16'' - (2 \times 1'') = 14''$.

Aluminum Beams   Aluma beams are extruded aluminum beams that (1) are shaped for maximum bending and deflection capacity, (2) provide a slot for a "nailer", and (3) allow for a slotted bolt to attach walers.

For these reasons, the beam is shaped as shown in Appendix 9. The pocket in the top is for a wood or fiberglass 2 × 2 that allows for a sheet of plywood to be nailed or screwed to it. The slot in the bottom is for a slotted bolt that attaches an aluminum, steel, or wood waler. Appendix 9 also includes design information for aluminum beams to be used for form design.

ALUMINUM BEAM DESIGN WITH MANUFACTURERS DESIGN CHARTS   On many occasions the manufacturer of a particular component provides the engineering data necessary to use its product. This data is easily obtained by contacting the local representative of the company that provides such a product. The data received will not always follow a typical protocol. For instance, the information in Appendix 9 (courtesy of Aluma Systems) is from an aluminum beam manufacturer. This chart utilizes maximum uniform load information based on some spacing shown on the chart. The product manufacturers have precalculated bending, shear, and deflection to determine the maximum allowable uniform load.

Before moving on to Example 9.4, the aluminum beam chart should be studied. The chart is broken up into an x and y axes. The x axis represents the type of span: simple, two spans, or three spans. Most of the time formwork components span three or more spacings. The y axis represents the span of the aluminum beam between supports, from 4 to 12 ft. The values within the table in Appendix 9 are the allowable uniform loads on a single beam at the given span. The fine print of the chart has some useful information. The letters designate which of these three criteria governed the spacing for the load:

R denotes that shear governs

M denotes that bending moment governs

D denotes that deflection governs using L/360 criteria

* denotes deflection governs not exceeding $\frac{1}{4}''$ deflection

FOS = 2.2 : 1.0

The first step to using this type of chart is to determine the spacing of the aluminum beams, which is based on the loading and the plywood thickness. Once the uniform load is known, the number of spans is estimated (single, two, three, or more). Now, the span length of the aluminum beam can be found on the y axis. This is the farthest the beam can span given the other conditions.

Example 9.4   Composite Form Design

Using a design pressure of 1100 psf and the wood design charts from Example 9.2, design a two-sided wall form using $3/4''$ HDO plywood in the strong direction (see Table 9.11), aluminum beams (Appendix 9) and steel double channels (Appendix 1).

TABLE 9.11   High-Density Overlay (HDO) Plywood Properties[a]

Stud Spacing (in)	$\frac{5}{8}''$ (plf)	$\frac{3}{4}''$ (plf)	$1\frac{1}{8}''$ (plf)
8	1970	2050	2230
9	1600	1800	1976
10	1200	1500	1660
12	745	1060	1380
16	350	505	1000
19.2	125	305	555

[a]HDO plywood used in the strong direction and complying with L/360 deflection limits.

Do not exceed the capacity of a $1\frac{1}{8}''$ to $\frac{7}{8}''$ diameter C7T taper tie (SWL = 24,000 lb). Assume one wall form measures $20'$ high × $24'$ wide.

Figure 9.11 shows the allowable loads for HDO plywood in graph form.

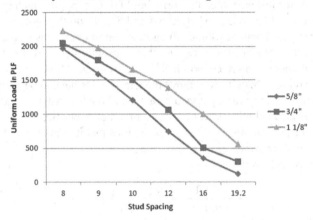

FIGURE 9.11 Allowable load graph for HDO plywood (Courtesy of Olympic Plywood).

Step 1: The concrete lateral pressure has been given at 1100 psf.

Step 2: Determine stud spacing using $\frac{3}{4}''$ HDO plywood.

Using the design load of 1100 psf and Table 9.12 for $\frac{3}{4}''$ HDO plywood, determine the stud spacing. The table indicates that the studs can be spaced at $10''$ OC:

Step 3: Determine double waler spacing with an aluminum stud at $10''$ OC:

$$W = 1100 \text{ psf} \times 10''/12''/\text{ft} = 913 \text{ plf}$$

The aluminum beam chart (Appendix 9) is now used to determine the spacing of the double walers. Aluminum beams are very strong and many times the spacing given is more than the rest of the form components can support. Using the column in Appendix 9 titled "3 Span (lbs./ft.)" and based on 913 plf, an aluminum beam can span 8.0 ft if L/360 is required and 7.5 ft if deflection is limited to $\frac{1}{4}''$. It will be assumed that the more stringent requirement is necessary; therefore, space double steel walers at 7.5 ft on center.

As mentioned above, this spacing will increase the loads on the waler and the ties. The waler has to be designed with a 7.5-ft tributary width or $W = 1100 \text{ psf} \times 7.5 \text{ ft} = 8250 \text{ plf}$ waler linear load. It will soon be determined whether or not this load makes the system economical. A large part of formwork costs are in the tie placement and removal. Therefore, if the tie quantity can be reduced, costs can typically be lowered. However, if this reduction in quantity results in a very large tie, the objective could be lost. This example will be completed taking these thoughts into consideration.

If the walers are spaced at 7.5 ft on center, $W = 8250 \text{ plf}$.

Step 4: Determine what the tie spacing would be using the prescribed 24,000-lb SWL taper tie.

$$\text{Maximum tie spacing} = 24{,}000\text{-lb tie}/8250 \text{ plf} = 2.91 \text{ ft}$$

If the ties for a 20-ft-tall wall form were spaced at 2.9 ft on center, then there would be approximately 7 rows of ties × 4 walers = 28 ties. If a 5 × 5 tie and waler pattern was employed, for instance, then the form would require approximately 20 ties, and the tie spacing would increase to between 4 and 5 ft. To do this, however, the walers would also be spaced between 4 and 5 ft and the aluminum beams would not be used to their maximum capabilities. However, this is usually the correct option because one additional waler is more economical than 8 more ties. It should be pointed out that besides having to install and remove the ties, the ties have to be prepped and the holes left by the ties have to be patched by a mason/finisher. After considering all the labor that is associated with each tie (3 to 6 man-hours per tie, $200–$500 each in labor and materials combined), it becomes obvious that it is important to reduce the quantity whenever possible.

Another approach would be

$$\frac{24{,}000\text{-lb tie}}{1100 \text{ psf}} = 21.82 \text{ SF per tie}$$

This creates an average combination tie and waler spacing of 4'8" by a tie spacing of 4'8" or other combinations that equal approximately 22 SF.

Taking all of this information into consideration and spacing the walers at 5'0" on center and the ties at 4'6" on center produces a layout of 6 walers with 20 ties. This layout is illustrated in Figure 9.12. One can then compare the cost between using 2 more walers versus 8 more ties.

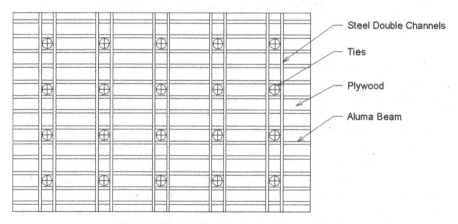

FIGURE 9.12   Panel layout for Example 9.4.

For two additional walers, purchase, install, and remove walers:

Material: 21 ft × 20 lb/ft (estimate) × 2 ea = 840 lb of steel at $0.60/lb = $546

Labor: 2 ea at 2 man-hours/each (estimate) = 4 man-hours × $50 = $200
Total: $746 for two additional walers

For additional ties, prepare, install, remove, and patch hole: Using the price range above and an average of $350 per tie, eight additional ties would cost 8 × $350 = $2800:

$$\$2800 - \$746 = \$2054 \text{ savings by using the 6-waler, 20-tie design}$$

Also take into account that the walers are placed one time during form fabrication, whereas the eight ties repeat as many times as the form is reused. If the form is used eight times, for example, the savings becomes over $21,000.

Step 5: Design the double waler using the waler and tie spacing above.

$$W = 1100 \text{ psf} \times 5' = 5500 \text{ plf}$$

Figure 9.13 shows a loading diagram of a double channel.

$M_{max} = 5500 \text{ plf}(4.5')^2/10 = 11,138 \text{ ft-lb}$
$S_{min} = (M/F_b)/2 \text{ channels}$
$S_{min} = [(11,138 \text{ ft-lb}) (12''/\text{ft})/21,600 \text{ psi}]/2 \text{ channels} = 3.09 \text{ in}^3$

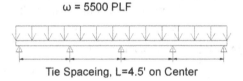

ω = 5500 PLF

Tie Spaceing, L=4.5' on Center

FIGURE 9.13   Loading diagram of a double channel.

Using American Standard Channels, the choices close to this minimum section modulus are listed in Table 9.12. The C6×8.2 appears to be the *best* of the three.

TABLE 9.12   Available Channel Sizes for Example 9.4

Channel Size	Section Modulus (S_x)	Weight per Foot	Conclusion
C5×6.7	3.0	6.7	Section too small
C5×9	3.56	9	OK, but heavy
C6×8.2	**4.38**	**8.2**	**OK, lightest, BEST**

The C6×8.2 also has a moment of inertia of 13.1 in⁴. This would also help in minimizing the deflection in the double waler because it also has a higher moment of inertia, which would lower the maximum deflection. The C5×6.7 channel could be used if the tie spacing was reduced a few inches. Contrary to that, since the C8×6.2 channel is overdesigned, the tie spacing also could be increased to maximize the waler size. However, this would contradict the goal to not exceed a 24,000-lb taper tie. All these points should be examined to make sure the design is the most economical.

Final Design Conclusion:

$3/4''$ HDO plywood

Aluminum beam studs spaced at $10''$ on center

Double C6×8.2 American Steel Channels spaced at 5 ft on center

$1\frac{1}{8}'' \times \frac{7}{8}$ " C7T taper tie (SWL = 24,000 lb) spaced at 4.5 ft OC

Figure 9.14 shows a taper tie being used with an all-wood form system. The tie lengths would have to be calculated to order the correct equipment. Here is how that would be accomplished for a taper tie system.

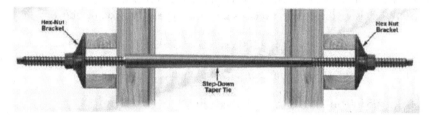

FIGURE 9.14 Taper tie cross section.

Assume the wall thickness is $16''$ reinforced concrete. Table 9.13 lists the measurements necessary for calculating the length of the she bolt.

TABLE 9.13 She bolt Length Calculation

Item	Length (in)
Concrete	16
Plywood × 2	1.5
Aluminum beams × 2	13
Double C − channel × 2	12
Plate washer × 2	4
Wing nut × 2	3
Excess threads × 2	6
Required taper tie L	55.5

Available taper tie lengths are $30''$, $34''$, $36''$, $42''$, $48''$, $54''$, and $60''$.

The $54''$ tie would work, and the excess thread availability would be reduced from $6''$ to $4.5''$.

The $60''$ tie will also work if the thread lengths are long enough that the wing nut does not "bottom out" (run out of threads). This tie has $10''$ of thread at the long end and $8''$ of thread at the short end. Based on this data, the wing nut would not run out of threads.

CASE STUDY: RADIUS FORMWORK, PLEASANT GROVE WWTP CONSTRUCTION

Project Overview

Waste water and water treatment plants are great projects to study in construction because they involve almost every type of construction including excavation, mechanical, electrical, underground, concrete structures, pile driving, and all aspects of architecture. The Pleasant Grove Wastewater Treatment Plant (WWTP) in Roseville, California, was no exception. This project was a brand new plant for the City of Roseville, making it the city's second WWTP. The growth of the city required increased capacity, but the original plant had already been expanded and was running out of real estate. Also, the new plant location was closer to the city's expanding areas thus added convenience.

A large portion of this project involved new concrete structures to transfer and treat water throughout the hydraulic flow process. These structures ranged in size and shape, from large rectangular structures to small, round, or tall structures. The focus of this case study is on three structure types that had curved, vertical walls. The first type included four clarifier tanks that had a 65-ft average radius and stood 16 ft tall. The second type involved three oval-shaped oxidation ditches that had curved ends measuring 60 ft in diameter and stood 16.5 ft tall. The third type included two digester tanks that had a 25-ft radius walls and stood 32 ft tall.

It was decided early in the form decision process that the digester, with a much different radius and wall height, would have a separate form system than the other two. The 32-ft height and the fact that there were only eight wall placements (four 90° pours per digester) directed the decision to rent a steel form. The project looked at EFCO and SYMONS. The two systems are very different. The EFCO system uses the "ready radius" panels that can be curved to a desired radius on site. The SYMONS system comes with precurved walers that are custom fabricated for the desired radius. After a cost analysis, the project decided to go with the SYMONS form (shown in Figure 9.CS1).

Like many decisions in construction, they are made with very detailed cost analyses, which include estimated production rates. However, since only one method is eventually used, the method not chosen is never tested. Therefore, a final, actual cost comparison can never be made so the contractor must accept the decision. At this point, all efforts should be focused on making the chosen option as cost-effective as possible.

Since the digester wall decision was made, now the clarifiers and oxidation ditch walls needed to be designed. The geometry between the two structures were almost identical and the number of wall placements made any decision potentially a big "money maker" or a big failure. All in all, there would be a

(continued)

CASE STUDY: RADIUS FORMWORK, PLEASANT GROVE WWTP CONSTRUCTION (*Continued*)

FIGURE 9.CS1   Steel forms used for a digester tank.

total of 44 wall placements of walls with 60- to 65-ft radii and 16- to 16.5-ft height. The similarities in these two structures had to be taken advantage of. The wall height was easy. The form engineers knew that a 16'9" panel would work with all 44 walls. It was the radius that was in question. Could a form panel be made that could be flexible enough to work with both the 60-ft radius and the 65-ft radius? Could a form be made with a 62.5-ft radius that would work for all 44 wall placements? The answer was yes. An aluminum vertical studded panel was made with horizontally curved double steel channels (for the 62.5-ft radius) that could open up to a 65-ft radius and close up to a 60-ft radius. The $\frac{3}{4}''$ BB-OES plywood had no problem making the radius and no provisions were necessary to weaken or wet the plywood for flexibility. Figure 9.CS2 shows this form being used on a clarifier wall.

This form is a custom-job-built radius form using plywood, aluminum beams, and rolled double steel channel walers. The same process used in Example 9.4 would be used to design this form. One difference would be that the plywood would be changed to BB-OES or MDO because the HDO would be difficult to bend into the necessary radius. Another difference would be that the aluminum studs would be turned in the vertical direction and the double steel channels would be placed horizontal and curved to the radius of the permanent structure geometry. These forms were lightweight, approximately 15 psf, and were built for the tanks that did not exceed 16.75 ft in height.

FIGURE 9.CS2   Aluminum and rolled steel channel radius forms.

Formwork Geometry Form design is not just about analyzing the forces from the concrete and stresses of the material. Structures have all sorts of shapes and sizes. Sometimes the more difficult task is designing forms that conform to the shape of the permanent structure. Many times, features have to be added in order to solve a problem that the geometry caused.

Corners, curves, and intersections are all characteristics of concrete structures that add difficulty to the design process. It is not always the fact that the wall is tall and the pressure from the concrete is great. Sometimes a small structure can have none of these characteristics except that the corners and intersections of wall create weak spots in the form design. At times, the form designer spends more time figuring out how one form panel is going to attach to another form panel, or how to tie a corner or intersection together when the ties do not have a proper termination point on the opposite side of the wall. Figure 9.15 illustrates this point for both a corner form connection and an intersection. The arrows represent unsupported sections (weak points) in the forms.

FORM DRAWINGS Once the calculations for a design are complete, the work has just begun, so to speak. Someone has to take the calculations, which are typically not-to-exceed values, and create the panel and layout drawings. These drawings show plan views and sections of the elements and also refer to accompanying individual panel drawings that detail the panel for size, materials, spacing of members, and tie locations and quantities at a minimum.

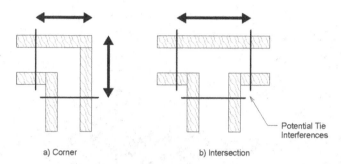

a) Corner   b) Intersection

Potential Tie Interferences

FIGURE 9.15 Formwork corners and intersections.

CASE STUDY: SOUND WALL CONSTRUCTION OVER BIDWELL PARK, CHICO, CALIFORNIA

Project Overview

When Highway 99 was widened over prestigious Bidwell Park in Chico, California, between 2011 and 2014, the contractor was faced with many challenges. Working in or over Bidwell Park brings the attraction of many groups and individuals that do not support change to the park. When John and Annie Bidwell donated this land to the City of Chico, the agreement was very specific to leave the park as is whenever possible. Years of planning, permits, and negotiations preceded the construction of the Highway 99 project.

Another challenge facing the contractor was protecting Big Chico Creek (shown in Figure 9.CS3), which runs in the middle of the narrow 3600+-acre park and under the Highway 99 overcrossing and eventually through downtown Chico and Chico State University.

The new piers for the widened bridge were designed to be constructed adjacent and sometimes within the creek low-water line. In the third year of the project, just when one thought all the challenges were behind the contractor, a final logistic/access issue had to be planned. The new sound wall, which is attached to the bridge barrier rail, is 30 ft above the ground on one side and is adjacent to two lanes of live traffic on the other side. The work access lane available to the contractor was less than 20 ft in width. A small hydraulic rough terrain crane could barely fit within these limits and would not allow any other access when being used. In addition, an operator would have to be on the crew continuously for the duration of the operation.

Furthermore, if a crane were placed on the ground level, the access along the park and the city streets would have made it very difficult, if not impossible, to set up a crane that could reach the portions of the wall that it needed to reach. Figure 9.CS4 illustrates the wall and the basic geometry of the structure.

FIGURE 9.CS3   Big Chico Creek.

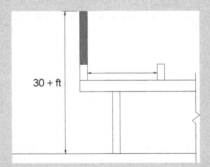

30 + ft

FIGURE 9.CS4   Schematic of sound wall.

Traveler

The contractor had to figure how to form the cast-in-place, two-sided wall from the bridge with just the right lane and shoulder closed. The contractor employed an EFCO form system and developed a traveling truss that rolled along the closed lane and supported both the inside and outside forms. Figure 9.CS5A shows the truss beam that is supporting the inside and outside forms.

Once the wall forms were placed on the truss beam, they never had to be hoisted with a crane again. From this point forward, the truss, which was mounted on casters, was moved with a forklift. The wall panel taper ties were

(*continued*)

CASE STUDY: SOUND WALL CONSTRUCTION OVER BIDWELL PARK, CHICO, CALIFORNIA (*Continued*)

FIGURE 9.CS5A Inside and outside forms supported by cantilevered beam.

FIGURE 9.CS5B Crane hoisting first form into place.

removed, the panels were pulled away from the wall and the truss was pushed 40–50 ft to its next wall section. Figure 9.CS5B shows the first wall placement from a section that the crane could access from the ground level.

Truss and Counterweights

EFCO designed a truss and counterweight system shown in Figure 9.CS6. The truss was made from EFCO equipment and components. The wall forms were their standard wall panel using plywood, E-beams, and EFCO double steel channels (super studs).

FIGURE 9.CS6   Traveling truss with counterweights.

The counterweights were precast onsite to custom dimensions and weights that fit the framework of the truss and kept the forms supported in position (see Figure 9.CS7).

Form Fabrication with Form Liner

In addition to the access complications, the wall design required form liner that cast a scenic imprint into the wall for aesthetic purposes. The contractor mounted a polyurethane form liner that usually costs between $30 and $60 per SF onto the form and repeated the pattern with each form use. Figure 9.CS8 shows the wall forms being fabricated with the form liner prior to the first placement.

(continued)

CASE STUDY: SOUND WALL CONSTRUCTION OVER BIDWELL PARK, CHICO, CALIFORNIA (*Continued*)

FIGURE 9.CS7 Counterweights (deadmen) on casting bed.

FIGURE 9.CS8 Prefabricated wall forms with form liner.

9.2.4 Estimating Concrete Formwork

Using Form Ratios to Accurately Verify a Concrete Estimate Reinforced concrete construction estimating requires a great deal of labor and temporary structure knowledge. When the finished product is delivered, one may not realize how much work went into the creation of the structure. Competently estimating this work involves a great deal of past project experience, and form ratios created by experts can be the link that brings past experience to the estimator.

Reinforced concrete estimating utilizes several categories of costs. Labor is one of if not the largest cost element. The formwork, the reinforcing steel, the placement of the concrete, the finishes, and some additional miscellaneous items are the other cost categories. The furnishing and placing of reinforcing steel is traditionally a subcontractor cost. Materials are also required in two categories: permanent and temporary. STS (small tools and services or supplies) is another cost category. Finally the cost of construction equipment, such as forklifts, cranes, and pickup trucks, must be included.

For consistency, examples shown in this book will use labor rates at $40.00 per labor hour. Other costs shown are actual project costs from the San Francisco Bay Area and the Central Sacramento Valley (northern California) that have been altered slightly to preserve confidentiality.

What Is a Form Ratio? A form ratio (FR) is developed by dividing the neat volume of concrete in cubic yards (CY) for a structural element into the total contact area of formwork in square feet of the same element as shown below:

$$FR = \text{total forms (SFCA)}/\text{total volume (CY)}$$

The formwork quantity should include any concrete surface that requires a form to support the plastic concrete during its curing process. This should include footing and slab-on-grade edge (SOG) forms, all wall surfaces, the formed area of all elevated concrete surfaces (excluding additional surfaces for access platforms), construction joints, miscellaneous curbs and equipment pads, and all other formed surfaces. The volume of concrete should include all concrete elements except fill concrete and other concrete not requiring a form. However, unformed concrete volume is still accounted for in the estimate because the unformed concrete must be purchased and there may be some incidental costs associated with this volume in the estimate. During actual construction, it is crucial that the project staff record actual costs in this same manner. This will be discussed more when we discuss the historical cost data.

Table 9.14 illustrates that the larger the element's thickness, the smaller the ratio. Also, if an element is supported by the ground (such as footings and SOGs), the ratio is lower because no formwork is required beneath the concrete. Wall concrete requiring forms on both sides generates higher ratios because twice the formed surfaces are required for formwork. Other surfaces, such as the top of the wall and top surfaces of slabs are not formed. These unformed areas require a "wet" finish, which is applied by cement masons who strike off and smooth the surface before the concrete takes final set. Between the elements with high and low ratios are supported slabs and soffit (bottom) beams, which must be supported by falsework.

TABLE 9.14 Formwork Ratios for Different Elements of Concrete Structures

Element	L (ft)	Width (ft)	HT (ft)	Forms (SF)	Concrete (CY)	Form Ratio
Bridge footings	18	18	6	432	72	6.00
Slab-on grade (SOG), with CJs 1[a]	100	40	1	280	148	1.89
Small walls	100	20	1	4100	74	55.35
Large walls	100	20	2.5	4100	185	22.14
Bridge piers	6	6	40	960	53	18.00
Small columns	20	5	5	250	19	13.50
Elevated slabs	100	20	1	2240	74	30.24
Misc. curbs & pads	20	20	0.5	40	7	5.40
Totals				12402	632	19.60

[a]CJs-Construction Joints.

Historical Cost Data Historical cost data are the actual costs or production rates achieved by a construction company self-performing similar work from project to project. Costs are recorded as each project is completed, and the final data is used for determining the project's financial success and for estimating similar projects in the future. Successful companies prefer to use their own historical costs over data published in cost manuals because they represent what the company has achieved on an actual project. However, published cost and production rates based on documented equipment and labor rates can be obtained from various sources. The data is based on an accumulation of records from companies throughout the industry doing the same type of work. The resulting rates may represent what is expected on an industry-wide basis, but do not necessarily represent what has been achieved by the estimator's own company. Therefore, it is usually preferable for a company to rely on its own historical cost data when it is available.

Historical costs should take into account safety ordinances, the capabilities and availability of a company's supervisory personnel and craft workforce, the area's available workforce, and how much similar work the company has previously performed. Historical costs also include the cost of daily shutdown periods such as water breaks, 10-min breaks, and bathroom breaks. The production rates generated by this data take into account all the above, which, when applied to accurate work quantity data, results in a good depiction of actual overall labor costs. These do not include interruptions by the owner or subcontractors or delays resulting from adverse weather, labor strikes, or other abnormal interruptions to the work. Published cost manuals that attempt to take into account all the above using multiplying factors are not company specific and, therefore, are not as accurate.

Costs Associated with Reinforced Concrete Construction Reinforced concrete construction requires performance of many different items of work resulting in several categories of costs. The major costs are permanent materials, formwork erection and fabrication, small tools and services or supplies (STS), furnishing and placing reinforcing steel, and placing and finishing of the concrete. Of these, the ones directly affected by the form ratio are the activities related to the formed contact area. These specific activities are the labor for erecting and removing formwork, the placement of the concrete and the dry finishes required to the concrete surfaces.

Formwork Labor Costs Formwork is the largest contributor to concrete labor costs. As the formwork ratio increases, the costs per cubic yard associated with formwork also increases. For example, if the cost of formwork per cubic yard of a structural element is $50 and the form ratio for that element is 15, one would expect the cost of an element with a form ratio of 25 to be higher. Table 9.15 and the graphic version shown in Figure 9.16 indicate how formwork labor costs increase with the higher ratios. When such graphs are used in analyzing costs for previously performed work, it is helpful to develop a linear trend line so that the approximate concrete labor costs can be easily interpolated from a graph similar to Figure 9.16.

TABLE 9.15 Formwork Labor Costs per CY for Different Elements of Concrete Structures[a]

Element Type	Form Ratio (SF/CY)	Average Labor Factor (man-hours/SF)	Cost per SF ($40/man-hour)	Cost per CY
Bridge footings	6.00	0.16	$ 6.40	$ 38.40
Slab-on-grade, with CJs	1.89	0.22	$ 8.80	$ 16.63
Small walls	55.35	0.14	$ 5.60	$ 309.96
Large walls	22.14	0.10	$ 4.00	$ 88.56
Bridge piers	18.00	0.08	$ 3.20	$ 57.60
Small columns	13.50	0.15	$ 6.00	$ 81.00
Elevated slabs	30.24	0.25	$ 10.00	$ 302.40
Misc curbs & pads	5.40	0.30	$ 12.00	$ 64.80
Totals (avg)	19.60		$ 7.00	$ 119.92

[a]Costs are for labor only; materials and equipment are not included.

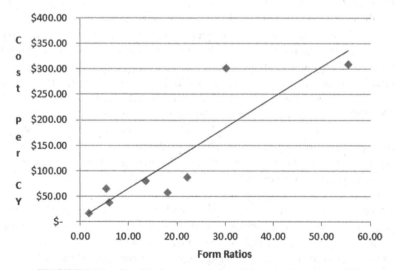

FIGURE 9.16 Graphical representation of formwork costs per CY.

In addition to the erect and strip labor costs, formwork fabrication labor and material costs need to be considered for every square foot of formwork area to be fabricated. As form ratios increase, formwork typically becomes more complicated; this increases the necessity for forms, which reduces the reuse factor. As the formwork ratio increases, the form reuse factor decreases. The reuse factor has to do with how many times one form can be used before it is thrown away or modified significantly (erect and strip quantity divided by form fabrication quantity), or simply the number of times a form is reused. If the form reuse factor is low, the project must produce more forms. For example, if the formwork material costs $5.00 per SF and they can be fabricated at a labor rate of 0.08 man-hour/SF (with labor @ $40.00/man-hour), the project would incur an additional cost of $8.20 per SF on form fabrication.

Another significant phenomenon is the trend that occurs in placing and removing forms by the square foot as the ratio changes. For small and detailed structures, when the ratio increases, the man-hours required per square foot decreases. In some sample past cost analyses of a form erect and strip operation, a form ratio of 16 yielded 0.195 man-hours per square foot. As the form ratio reached 34, the demand on erecting and removing the forms reduced to 0.148 man-hours per square foot. This is caused by the amount of area produced by the forms being different for the same amount of effort.

Concrete Placement As mentioned earlier, low form ratios are typically associated with slab-on-grades and thick elements, particularly those that are formed on one side only. When concrete is placed in these elements, more concrete can be placed per hour for the same size placing crew. The more concrete placed per hour, the less the labor hours required per cubic yard. As a comparison, high form ratios are associated with thin walls and elevated slabs and beam/girder elements (falsework supported). Thin walls have to be poured slowly, therefore, for every vertical foot of concrete, fewer cubic yards are placed. This can be compared to wider walls, which require much more volume for every vertical foot of placement. If a wall thickness doubles, the ratio and the cost per cubic yard are reduced to half. Elevated slabs and beam/girder elements produce slower concrete placements because they are typically thinner than on-grade element, and they sometimes have to be placed in a particular sequence to satisfy falsework design requirements. All forms are above-grade designed with specific pour rates, which must be followed to avoid form failures.

Concrete conveyance costs go hand in hand with placement costs. An example situation in which the concrete is conveyed to the forms by a concrete pump/boom system costing $180/h and is placed at a rate of 72 CY/h would require adding $2.50 per CY (180/72) to the placing costs. The longer a pour takes, combined with how many cubic yards are actually being placed per hour, greatly affects the placement cost. Consequently, if the pump is on site longer, the placing crew is also there longer. The setup and cleanup would be the same at the beginning and end of the pour, regardless of the time required to place the concrete.

Concrete Finishes Finishes in concrete work have two common classifications: wet and dry. Wet finishes are applied shortly after the concrete is placed but prior to the initial concrete set. Dry finishes are required on formed surfaces after the forms are removed in order to achieve the requirements of the specifications for smoothness and other aesthetics. Wet finish costs do not fluctuate with form ratios as dry finishes do. Dry finishes have a direct relationship to the formed surfaces. Formed surfaces will require some type of dry finish to meet the specification. Almost all specifications require either a point and patch or sack finish. Surfaces that will not be exposed to view when the structure is put into service generally require a much less stringent dry finish requirement. These types of finishes would also apply to surfaces that are backfilled and buried underground. The labor costs for these finishes range from 0.008 to 0.015 man-hours per square foot, depending on the specification requirements. On the other hand, concrete surfaces exposed to view, such as the outside surfaces of structures and public area exterior surfaces, will produce labor costs from 0.02 to 0.06 man-hours per square foot. Therefore, when the form ratio fluctuates from a 5 to a 10, the finish costs will range from $1.00–$10.00 per CY ($40.00/labor-hour).

Architectural finishes that are formed typically require a more expensive form and dry finish. There are also additional STS costs for the form liner material to produce the different patterns in the concrete. These form liners are attached to the face of the formwork in order to achieve the architectural design surface. Various patterns available on the market cost between $5.00 and $40.00 per SF. The inexpensive material is hard PVC and the higher quality, more expensive material is made from an elastomeric and can resist wear from multiple concrete placements. Form liner costs vary directly with the surface area and are related to the form ratio.

Small Tool, Services, and Supplies (STS) In addition to the labor costs for reinforced concrete estimates, small tools and services required to support formwork materials must be considered for purchase. This category includes formwork placement tools and incidentals and concrete placement and finishing tools. Formwork STS includes plywood, dimensional lumber, form ties, and hand power tools. Concrete placement STS includes vibrators, shovels, curing hoses, fittings and blankets, rubber gloves and boots, and construction joint preparation equipment such as sandblast pots, nozzles, hose, respirators, and goggles. Finishing STS include trowels, floats, cement and sand for dry patch, and personal protective equipment.

Formwork ratios fluctuate up and down, which affects the items associated with SFCA. Of the above-mentioned items, certain ones are directly related to the square foot of contact area (SFCA). These are formwork fabrication, placing, and removing formwork, dry finish, and wet finish. Typical costs for these items are:

Form fabrication: $1.50–$30.00/SF of fabricated forms

Form erect and strip: $0.30–$0.40/SF for all forms that are placed and removed, regardless of reuse

Concrete placement: $1.70–$2.00/CY of concrete placed

Dry finish concrete: $0.10–$0.15/SF of concrete, which requires some type of dry finish

Higher form ratios produce higher STS costs per cubic yard and labor hour. For example, a ratio of 5.0 would result in $1.50–$2.00/SFCA of formwork STS ($0.30/SF × 5.0 or $0.40/SF × 5.0). On the other hand, a form ratio of 20.00 would produce $6.00–$8.00/SF of formwork erect and strip, respectively ($0.30/SF × 20.0

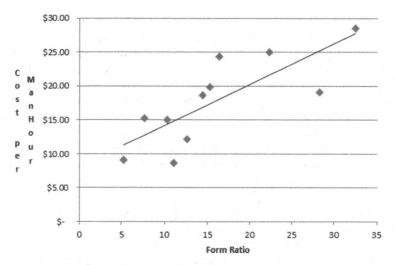

FIGURE 9.17   STS cost as related to form ratio.

or $0.40/SF × 20.0). Figure 9.17 shows the relationship between STS costs and concrete labor man-hours. The trend line indicates the increase in STS costs with an increase to the form ratio. The data points represent past project costs.

Reinforcing Steel   Reinforcing steel can vary by structure type and project. However, when concrete ranges from 8 to 60 in in thickness, the reinforcing steel ratio (pounds of steel per cubic yard of concrete) can vary in a similar fashion to the form ratios. For instance, a 12-in wall will typically have reinforcing steel at both surfaces ($1\frac{1}{2}''$ to $2''$ from the surface). In comparison, a $5'0''$-thick wall or column would also have two planes of reinforcing steel, except this larger element would have at least $4'0''$ of unreinforced concrete between the two layers of reinforcing steel. The 12-in wall would have a much greater weight of reinforcing steel per cubic yard than the larger element. If it is assumed that these two elements have a form ratio of 5–20, then it would be expected that the reinforcing weight per cubic yard would increase with the ratio. As an example, a 6-ft^2 column that is 20 ft tall and reinforced with 5-lb bars 2 in from the face, all around, and 12 in on center in each direction would produce 36 lb of rebar for every cubic yard of concrete. However, a 12-in-thick wall, 6 ft long and 20 ft tall with the same reinforcing layout produces 108 lb of rebar for every cubic yard of concrete. Structures that are more heavily reinforced would achieve higher ratios proportionately.

When a concrete quantity takeoff is complete, an estimator can "plug" a dollar amount for the reinforcing subcontractor. To do this, he often relies on the volume of concrete and a review of the size and spacing of the reinforcing steel. The size and spacing is going to indicate to an estimator approximately how many pounds of reinforcing steel may be needed for every cubic yard of concrete, similar to the above example. Once the weight is estimated, the cost (plug) can be obtained by reviewing historical project data that shows costs ranging from $0.75 to $1.25 per pound of placement ($1.10 to $1.75 if the reinforcing steel is epoxy coated). Since reinforcing is traditionally performed by a subcontractor specializing in furnishing and

TABLE 9.16 Portions of Concrete Costs Related to the Form Ratio[a]

Element Type	Form Ratio (SF/CY)	Erect & Strip Labor	STS	Place & Finish Labor	Total Cost per CY
Bridge footings	6.00	$120.00	$11.50	$24.00	$155.50
Slab-on-grade	1.89	$140.00	$5.34	$7.56	$152.90
Small walls	55.35	$320.00	$85.53	$221.40	$626.93
Large walls	22.14	$180.00	$35.71	$88.56	$304.27
Bridge piers	18.00	$160.00	$29.50	$72.00	$261.50
Small columns	13.50	$360.00	$22.75	$54.00	$436.75
Elevated slabs	30.24	$400.00	$59.86	$120.96	$580.82
Misc curbs & pads	5.40	$375.00	$10.60	$21.60	$407.20
Totals (avg)	19.60				$365.73

[a]Margin and indirect costs are not included in the above estimates.

installing reinforcing steel, these costs are left out of the overall labor costs normal to the concrete labor estimate.

Overall Costs The costs mentioned so far are estimated as individual components. Once they are analyzed individually, they can be combined into total costs. Table 9.16 shows some sample costs for the total concrete estimate. The total costs are an accumulation of the costs previously mentioned that fluctuate with form ratios. For instance, the purchase of the actual concrete has been left out since the volume of concrete is not affected by the form ratio by itself. Margins (company profits) and indirect costs were also left out of these costs. In other words, the costs shown are an estimate of only the direct labor costs that are affected by form ratios.

Types of Structures Structure types vary between the different uses of the structures. The geometry, size, and shape of each structure determine the form ratio. The following are categories of concrete structures and their elements:

Highway Bridges: footings, abutments, piers, and superstructure

Water/Sewage Facilities: slabs-on-grade, walls, and elevated slabs

Commercial and Industrial Buildings: slabs-on-grade, columns, walls, and elevated slabs

Bridges in California, as in many other western states, are typically the concrete box-girder type. The substructure (footings and columns/piers) and the superstructure (spans from pier to pier) are all cast-in-place reinforced concrete with posttensioning strand. The substructure form ratio is below 12. The superstructure form ration is above 20. Water treatment structures and commercial and industrial buildings have form ratios greater than 18. The foundations and walls of the latter are less thick and exhibit elements that do not have to support high dead loads; therefore, they consist of smaller elements. Refer back to Table 9.16 for typical form ratios as they apply to these elements.

When estimating different types of structures, the estimator should categorize the types of elements within the project type and keep these cost estimates separate. This would eliminate the chances that costs for the low ratio structures are not mixed with cost of the higher ratio structures. Once the elements are categorized, the quantity takeoffs should remain separate and only combined when intending to represent the overall project.

Estimating with Formwork Ratios

The Work Quantity Takeoff and Labor Work-Hours   The construction specifications will assist the estimator in determining how to categorize the takeoff in the cost-estimating process. A quantity takeoff is performed for all the individual work items that involve labor and materials. It is recommended that the structure types and the elements of the structures be categorized prior to the actual takeoff. Once it is determined which categories contain which structures and their specific elements, the estimator can begin quantifying the square feet of form contact area, the cubic yards of concrete, the dry and wet finishes, and the like. Spreadsheets are often used to break the elements down into components within an element. The spreadsheet (whether done by hand or with a computer) should show subtotals for each element, and for each structure type. It is very common for a company to require two separate quantity takeoffs, performed by two different takeoff engineers, when projects are bid by lump sum. This assures accurate quantities and reduces the risk exposure to monetary loss.

When the structure takeoff is complete, the form ratio for each of the work activities can be calculated. Then the estimator must decide on the man-hour production rate for each of the work activities that must be performed for each structural element. Table 9.17 shows an example of tabulated slab-on-grade historical costs data sheet.

If an estimator was pricing slab-on-grade formwork, data would be tabulated as shown in Table 9.17. Three past projects are listed next to an actual work quantity for the project being estimated. Historical production rates of 0.185, 0.124, and 0.234 man-hours per square foot are shown in the table as an example. The estimator must then decide what production rate to apply to the present project slab-on-grade activity. The three past projects can be used to determine the appropriate production rate. One way to do this would be to average the three past project production rates. However, this method would not consider the differing work quantities of the three projects. For instance, if the first project required 2000 ft^2 of forms, the second

TABLE 9.17   Example of Completed Project Quantity and Labor Hour Factors

Description	This Estimate				Past Project 1		Past Project 2		Past Project 3	
	QTY	UNIT	Man-hour/ SF	TOTAL Man-hours	QTY	Man-hour/ SF	QTY	Man-hours/ SF	QTY	Man-hours/ SF
Erect and strip Slab-on-grade forms	3200	SF	0.400	800	2000	0.185	5500	0.124	350	0.234

5500 ft^2, and the third 350 ft^2, an average would not reflect the effects of their differing quantities.

What is the effect of differing quantities? Larger work quantities such as those for the project 2 with 5500 ft^2 typically produce a better production rate because the operation generally continues to improve with time. On the other hand, project 3 with 350 ft^2 did not experience enough repetition to refine the process and produce better unit rates. In this case, it is not a surprise that the unit rate of 0.234 exceeds the others. The 2000-ft^2 project 1 falls somewhere between the other two.

If the project being estimated has a slab-on-grade quantity of 3200 ft^2, and the estimator has no reason to believe that the project elements are much different from the projects in the past cost summary, then the estimator would look at the two larger quantities (5000 ft^2 and 2000 ft^2) and choose a production rate somewhere in between. In this case, it would be appropriate to interpolate the unit rate between the other two, resulting in a rate of 0.163 man-hours per square foot.

When this is done, the results for each structure element can be totaled to represent the entire project. These totals will include the total cubic yards of concrete and the total square feet of contact form area resulting in the total labor hours per cubic yard of concrete for a given form ratio. The corresponding information for the present project can then be plotted for a comparison between the historical rates and anticipated performance. If the plots for the present project fall above or below the "trend line," then it is evident that the estimate is either high or low when compared to historical data.

When the activities have all been compared to past costs individually, and their production rates have been determined, another analysis should be done. Since we know the form ratio for the current estimate and the total concrete quantities, we should be able to line them up with historical costs of other projects and their totals. By doing this, similar projects are being compared. It would not make sense to compare costs between a project with a form ratio of 12 to a project with a form ratio of 22. This would defeat the purpose of using these ratios and comparing job totals. The form ratios of the project being compared should be within 10–15% higher or lower than the trend line of similar projects. Greater variations are cause for concern with regard to the present estimate. It should be noted that the above analysis pertains only to cost of labor and associated STS. A similar analysis can be performed for other concrete costs.

Tracking Project Costs Everything that has been discussed so far would be irrelevant if the project staff did not do an accurate job of tracking costs at the project level. A system should be in place to record quantities completed, labor hours spent, and total activity costs including equipment and materials. A system like this is almost always used so that companies can document the actual costs as compared to the estimated budget.

In addition, it is very important that the project staff accurately code the cost of individual activities. Project personnel are sometimes tempted to make an activity seem more cost effective than it is or claim quantities for work not actually performed, thus producing inaccurate results. The results, if used on future estimates, would generate an unrealistic estimate and eventually jeopardize the financial health of the company. The project reporting system needs to match the estimating system by category. These categories are similar to the estimating activities mentioned earlier.

9.3 CONCLUSION

Many items were covered in this chapter between formwork design and formwork costs. Since formwork can make up to 75% of the concrete costs, it cannot be emphasized enough how important the design of the forms are to these total concrete costs.

Concrete estimates (including formwork) are performed in similar manners from company to company. The student of this subject should leave academia with a solid understanding of the components between design and costs. Then, when putting the studies to practice, learn the actual methods used by the company in which the student is employed.

CHAPTER 10

FALSEWORK DESIGN

Chapter 9 set the stage for basic formwork calculations and the three standard design checks: bending, shear, and deflection. As we learned with formwork, concrete structures leave a permanent impression; and in many cases the final product is viewed by the public, Therefore, in its temporary nature, concrete not only has to comply with bending and shear issues, but concrete cannot deflect or undulate more than the human eye can see. Falsework is similar, but it is easier to see these undulations in vertical construction (walls and columns) than it is to see in horizontal constructions. With that said, let's bring to the table the concepts of risk and catastrophe. It is one thing to build a product that is not visually acceptable or pleasing to the eye. However, when constructing elevated, horizontal concrete segments, we must contend with gravity, and a whole other element comes into play.

10.1 FALSEWORK RISKS

Falsework supports elevated horizontal concrete elements such as building slabs, bridges, and beams. For years, owners and engineers have recognized falsework as being a temporary process to construction that needs to be engineered by a licensed, professional engineer. For years, vertical construction was not in this category, but in recent years, even vertical formwork has fallen under the scrutiny of the owner and his engineer. No owner wants any sort of incident associated with the project, no matter how large or small. Chapter 11 will get into design issues as they pertain to pipe falsework; and it can be assumed that these same issues would also be of concern for smaller horizontal and vertical concrete works. Figure 10.1, for example, shows falsework over a river using pipe post falsework. The risks of this type of work are illustrated from fall risks, equipment risks, and environmental risks at a minimum.

229

FIGURE 10.1   Pipe post falsework.

The intent of this book is not to prepare the student for designing and building formwork and falsework to construct the new San Francisco/Oakland Bay Bridge, the Tappan Zee Bridge in New York, or a 100-story downtown office building. The intent here is to educate the student for designing and building safe systems that are economically efficient and within the project's budget. The larger projects will be experienced eventually when the student builds confidence and is surrounded by a team of experts in their field.

10.1.1 Falsework Accidents

It seems that a year never goes by in the United States without a report of a false-work accident. These accidents are not just the result of design problems but are often caused by human error or bad decisions made by field supervisors from super-intendents to foremen and below. Even though methods and procedures are just as important, if not more important than design factors in temporary structures, the focus of this book will remain on design. In any case, whatever their cause, accidents are always unfortunate; and the only good that comes from them is the inherent lesson to be learned.

CASE STUDY OF A FALSEWORK ACCIDENT

Project Overview

Soon after the turn of the 21st century, a highway project was constructed in northern California by a joint venture company for the State of California (owner agency). The project involved road widening, overhead bridge ramps, and accompanying drainage on Highway 149 in Butte County. Figure 10.CS1 shows typical frames (bents) and stringers (beams) in place on this project for the typical pipe falsework system.

FIGURE 10.CS1   Typical falsework for this project.

The completed project would eliminate the need for surface intersections with stop signs or signals. On July 31, 2007, an overpass that was under construction collapsed.

Fortunately, no one was killed as a direct result of the collapse. However, one fatality resulted from traffic buildup, and one construction worker and one person driving underneath the collapse were injured. This relatively fortunate outcome was not the result of good decision making or good planning, but was pure luck. The contractor and the State of California dodged a deadly and costly bullet.

Forensic studies were performed, as one would expect, by Cal OSHA. However, the cause of the accident was not ultimately decided by a judge. No one served any time in prison. The injured persons settled with the contractor and insurance companies; and by 2011, one would have thought nothing had ever happened. So what actually caused the collapse of the falsework?

The Cause

A falsework bent consisting of pipes and beams collapsed onto a passing truck, nearly crushing the driver and causing a construction worker to fall, when a

(continued)

CASE STUDY OF A FALSEWORK ACCIDENT (*Continued*)

guying cable was removed. The frames and beams were placed during a previous night shift with the highway closed in both directions with traffic detoured. During this shift, the frames were braced with wire rope guys to temporary deadmen (counterweights). Before the wire rope guys could be transferred to the adjacent bent for a more permanent attachment, the shift had ended and the highway was reopened to traffic. The following morning shift began, and the crew and owner agency realized the guying was not complete. Instead of setting up another road closure the following evening and correcting the situation without live traffic, the decision was made by the contractor and owner agency to relocate the guys that morning.

When the guy was released, so much tension had built up that the frame moved enough to generate momentum that toppled the frame and stringers it supported. The falsework bent failure was caused by human error. Had the plan been followed, the accident would probably have been avoided. However, had a better bracing system been designed and installed initially, this could have prevented the accident.

Corrective Measures

The project was delayed for several weeks. Once the investigation allowed the contractor to continue, a new bracing system was put into place. Figure 10.CS2 shows the new pipe bracing system in place.

FIGURE 10.CS2   Push–pull braces added to falsework.

This bracing was added to the new falsework bents that were erected after the accident. They went to a push–pull system to take the wire rope tension out of the equation and to allow bracing on only one side of the bent since traffic was on the other side.

The parties involved spent several years, after the completion of the project, in settlement negotiations. The settlements were never made public.

The next section will discuss the measures that are taken by falsework contractors and owner/engineers to assure that accidents due to either design or to human error or incorrect methods are avoided.

10.1.2 Falsework Review Process

Two types of falsework review procedures are common in the construction industry. These review procedures are engineering checks and balances and they are (1) in-house reviews and (2) owner/engineer reviews.

In-House Reviews In-house reviews occur within the company performing the falsework erection and removal. This part of the process can, and frequently does, involve consulting engineers when the contractor does not have an engineer on his payroll or its engineering staff is too busy. Regardless of the engineering component, this group represents the contractor.

Some contractors develop written procedures that are followed during the design process, prior to submitting it to the owner/engineer. These written procedures contain some of the following information:

Type of falsework risk
Eligibility of persons involved
Redundancy
Risk avoidance

The contractor must categorize the risk somehow. An example would be to rate the falsework type as low, medium, or high risk. Once the levels have been decided, each risk level would represent a specific type of falsework. In other words, the lower risk falsework would be used for the most straightforward, easier systems; higher risk falsework would be used for projects spanning sensitive waters, busy highways, or in very tall systems. Table 10.1 has been prepared as one of many options for a review policy matrix. The dimensions used are recommendations or estimates. Each company should consult its legal department before preparing a similar policy.

Each company should develop a policy, communicate that policy to the employees, and manage the policy. Obviously, some terms would still have to be carefully defined in the document. For instance, the word "minimal" could mean different things to different people. The policies cannot have vague wording or contradictions.

TABLE 10.1   Example of Falsework Risk Categories

Risk Category	1. Designer and 2. Check Engineer	Falsework Type
Low	1. In-house engineer, consultant engineer 2. In-house engineer, consulting engineer (not the same)	Elevation less than or equal to 20 ft in height, beam spans less than or equal to 30 ft in length. No risk to public; no risk to environment.
Medium	1. Consultant engineer 2. In-house engineer, consulting engineer (not the same)	Elevation less than or equal to 30 ft in height, beam spans less than or equal to 40 ft in length, minimal risk to public, some minor roadways, minimal risk to environment.
High	1. Consultant engineer 2. Additional consultant engineer	Elevation greater than 30 ft in height, beam spans less than 40 ft in length, high probability there is risk to public, such as highways, railroads, and private property, high probability there is risk to environment, both which would mean substantial financial company losses.
Extremely High Risk	1. Consultant engineer 2. Additional consultant engineer	Permanent Design/Temporary Support (Chapter 13)

Persons who are eligible to design and check the design should be clearly specified as well. In no case should these two responsibilities be performed by the same person. Also, in no case should these two responsibilities be performed in-house. In some cases (high risk), the company may not want any in-house involvement. The risk, in this case, would be 100% subcontracted to a consulting firm. Each company should evaluate its own risk and write a specific policy that addresses the most critical management concerns. An additional column could also be added to Table 10.1 titled "Inspection Responsibility"

Owner/Engineer Reviews   On the other side, that of the owner/engineer, similar concerns are shared for possibly different reasons. No owner/engineer wants any problem of any kind to occur during the construction process. Since falsework comes with additional risks from normal construction activities, it is scrutinized to a higher degree. Falsework approvals are, in fact, so important to CalTrans that the CalTrans Standard Specifications manual contains a separate specification section for falsework. Their specifications cover anything that has or could possibly go wrong on a highway project.

Owner/engineer reviews can contain a great deal of requirements. The following list is just some items that might be required in a falsework submittal:

Material compliances and specifications

Letters of certification

Welding certifications for previously welded splices

Calculations for all elements

Shop drawings, separate submittals for each bridge and each frame

Erection and removal procedures, with sequencing

Material grades and species

Anticipated settlements

Required experience of designer when high risk (minimum 3 years)

10.1.3 Falsework Design Criteria

The basic design checks that apply to both formwork and falsework are:

Bending stress for plywood and dimensional lumber, $l = 10.95\sqrt{F_b S/w}$

Shear stress for plywood, $L = F_v(Ib/Q)/0.6w$

Shear stress for dimensional lumber, $L = [F'_v(b)(d)/0.9w] + 2d/12$

Deflection for plywood and dimensional lumber, $l = 1.69\,3\sqrt[3]{EI/w}$, l is clear span in inches, and L is clear span in feet.

If a form designer is using plywood and dimensional lumber for all falsework components, these formulas will be instrumental. When other materials are used, as will be the case later in this chapter, other formulas and charts will be necessary.

Falsework loads, even though originating from material weight and gravity, are actually fairly light compared to hydrostatic pressure. In the last chapter, the loads from hydrostatic pressure were in the range of 800–1200 psf. Many falsework loads, except for large public works and transportation projects, are in the range of 200–500 psf. As mentioned in Chapter 8, falsework loads include dead loads and live loads and self-weight of stringers. However, even with these additional forces, most falsework loads are still lighter than formwork loads. As contractors, we need to be continuously reminded of the risk in construction of elevated horizontal concrete works and always perform the checks and balances that go along with this type of work that were detailed earlier in this chapter.

This chapter will show three examples of different types of falsework in construction. The first two will be standard wood falsework supporting elevated reinforced concrete slabs. One will use the conventional method for calculating loads, and the other will use the Appendix 7 formwork charts used in the previous chapter. The first example will also incorporate the load calculations learned in Chapter 8 so the student can continue to master this process. Example 10.3 will utilize aluminum beams and 15-k shoring frames.

As mentioned in Chapter 8, elevated slab loads are a combination of the concrete and material dead load (DL) and the live load (LL) from workers, equipment, tools, and the weight of the falsework beams. The wood and plywood weight are included with the DL. The formula given was P_{max} = DL (concrete, rebar and forms) + LL (workers, equipment, tools and steel beams).

DL = Slab thickness (feet) X 160 pcf (160 = concrete unit weight including rebar and forms)

LL = Depends on the type of construction (ranges from 25 to 75 psf)

Light duty = 25 psf (painters and cleaning)

Medium duty = 50 psf (carpentry)

Heavy duty = 75 psf (concrete, masonry, demolition, and steel)

FIGURE 10.2   Water tank falsework using aluminum system.

10.1.4   Load Paths for Falsework Design

The following paths and component terms will be used in this chapter, depending on the type of construction. Following the example types, the load paths described will be flat slab with wood posts; flat slab with dropped beams using aluminum joists, beams, and frames; and bridge falsework using wood and steel beams.

1. Elevated slab support with all wood components—plywood, joist, cap beam, post, ground, or lower slab
2. Elevated slab support with aluminum components—plywood, aluminum joist, aluminum cap beam, aluminum 15-k frames, ground or lower slab (Figure 10.2)
3. Bridge falsework with a combination of wood and steel beams—plywood, wood joists, steel stringers, steel cap beams, posts, lower cap beams, sandjacks/ wedges/corbels, pads, or ground slab

Example 10.1   Flat Slab Falsework
Design falsework to support a 25″ slab with heavy-duty live load. In order to save time, deflection will be ignored in these first two examples. Use Douglas fir–Larch, 4×4 joists, and 4×8 cap beams.

Step 1: Determine the loading from the DL + LL (see Figure 10.3).

DL = 25″/12″/ft X 160 pcf = 333 psf
LL workers/equipment = 75 psf
Total = 408 psf (use 410 psf)

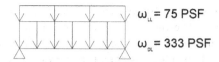

$\omega_{LL} = 75$ PSF

$\omega_{DL} = 333$ PSF

FIGURE 10.3   Load diagram of example 10.1.

Step 2: Select the plywood and determine the joist spacing.

Since the plywood will be placed by hand and the loading is relatively light compared to wall forming, use $\frac{5}{8}''$ BB-OES in the strong direction. The plywood has an allowable bending stress value of 1600 psi and allowable rolling shear value of 60 psi.

$$W = 410 \text{ psf} \times 1 \text{ ft} = 410 \text{ plf}$$

Bending

$$l = 10.95\sqrt{(F_bS/w)}$$

where
l = clear span (in)
F_b = allowable bending stress of material (1600 psi)
S = section modulus (0.339 in^3)
W = uniform load on span (410 plf)
10.95 = constant
$$l = 10.95\sqrt{(1600)(0.339)/410} = 12.59 \text{ in}$$

Shear-Plywood

$$L = \frac{F_v(Ib/Q)}{0.6\,w}$$

where
L = clear span (ft) divide by 12 to compare in inches
F_v = allowable shear stress of material (60 psi)
Ib/Q = rolling shear constant (5.824 in^2)
W = uniform load (410 plf)
0.6 = constant

$$L = \frac{(60)(5.824)}{0.6(410)} \times 12''/\text{ft} = 17.04 \text{ in}$$

Deflection is ignored for this example because it only controls long spans with light uniform loads.

$$l = 1.69\sqrt[3]{(EI/w)}$$

where
l = clear span (in)
I = moment of inertia (in^4)
W = uniform load on span (plf)
E = modulus of elasticity
1.69 = constant

Ignoring deflection, the joist spacing can't exceed 12.6 in of clear span, which is governed by *bending*. The on-center spacing of the joists will be greater, and the amount depends on the width of the joist. If a Douglas fir–Larch (DFL) 4 × 4 (b = 3.5″) joist is used, then the on-center spacing would be 12.6 + 3.5″ = 16.1″, say 16″ OC. Figure 10.4 shows this spacing between joists.

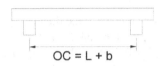

OC = L + b

FIGURE 10.4   Joist spacing for example 10.1.

Step 3: Determine the joist size and cap beam spacing.

In step 2, the joist size was selected to be a DFL 4 × 4. This joist size will be used to determine the cap beam spacing. First the uniform load on one joist must be recalculated.

$$W = 410 \text{ psf} \times (16''/12''/\text{ft}) = 546 \text{ plf}$$

along each joist. With this uniform load and the properties of the DFL 4 × 4, determine the cap beam spacing. Perform the two checks, ignoring deflection.

Bending

$$l = 10.95 \sqrt{(1000)(7.15)/546} = 39.63 \text{ in}$$

Shear—Dimensional Lumber

As a reminder, the dimensional lumber calculation for shear is different than plywood:

$$L = \left(\frac{F'_v bd}{0.9w} \right) + \left(\frac{2d}{12} \right)$$

where L = clear span (ft), divide by 12 to compare in inches
F'_v = twice the allowable shear stress (2 × 95 psi = 190 psi)
b = base dimension of wood section (3.5 in)
d = depth dimension of wood section (3.5 in)
w = uniform load (546 plf)
0.9 = constant

$$L = \frac{190 \times 3.5 \times 3.5}{0.9(546)} + \frac{2 \times 3.5}{12} = 5.32 \text{ ft}$$
$$l = 5.32 \times 12''/\text{ft} = 63.8 \text{ in}$$

Deflection is ignored for this example because it only controls long spans with light uniform loads.

Ignoring deflection, the cap beam spacing can't exceed 39.6 in of spacing, which is governed by *bending*. Space the cap beams at 39″ OC because 39″ is 3.25 ft, and this will normally work out better for the overall layout of the falsework. Figure 10.5 shows the cap beam spacing.

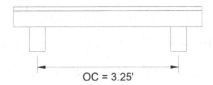

OC = 3.25'

FIGURE 10.5   Cap beam spacing for example 10.1.

Step 4: Determine cap beam size and post spacing.

In step 3, the cap beam size was selected to be a DFL 4 × 8. This cap beam size will be used to determine the post spacing. Once again, the uniform load on one cap beam must be recalculated.

$$W = 410 \, \text{psf} \times (39''/12''/\text{ft}) = 1333 \, \text{plf}$$

along each cap beam. With this uniform load and the properties of the DFL 4 × 8, determine the post spacing. Perform the two checks, again ignoring deflection. Be sure not to double the cap beam properties as was done in formwork using double walers. Cap beams are single beams, so the properties are left single.

Bending

$$l = 10.95\sqrt{(1000)(30.66)/1{,}333} = 52.51 \, \text{in}$$

Shear

$$L = \frac{190 \times 3.5 \times 7.25}{0.9(1333)} + \frac{2 \times 7.25}{12} = 5.23 \, \text{ft}$$

$$l = 5.23 \times 12''/\text{ft} = 62.8 \, \text{in}$$

Deflection is ignored once again.

Ignoring deflection, the post spacing can't exceed 52.5 in of spacing, which is governed by *bending*. Space the cap beams at 52″ OC. Figure 10.6 shows the cap beam spacing.

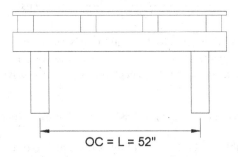

OC = L = 52"

FIGURE 10.6   Post spacing for Example 10.1.

Step 5: Determine post load and sizing.

The post load can be determined by distributing the P_{max} load over the tributary area of one post.

$$P_{post} = P_{max} \times (\text{cap beam spacing} \times \text{post spacing})$$

$$P = 410 \text{ psf} \times (39''/12''/\text{ft}) \times (52''/12''/\text{ft}) = 5,774 \text{ lb}$$

The tributary area per post is 14.08 ($39'' \times 52''/144$) ft^2. A post must be designed that can support 5774 lb. For this example, it is assumed that the post height (from ground to bottom of cap beam) is 13 ft. This represents L_e = effective post length. With this information, a 4×4 DFL post can be checked. If this post does not work, a larger size should be checked or the post spacing can be decreased to lower the load per post. Typically, L_e can't be changed because the elevated slab must remain at the same elevation per the contract plans. However, L_e can be reduced if lateral bracing is added. For this example, it is assumed that lateral bracing will not be used, and the L_e will remain at 13 ft.

To check the post size, perform the following check:

$$\frac{P}{A} > \frac{0.3E}{(L_e/d)^2}$$

where P/A = normal axial stress on the post

E = modulus of elasticity of DFL

L_e = effective length of the post in inches

d = least dimension in a cross section of the post

Normal stress on the post would be

$$\frac{P}{A} = \frac{5774 \text{ lb}}{(3.5)(3.5)} = 471 \text{ psi}$$

This is the normal stress at the end of a 4×4. The allowable compression stress parallel to the grain for DFL is 1600 psi, which is well below the crushing limits. Buckling is the main issue when a 4×4 is required to support a load as a 13-ft-long post. To do this, use the formula above to determine the allowable buckling stress and ultimately the allowable load the post can support:

$$\frac{P}{A} > \frac{0.3(1,500,000)}{(13' \times 12''/\text{ft})/(3.5'')^2} = 227 \text{ psi}$$

The allowable buckling stress for this condition is much less than the normal stress; therefore, the buckling stress dictates the design of the post. To determine the allowable post load, plug the buckling stress back into $f_{bs} = P/A$, solve for P, and compare this to 5774 lb.

$$P_{all} = 227 \text{ psi} \times (3.5 \times 3.5) = 2781 \text{ lb} < 5,774 \text{ lb} \quad \text{(does not work)}$$

Since the allowable load due to buckling is not equal to or greater than the actual load of 5774 lb, a larger post must be considered. Before going to a 6× (which is very expensive), the designer should try a 4×6, 4×8, and so forth. The largest $4\times$ that should be considered is the size just below the cross-section area of a $6 \times 6 = 36$ in^2 (which would be the first $6\times$ to be tested). A 4×8 is 32 in^2 and a 4×10 is 40 in^2; therefore, if a 4×8 does not work, then the 6×6 should be considered. Since the d dimension of both a 4×6 and 4×8 are 3.5″, the allowable buckling stress can be multiplied by the cross-sectional area to quickly calculate the allowable load on each. In other words, there is no need to recalculate the allowable buckling stress, as it will not change as long as the least dimension, d, does not change:

P_{all} for $4 \times 6 = 227$ psi $\times (3.5'' \times 5.5'') = 4370$ lb < 5774 lb   (does not work)

P_{all} for $4 \times 8 = 227$ psi $\times (3.5'' \times 7.25'') = 5760$ lb < 5774 lb   (does not work, but very close)

By observation, it appears a 4×10 would work, but for economy purposes, it was pointed out that a 6×6 should be used if a 4×8 did not work. Unfortunately, now, the buckling stress must be recalculated due to the increased d dimension changing from 3.5″(4×) to 5.5″(6×).

$$F_{bs} = 0.3(1{,}500{,}000 \text{ psi})/(13 \times 12''/\text{ft}/5.5)^2 = 559 \text{ psi}$$

$$P_{\text{all}} = 559 \text{ psi} \times (5.5 \times 5.5) = 16{,}910 \text{ lb} > 5774 \text{ lb} \quad \text{(OK)}$$

Usually when the post goes up to the next minimum dimension, the allowable capacity doubles or triples. Also, if the designer really wanted to use the 4×4's, he could go back to the beginning, then add lateral bracing and decrease the L_e. Another way to make the 4×8 work would be to slightly decrease the spacing between either the posts or cap beams, thus reducing the tributary area per post and eventually the post load. At this point, a cost comparison should be done between the 6×6 and the labor and material cost for the lateral bracing (lacing). Finally, the savvy student has also recognized that since we used a 6×6, the b dimension used in step 4 should be increased to 5.5 in, which changes the post load. Technically, this slight change can be ignored because, by using the 6×6, we built in an additional factor of safety = $16{,}910/5774 = 2.93$ FOS.

Bearing Capacity

The crushing stress at the end of the 4×4 was discussed, but the crushing between the 4×8 cap beam and the 6×6 post has not been checked. This is more critical because this crushing is considered perpendicular to the grain of the 4×8 cap beam, which is more sensitive than the end of the post (parallel to the grain of the post). Figure 10.7 illustrates the bearing area in question.

$$f_{c\perp} = P/A(\text{rea})$$

$$f_{c\perp} = 5774 \text{ lb}/19.25 \text{ in}^2 = 300 \text{ psi}$$

The allowable crushing stress perpendicular to the grain for DFL is 625 psi > 300 psi (OK).

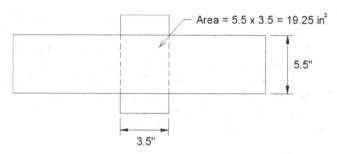

FIGURE 10.7 Bearing area between the post and cap beam.

Final Design:

5/8″ BB-OES plywood used in the strong direction

DFL 4 × 4 joists spaced at 15″ OC

DFL 4 × 8 cap beams spaced at 42″ OC

DFL 6 × 6 × 13 ft long

No lacing for posts, crushing in both directions (OK)

10.1.5 Falsework Design Using Formwork Charts

The next example will utilize the formwork charts in Appendix 7. The charts selected in this chapter are from the *Williams Form Catalog*. These particular charts use fairly liberal shear values, hence all spacings are governed by either bending or deflection.

Example 10.2 Flat Slab Falsework
Using the formwork tables in Appendix 7, design falsework to support a 21″ slab with medium-duty live load. Deflection will still be ignored.

Design a falsework system using the charts in Appendix 7 and following these parameters:

DL = 21″/12″/ft X 160 pcf = 280 psf

LL workers/equipment = 50 psf

Total = 330 psf (Figure 10.8)

$\omega_{LL} = 50$ PSF

$\omega_{DL} = 280$ PSF

FIGURE 10.8 Slab load diagram.

Step 1: Plywood selection and joist spacing.

For this example, $3/4''$ BB-OES plywood will be used. Based on this plywood and the charts in Appendix 7, determine the joist spacing. This appendix is using material that has the following characteristics:

Max deflection L/360, not to exceed $\frac{1}{16}''$

$F_v = 72$ psi

$F_b = 1930$ psi

$E = 1,500,000$ psi

From the previous calculation, $w = 330$ psf

Based on the plywood being used with the face grain parallel to the span (left side of chart) and $w = 330$ psf, the joist spacing should not exceed 17″ OC. In order to make the fabrication and erection most productive, the joists will be spaced at an even 16″ OC. Like the charts in the previous chapter, the values given are the center-to-center spaces.

Step 2: Joist selection and cap beam spacing.

$$\text{Uniform joist load: } w = 330 \text{ psf} \times (16''/12''/\text{ft}) = 440 \text{ plf}$$

The next chart in Appendix 7 that is used is intended for all single-beam (studs, joist, and others) members. Also, this chart represents beams that are supported by three or more supports ($M = wl_2/10$), not simply supported ($M = wl^2/8$.).

In this example, a 2×6 joist will be used. So with a new uniform load on the joist of 440 pfl, determine the maximum spacing of the cap beams, which is the maximum span of the 2×6.

Some interpolation has to be used on the charts when the uniform loads are in between the values shown. Always go to the more conservative number. Taking this into account, cap beams should not be spaced farther than 56″ OC. If the designer wanted to make the layout of this system go without the potential of error, 54″ could be selected as an even 4′6″ OC.

Step 3: Select a cap beam size and determine the safe spacing between the posts.

A 4×8 has been selected for a cap beam. The chart selected for this step is the same as the single-beam chart because a cap beam is a single beam. Be cautious not to accidentally use the double waler chart as was used in the Chapter 9 examples.

The new uniform load on a single 4×8 cap beam is determined as follows:

$$W = 330 \text{ psf} \times 4'\text{-}6'' = 1481 \text{ plf}$$

The single-beam chart indicates that a 4×8 cap beam can span 60 in between posts. Therefore, post spacing should not exceed 60″ (5′0″).

Step 4: Determine the post load on a single interior post (post design will be skipped in this example, so refer to Example 10.1).

$$P_{all} = P_{max} \times (\text{cap beam spacing} \times \text{post spacing})$$
$$= 330 \text{ psf} \times (4.5' \times 5') = 7425 \text{ lb}$$

Final Design:

$\frac{3}{4}''$ BB-OES plywood used in the strong direction

2×6 joists spaced at 16″ OC

4×8 cap beams spaced at 4′6″ OC

Space posts at 5′ 0″ OC (post size not determined in this example)

Post load $= 7425$ lb

Example 10.3   Elevated Slab with Dropped Beams

Design an elevated slab and dropped beam falsework using an aluminum beam and shoring frame system.

This example will utilize standard form plywood, aluminum joists, and cap beams supported by 15 k per leg capacity aluminum frames. All components except the plywood are premanufactured metal products. Appendix 7 will be used for plywood properties and Appendix 9 will be used for the aluminum beam properties.

The sketch in Figure 10.9 shows a typical section through the slab and dropped beams. This type of falsework is more complicated because the dropped beams and the slab are at different elevations. Therefore, two levels of decking are required.

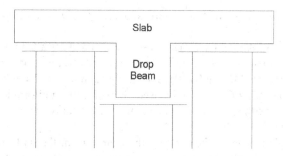

FIGURE 10.9   Slab and dropped beam.

The design load including live load and material will be given as the following:

Elevated slab section = 220 psf
Dropped beam = 480 psf

As mentioned earlier, the aluminum beam charts from the previous chapter will be sufficient for the joists and cap beams; and each post of the modular frames can support 15 k, which already includes a 2.2 : 1.0 factor of safety. The frames come in a variety of widths and heights. Figure 10.10 shows an arrangement of shoring frames for an elevated slab and beam system.

Step 1: Determine the joist spacing for both the slab and dropped beams using the same $\frac{3}{4}''$ plywood as Example 10.2. Use Appendix 7 for aluminum beam spacing.

Elevated slab = 220 psf, maximum joist spacing = 19" OC
Dropped beams = 480 psf, maximum joist spacing = 15" OC

Stage 1: Elevated Slab

Step 2A: Determine the maximum spacing of the slab joists using 19" spacing and the chart for aluminum beams in Appendix 9.

$$W = 220 \text{ psf} \times (19''/12''/\text{ft}) = 348 \text{ plf}$$

The aluminum beam chart in Appendix 9 determines the maximum span of the aluminum joist supporting the elevated slab. It will be assumed that these joists will

FIGURE 10.10   Slab and dropped beam.

span three cap beams (two spans). Therefore, the two-span section of the chart will be used. In the two-span column, go down to 348 pfl and the beam span will be indicated on the left, y axis.

Using the $\frac{1}{4}''$ maximum deflection (started values in Appendix 9) the beam can span 10.5 ft. This is a very long span, but at the same time, the load is very light. At this point a decision has to be made on whether to space the joists at 10.5 ft or less to change the cap beam load. To make this decision, list the different cap beam loads that apply to some various joist spans.

Table 10.2 shows span options for the aluminum beams and cap beams.

TABLE 10.2   Aluminum Beam Spans with Cap Beam Spacing from Appendix 9

Joist Span (ft)	Cap Beam Uniform Load (plf)	Spans (ea)	Cap Beam Spacing (ft)
10.5	2310	2	4.0
9	1980	2	4.5
8	1760	2	5.5
6	1320	2	7.0
5	1100	2	7.5

Another consideration is the size of the slab sections in plan view. If the slab lengths are 15 ft, for instance, then half of the section would be 7.5 ft. The joists can't span the whole 15 ft; therefore, it would make sense to split the distance in half and span the joists only 7.5 ft. In addition to the geometry of the concrete segments, the shoring frames come in certain sizes that range from 4-ft dimensions to 7-ft dimensions as shown in Figure 10.11. With this said, it would not make sense for either span to be more than 7 ft.

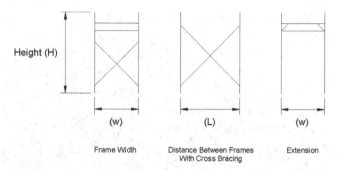

FIGURE 10.11   Shoring frame options.

Looking at Table 10.2, many options are available; but if we look at the shoring frame geometry, the spacing between the frame legs can dictate the beam spacing. Sometimes in temporary structure design, the beams allow greater spacing than is practical for the rest of the design. This happens frequently with aluminum components because aluminum is so efficient that the rest of the supports can't match its strength; therefore, the aluminum gets underutilized to compensate for the system. This does not mean that it is not economical. Even underutilized, aluminum systems have great value.

Table 10.3 shows some examples of the available frame sizes that can be purchased or rented.

TABLE 10.3   Some Available Shoring Frame Sizes

Frame Options	(W) Width of Frame (ft)	(L) Length with Cross Brace (ft)	(H) Height of Frame
1	4	8	6 ft
2	4	10	6 ft
3	2	8	5 ft
4	2	10	5 ft
1 Extension	4	8 or 10	5'4"
2 Extension	2	8 or 10	4'4"

In this example, we will space the aluminum beams to match an efficient aluminum frame system shown in the charts above. The 4-ft-wide frames are the most common and should be used whenever space allows. These frames will also maximize the system by using fewer frames.

Figure 10.12 shows the plan view layout of the shoring frames. The dropped beams are shown with double lines and the flat slabs are in between.

Stage 2: Drop Beam

Step 2B: Determine the maximum spacing of the drop beam joists using 15" spacing and the chart for aluminum beams.

$$W = 480 \text{ psf} \times (15''/12''/\text{ft}) = 600 \text{ plf}$$

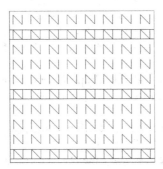

FIGURE 10.12 Plan view layout of shoring frames.

The aluminum beam chart in Appendix 9 determines the maximum span of the aluminum joist supporting the drop beam. It will be assumed that these joists will span three cap beams (two spans). Therefore, the two-span section of the chart will be used. In the two-span column, go down to 600 pfl and the beam span will be indicated on the left, y axis. As before, the span is very long and the geometry of the structure may ultimately govern the actual beam spans. The chart in Appendix 9 indicates a maximum span of 9 ft using the $\frac{1}{4}''$ max deflection. However, like the slab, this spacing may not meet the geometry of the structure. The beams are less than 2 ft wide so only one row of frames per beam will be necessary. If a 4-ft frame is used, the beam will load a single row of frames. Figure 10.13 shows a single row of frames under a single cap beam.

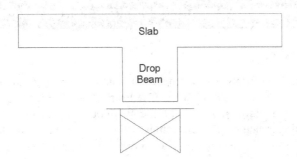

FIGURE 10.13 Beam shoring.

Stage 3: Shoring Frames

Step 3: Shoring tower capacities are rated based on what one leg can support with a 1.5–2.5 : 1.0 FOS. The design of the shoring towers involves calculating how much load (tributary area) is supported by one individual leg. Flat slab reinforced concrete decks are easily calculated for each shoring leg because the unit weight per square foot is the same throughout the system. This dropped beam scenario, however, is more complicated because of the two different thicknesses of the top slab (220 psf) and the dropped beam (480 psf).

If the top slab shoring creates a 4-ft × 7-ft pattern, the load per shoring leg is as follows:

$$220 \text{ psf} \times 7 \text{ ft} \times 4 \text{ ft} = 6160 \text{ lb} < 15,000 \text{ lb} \quad \text{(OK)}$$

Using these same frames under the dropped beam as in Figure 10.13 would result in the following shoring leg load:

$$480 \text{ psf} \times 2 \text{ ft wide beam} = 960 \text{ plf of beam}$$

If the same pattern is used $(4' \times 7')$, then the load for one leg would be as follows.

$$960 \text{ plf}/2 \text{ sides} \times 7 \text{ ft} = 3360 \text{ lb}$$

Or

$$480 \text{ psf} \times 7 \text{ ft} = 3360 \text{ lb}$$

Most aluminum systems are rated for 10–15 k per leg, which includes the applicable FOS. These shoring towers would be more than acceptable for this example; and one might even consider a lighter weight steel system (10–11 k).

The case study to follow is an example of how a building contractor decides to form multiple-story slab decks and the methods used. This case study is purposely not similar to the chapter examples so diversity can be illustrated.

CASE STUDY: BUILDING FALSEWORK AND FORMWORK—CHARLES PANKOW BUILDERS, LTD.

Background

Selecting optimal formwork systems for a concrete framed building can be a daunting task. For the past 50 years, Charles Pankow Builders, Ltd. has been a leader in the building industry, specifically in buildings that utilize concrete as the building structure. This case study will provide insights on why Pankow selected the forming systems they did on two different projects.

Criteria

The following criteria are reviewed by Pankow's preconstruction group, superintendents, and foremen, when selecting form work systems. One of the first and foremost criteria is selecting forming systems that reduce on-site labor, the riskiest part of the construction process. Other criteria include:

- Labor productivity
- Familiarity of forming system
- Form reuse

- Prefabrication and rental costs
- Schedule
- Site constraints

In general, Pankow does not own its own formwork and typically rents from formwork suppliers, which allows the company to be flexible in selecting the right formwork for the job.

Case Study: 20-Story Apartment Building, Los Angeles

Statistics

- 18 typical residential floors + 2 garage levels
- 8″ thick mild steel and post-tensioned slabs
- 9′8″ floor-to-floor height
- 154,000 ft^2 per floor, one slab pour per floor
- Rectangular concrete core
- 2 shear walls outside of core
- 15 concrete gravity columns, regularly spaced and in-line
- Hoisting via tower crane (30,000-lb capacity)

For this case study, we will concentrate on the typical floor forming systems shown in Figures 10.CS3 and 10.CS4.

FIGURE 10.CS3 Twenty-story apartment building, Los Angeles, California (courtesy of Charles Pankow, Ltd and Doka Forms).

(*continued*)

CASE STUDY: BUILDING FALSEWORK AND FORMWORK—CHARLES PANKOW BUILDERS, LTD.
(Continued)

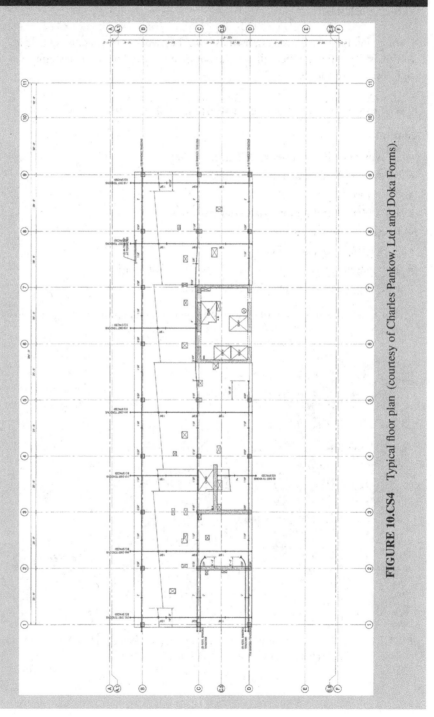

FIGURE 10.CS4 Typical floor plan (courtesy of Charles Pankow, Ltd and Doka Forms).

Floor Slab Forming System

Given the size of the floor plate, the regularly spaced and aligned columns, and the number of reuses of the forms, a column-hung, large fly table system (rented from Atlas Construction Supply, Inc.) was selected, a system that is Pankow's first choice for a project such as this. A fly table slab-forming system (Figure 10.CS5) provides Pankow good labor productivity, is fast setting, is a system Pankow's crews are familiar with, and reduces on-site labor when compared to other systems, for example, handset. The fabrication costs for these fly form systems are much higher than handset systems; therefore, a high reuse of fly forms is imperative for these systems to be economical for a project. A rule of thumb in the Pankow forming "handbook" is that fly tables will make economic sense when they are regular and can be reused at least 14 times.

FIGURE 10.CS5   Fly table (courtesy of Charles Pankow, Ltd and Doka Forms).

When comparing fly table systems, one aspect to consider is column-hung supports or truss/shored supports. The advantage of the column-hung systems over truss/shored systems is they do not require reshores. For truss/shored fly table systems, reshores are required to be installed three floors below the floor being placed in order for the wet concrete load to be transferred safely to the floors below. These shores get in the way of trades working below the concrete deck. The column-hung fly table system does not require reshores and allows the drywall and MEP trades to begin layout and installation immediately without reshores being in the way. This helps the overall construction schedule and

(continued)

**CASE STUDY: BUILDING FALSEWORK AND
FORMWORK—CHARLES PANKOW BUILDERS, LTD.
(*Continued*)**

keeps work flow closely behind the concrete operations. Figure 10.CS6 shows
a fly table being installed.

FIGURE 10.CS6   Fly table installation (courtesy of Charles Pankow, Ltd and Doka
Forms).

Concrete Core Forming System

A crane set, jump form system (Figure 10.CS7) was selected to form the concrete core. Because the core walls formed a rectangular box, casting these walls
ahead of the slab was very feasible. Like the flying form systems, jump form
wall systems cast ahead of the slabs have high first costs when compared to
other systems; however, the savings in labor and time makes up for this higher
initial cost.

One convincing reason to cast the core walls ahead of the decks was
schedule. This system allows the core walls and decks to be formed simultaneously, thereby reducing the floor-to-floor schedule by at least 3 days,
when compared to casting the walls and decks concurrently. When extended
over the entire duration of the project, this equates to nearly 2 months of
construction time and provides the owner the building early for 2 months of
additional rent.

FIGURE 10.CS7   Jump form system (courtesy of Charles Pankow, Ltd and Doka Forms).

A self-climbing, jump form system was also evaluated for this project; however, the project team felt the added cost for the hydraulic system associated with a "self-climber" was not worth it. As long as there is available tower crane time for raising a crane-set jump form system, a self-climber is not needed. In its spare time, the crane can also service the reinforcing steel subcontractor.

Also, if the jump form system is no more than three floors above the working deck, a personnel hoist does not need to be installed. The jump form system is accessed via ladders from a platform set inside the core.

Concrete Gravity Columns and Shear Walls

The column and wall forms (Figure 10.CS8) were set off of the previously cast slab. In order to meet the schedule, a full floor of column and wall forms were fabricated, and they had to be set and concrete placed in a single day. All of these forms were stripped the following day. The column forms selected utilized hinged forms that allowed the forms to be easily set and stripped by the carpenter crew. All the column and wall forms were flown down to the ground, requiring adequate storage space. The project site was tight; however, the plaza level at the ground floor provided enough space for these forms to be stored during the deck-forming operation.

(*continued*)

CASE STUDY: BUILDING FALSEWORK AND FORMWORK—CHARLES PANKOW BUILDERS, LTD.
(Continued)

FIGURE 10.CS8 Concrete gravity columns and shear walls (courtesy of Charles Pankow, Ltd and Doka Forms).

Typical Floor-Floor Schedule (5-day + Sat):

- Day 1: Finish stripping fly tables from floor below, set on column-hung brackets at floor above (slab was stressed previously).
- Day 2: Install slab rebar and Post Tensioning (PT), set MEP blockouts, set edgeforms.
- Day 3: Install slab rebar and Post Tensioning (PT), set MEP blockouts; place concrete core concrete.
- Day 4: Place slab concrete, set column rebar cages, and crane-up column forms at end of day.
- Day 5: Set column and wall forms, place column, and wall form concrete.
- Day 6: Strip column and wall forms and crane down to ground floor.

NOTE: A minimum of 2 h of overtime (OT) is worked every day to meet this schedule.

Case Study: 11-Story Apartment Building, San Francisco

Consider the Mission Street apartment building (Figure 10.CS9).

FIGURE 10.CS9 1321 Mission Street apartment building (Courtesy of Charles Pankow, Ltd and Doka Forms).

Statistics

- 11 typical floors above grade, one garage level
- 6.5″ thick mild steel and post tensioned slabs
- 9′6″ floor–floor
- 9000 ft³ per floor, one slab pour per floor
- Nonrectangular (J-shaped) concrete core
- 2 straight shear walls outside of core
- 31 concrete gravity columns, regularly spaced and in-line
- Hoisting via tower crane (13,000-lb capacity)

The typical floor plan is shown in Figure 10.CS10.

Floor Slab Forming System:

A large fly table system is Pankow's first choice to form these types of slabs; however, for this project, they did not make economic sense. First of all, the height of the building only allowed 10 reuses. Second, the shape of the building and irregular column spacing did not lend itself to large fly tables. And lastly, high-voltage electrical lines for the Muni bus line on Mission Street did not allow for the flying of large tables at the lower floors. In this case, Pankow elected to utilize a "handset" system that resembled "Ellis Shore" style forms but with a modern twist—Doka's Doka Flex System (Figure 10.CS11).

(continued)

CASE STUDY: BUILDING FALSEWORK AND FORMWORK—CHARLES PANKOW BUILDERS, LTD.
(Continued)

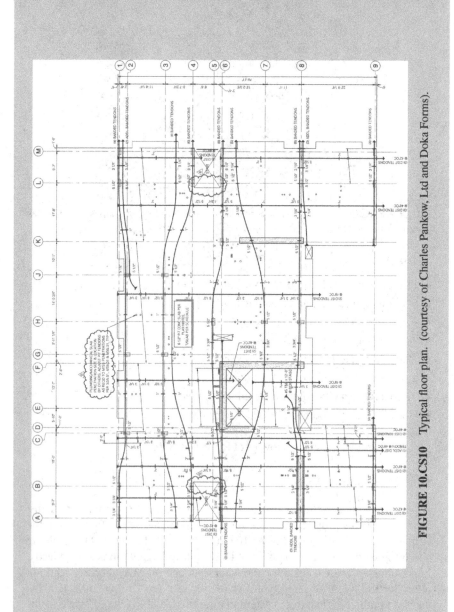

FIGURE 10.CS10 Typical floor plan. (courtesy of Charles Pankow, Ltd and Doka Forms).

FIGURE 10.CS11   Doka form system (courtesy of Charles Pankow, Ltd and Doka Forms).

Pankow did evaluate a "mini" table system for this project. An advantage of these mini tables is the preinstalled perimeter safety rail on the tables. Similar to the larger fly tables, the perimeter rail is preinstalled at the perimeter and, therefore, reduces leading edge safety risks. Although the tables have to be transported to the perimeter for the fly out, this is done very effectively via an electric transporter, affectionately called the "turtle." However, in the end, Pankow elected to go to a handset system because:

The handset system could be utilized from the ground floor slab through the roof and not require the crews to switch systems to form level 3 slab. The mini tables could not be utilized to form the ground floor and level 2 slab, a handset system would need to be used anyway. Switching systems would add another learning curve for the crews.

The handset system is not as dependent on crane time as the mini tables. The mini tables would require use of the crane for the entire day they are set and stripped. The handset system is broken down into small bundles that can be flown up to the floor above in a short amount of time.

The irregular column layout and shape of the building lent itself to a more flexible handset system.

(*continued*)

Many of the crews in the Bay Area are familiar with the Ellis Shore system, and the Doka Flex System is similar to Ellis Shore, but is quicker and easier to install. For this reason, the crew's learning curve with the system would be shortened.

This system requires reshores 3 floors below the formed slab. However, since the building is only 11 stories tall, the overall impact of reshores to the schedule is minimal.

Concrete Core Forming System

Unlike the Los Angeles project case study, the concrete core for the 1321 Mission project does not lend itself to be cast ahead of the slab. One reason is this project's core wall is J shaped and fully enclosed, that is, rectangular. A fully enclosed jump form "works" better than one that is not fully enclosed. Second, because there are so many columns (31 each) on the floor and because there are two additional shear walls outside the main core wall, the time saved by a jump form is reduced. Third, the added cost of a jump form as compared to a gang form set off the slab is difficult to amortize over just 11 floors. And lastly, even if time can be saved with a jump form (say 1 day per floor), that equates to about 2 weeks on the overall schedule, which is not worth the added cost (Figures 10.CS12–10.CS14).

The decision Pankow made was to utilize a crane set, gang form system that is cast concurrently after the slab is cast. This system requires the forms to be flown to the ground while the decks are being formed, and on this site, the storage area was very limited. In order to decrease the required storage space, the forms on the inside of the elevator were set on a platform covering the elevator shaft and did not need to be flown to ground.

Concrete Gravity Columns and Shear Walls

Pankow selected similar forming systems for these elements as the Los Angeles project. Hinged column forms and gang wall forms (Figure 10.CS15) were used. Due to the large number of columns and walls, a 3-day schedule was established for the forming and casting of these elements.

Typical Floor–Floor Schedule (8-day with Sat)

- Day 1: Strip deck forming system from floor below (slab was stressed previously), set on slab above.
- Day 2: Finish setting deck forming system, install slab rebar and PT, set MEP blockouts, set edge forms.

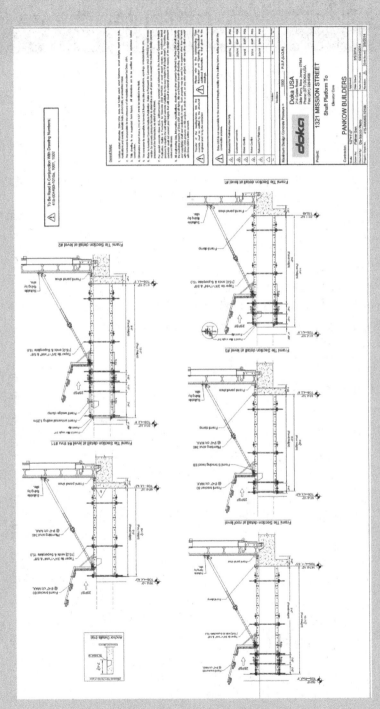

FIGURE 10.CS12 Column forms (courtesy of Charles Pankow, Ltd and Doka Forms).

(continued)

CASE STUDY: BUILDING FALSEWORK AND FORMWORK—CHARLES PANKOW BUILDERS, LTD.
(Continued)

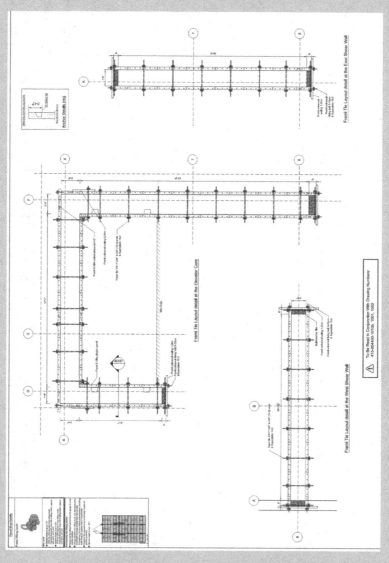

FIGURE 10.CS13   J wall (courtesy of Charles Pankow, Ltd and Doka Forms).

FIGURE 10.CS14 Column forms (courtesy of Charles Pankow, Ltd and Doka Forms).

FIGURE 10.CS15 Wall forms (courtesy of Charles Pankow, Ltd and Doka Forms).

(continued)

CASE STUDY: BUILDING FALSEWORK AND FORMWORK—CHARLES PANKOW BUILDERS, LTD.
(Continued)

- Day 3: Install slab rebar and PT, set MEP blockouts.
- Day 4: Install slab rebar and PT, set MEP blockouts, set column and wall rebar cages, raise core wall platform.
- Day 5: Place slab concrete, set column and wall rebar cages, crane-up column forms at end of day.
- Day 6: Set column and wall forms on previously placed slabs and strip edge form (reduced crew size).
- Day 7: Pour $\frac{1}{2}$ of columns, pour $\frac{1}{2}$ of shear walls, set wall forms, Stress slab PT.
- Day 8: Strip and set columns, pour second $\frac{1}{2}$ of columns, pour $\frac{1}{2}$ of shear walls.
- Day 9: Strip columns and walls, hoist the forms to the ground.

10.1.6 Bridge Project

Your professor will go over the design of a basic box girder falsework system.

Box-Girder Falsework Design A box-girder bridge is a very common bridge design used by various county, state, and federal agencies and is predominantly reinforced concrete. The foundations are the same as any other type of bridge using driven or drilled piles, concrete footings, and concrete piers. The cap beams spanning the piers are typically incorporated with the superstructure. In other words, the cap beams are within the depths of the superstructure and not between the superstructure and the top of the piers. The superstructure is a series of open boxes consisting of a top slab (driving surface), vertical stem walls, and a bottom slab. The number of boxes depends on the width of the bridge and the height of the box depends on the bridge span lengths. The longer the spans, the deeper the bridge is (more moment of inertia). The stem walls contain post-tensioning ducts with high-strength stands that are tensioned after the top deck is placed and cured.

The design of a box-girder falsework system is usually very complicated. However, the design can be simplified if all the conditions are reduced to basic scenarios such as:

Simple box-girder geometry

Not spanning over water, live traffic or environmentally sensitive areas

Heights less than 30 ft

If these conditions are favorable, then a fairly simple step-by-step list can be developed in order to produce an economical falsework system. Figure 10.14 shows a very common, but simple box-girder bridge cross section.

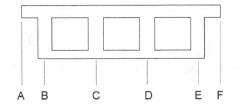

FIGURE 10.14 Box-girder bridge cross section.

Step 1: Stringer loads

Assume a stringer will be at each location A through F line (6 stringers). The two side stringers (A and F) are under the edge of deck (EOD) and the rest of the stringers are under the centerline of each outside and inside stem wall (B, C, D, and E). Determine the associated concrete area of each line. For example, stringer B will have to support half the overhang to line A and half the box to line C. Therefore, line B will take a shape like Figure 10.15.

Draw each section A–F and calculate the concrete volume (CF) of that section. Convert the volume into a weight using 160 pcf. Consider this section as a 1 ft-long bridge section and multiply by 1 ft.

$$\text{Weight} = \text{Area} \times 160 \text{ pcf} \times 1 \text{ ft}$$

Add a live load of 50 psf over the shadow area of the section.

Add 100–150 plf of bridge for the self weight of the stringer, which is not known yet.

Step 2: Soffit Design

Find the widest box width (stem wall to stem wall) and calculate the linear load from the bottom slab. Assume plywood and a joist has to span this distance and support the bottom slab load. Figure 10.16 shows this section.

$$W = (\text{thickness}/12''/\text{ft}) \times 160 \text{ pcf} \times 1 \text{ ft}$$

Add live load of 50 psf for live load.

Use the charts in Appendix 7 to determine the best plywood and joist size based on this load.

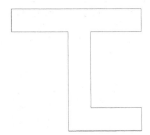

FIGURE 10.15 Stringer line B tributary area.

FIGURE 10.16 Bottom slab spanning stem to stem.

Step 3: Stringer Design

Determine the longest span of the bridge. Using the uniform loads per linear foot of bridge for each stringer line, determine the minimum size of each stringer A–F. For example, if line C and D weigh 3000 plf of bridge and the longest span is 34 ft, the minimum stringer moment and section modulus would be as follows:

$$M = (3000 \text{ lb})(34')^2/8 = 433,500 \text{ ft-lb}$$

$$S_x = (433,500 \text{ ft-lb})(12''/\text{ft})/21,600 \text{ psi} = 240.8 \text{ in}^3$$

Perform this calculation for all lines that are different. Notice that A and F are the same, B and E are the same, and C and D are the same. Therefore only three of these calculations should be run. A table like Table 10.4 should be used in order to keep the work organized. Only C and D are filled in using the $w = 3000$ plf and $L = 34$ ft.

From this information in the table, determine how many braces each stringer would need and decide which of the three stringers are best for each line with a 34-ft span.

Step 4: Top Cap Design

Determine the minimum size of the top cap beam supporting the stringers. Select how many posts will be needed. In the bridge section used in this example, three to four posts would be adequate. Using the linear loads from A–F and the 34-ft span, apply each total to the top cap beam and set three to four posts as supports. Figure 10.17 shows this FBD.

This is an indeterminate structure with four posts (supports). Using a beam program would be the most accurate and less complicated. Whichever method is used, determine the lightest top cap beam that will support this loading condition. The loads are set by the stringer layout and cannot be changed. However, the posts can be moved left and right to establish the lowest moment in the beam, thus, economizing the design. Finally check for web yielding over each post and determine if stiffeners are required.

TABLE 10.4 Stringer Design

	Lines A and F	Lines B and E	Lines C and D
Linear load (W = plf)	?	?	3,000
Length (ft)	34	34	34
M_{max} (ft-lb)	?	?	433,500
Minimum section (S_x)	?	?	240.8
Beam option 1	?	?	W18 × 130
Beam option 2	?	?	W21 × 111
Beam option 3	?	?	W24 × 103
Lateral braces required (#)	?	?	lb = $20,000/F_y(d/A_f)$

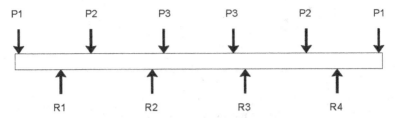

FIGURE 10.17 Stringers loading top cap beam.

Step 5: Post Design
Determine the load on the heaviest loaded post using that line's longest span and heaviest linear load. Assume that lines C and D have the heaviest linear load and the span of 34 ft is the longest.

$$P = W \times L$$

$$P = 3000 \text{ pfl} \times 34 \text{ ft} = 102,000 \text{ lb } (102 \text{ k})$$

Now determine the longest post length. In order to do this, elevations from the drawings are used and the elevation from the top of bridge to the ground needs to be known. The post length then becomes as follows:

Bridge top deck elevation (elev) − Gound elevation (elev)

= gross dimension (ft) − decking material (ft)

   − tallest stringer and top cap (ft)

   − pads, corbels,wedges, and sandjacks and bottom cap (ft)

= post length (ft)

With the maximum post load (P) and the maximum post length (L) calculate the minimum (square) post required:

$$P = \frac{(0.3)(E)}{(l/d)^2} \times \text{area} \leq P_{\text{act}}$$

Step 6: Lateral bracing load of the heaviest loaded falsework bent.
Determine the total bridge weight at the heaviest loaded bent. This would be the P_1, P_2 and P_3 loads all added up from step 4.

$$F = (2 \times P_1) + (2 \times P_2) + (2 \times P_3)$$

Apply 2% (minimum load) laterally to the top cap beam of the bent as shown in Figure 10.18.
Step 7: Stem wall formwork design.
Design the wall formwork for one of the internal stem walls (lines C or D). Assume full liquid head pressure for $P, \frac{5}{8}''$ plywood, all 2×4 studs and walers, and snap ties

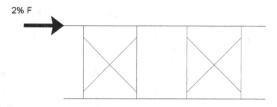

FIGURE 10.18   Lateral force.

with max capacity of 3500 lb. The wall should be formed from the top of the bottom slab to the bottom of the top slab. In other words, the wall height is equal to the bridge thickness minus the top and bottom slab.

Step 8: Top Deck Formwork Design

Design the formwork for the top deck slab from stem wall B to C or C to D, whichever length is greater. By choosing the widest box, the worst-case scenario is used. Use $\frac{5}{8}''$ plywood, 2 × joists, 4 × cap beams, and 4 × 4 posts.

CHAPTER 11

BRACING AND GUYING

Guying and bracing of temporary structures has become a much talked about and scrutinized subject in construction. The failure of such systems can not only be catastrophic but can prolong the misperception that construction is an unsafe industry. In addition, guying and bracing is a very visible aspect of construction; it is often what the public sees daily on their commute to work and hears about when accidents happen. The case study accident in Chapter 10 is an example of a guying and bracing accident. It was placed in Chapter 10 because it involved falsework.

Tilt-up construction, formwork placement, and rebar cage guying and bracing operations have indeed resulted in several high-profile accidents in the last 20 years. Iron workers, carpenters, laborers, and operators are among the tradesmen who have been severely injured or killed in these accidents. When these accidents occur, they are usually not "reportable" OSHA cases; they are more often "recordable," "lost-time," and, unfortunately, "fatality" OSHA cases. Reportable accidents are accidents that occur and do not require a doctor or prescription medication. A recordable accident, however, is an accident that does require medical attention and/or prescription medication. A lost-time accident means the injured person misses his next scheduled day of work.

These injuries and fatalities are not always caused by the design of the actual supports, but many times are the result of human mistakes and bad judgment. It is therefore crucial that designers and builders use the utmost caution when providing support for their temporary structures. Figure 11.1 shows a pier table form and reinforcing steel being supported by custom pipe braces.

FIGURE 11.1   Custom pipe bracing.

11.1   REBAR BRACING AND GUYING

Rebar bracing and guying are operations more likely to fail than others mentioned above. The reasons are multiple, but the size, weight, and the inherent unstable characteristics of rebar cages are primary reasons. Rebar is typically connected by tie wire, and the engineer of record for the permanent structure does not always take into account the tripping (standing), supporting, and guying of the vertical, unstable members. This responsibility is borne by the contractor and his engineer, who typically works for a different firm than the designer of record for the permanent structure. State agencies have begun to issue more guidelines for the contractor to stabilize rebar assemblages. For instance, the specifications on bar reinforcing cages measuring 4 ft in diameter or greater may have the following requirements:

A minimum number of bars, depending on the bar size, of each cage, equally spaced around the circumference, shall be tied to a minimum number of reinforcement intersections by double tie wire. The minimum number of bars would be selected by the designer and approved by the owner/agency.

A minimum percentage of remaining intersections may have to be tied with single tie wire connections. Staggering may be required.

Internal bracing may be required to avoid collapse during assembly, transportation, and hoisting.

Such requirements are becoming the standard for state and federal highway projects. It should be only a matter of time before the rest of the construction industry adopts such practices.

Rebar guying, if properly planned, does not have to be limited to external guy cables or stiff-leg bracing. Internal bracing can be added to the inside of the unstable cages that add rigidity during the lifting and guying stages. This text will not provide examples of this internal bracing, but it should be noted that this practice will lessen the stresses on the tie wire and the vertical and hoop rebar during the different phases of cage guying. In addition, in most cases, the planning and engineering that goes into the bracing of the rebar cage also translates to a form guying and bracing plan as well.

The first example in this chapter will cover a wall form or panel being braced to resist a typical wind force. The wall panel will be braced with a pipe brace in tension and compression, and a concrete anchor block called a deadman. Typical pipe braces consist of a smaller pipe inside a larger pipe and is designed to the smaller pipe. The ends of the pipe braces usually have adjustment screw jacks and plate steel for attachment. The wind chart from Chapter 3 has been included in this chapter as well for quick access to loading forces converted from mph to psf. The second example will be geared toward highway, bridge work, and reinforcing steel cages with wire rope guying systems.

11.2 FORM BRACING WITH STEEL PIPE AND CONCRETE DEADMEN

Example 11.1 Wall Form/Panel
Determine the loading on a pipe wall brace from the wind forces given, then size the pipe braces and the deadman (concrete anchor block) including the required factors of safety of $1.5 : 1.0$. The wind force is 80 mph and the braces and deadmen are at 15-ft centers. Design the braces for tension and compression, and design the anchor blocks to resist sliding in both directions as well. Use Figure 11.2 for additional information and dimensions.

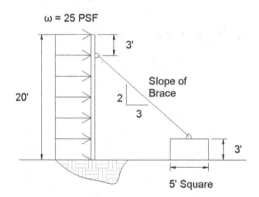

FIGURE 11.2   Example 11.1 sketch.

Steps to designing a wall panel bracing system:

1. Calculate the wind force on a 12″ strip of wall panel.
2. Distribute the force to a single brace based on the brace spacing and convert to force to a single resultant force (R).

3. With statics, and using the slope of the brace, determine the force in the brace and anchor block.

4. Determine the pipe brace length and size the pipe as a single pipe (smallest, inside pipe) to resist buckling (compression) and tension.

5. Transfer the brace loads to the anchor block and determine the minimum block weight to resist sliding.

Table 11.1 can be used for the forces generated from local wind speeds. Lets assume a 20-ft wall with wind speed of 80 mph.

TABLE 11.1   Wind Speed and Force

Speed (mph)	Force (psf)
10	0.27
20	1.08
30	2.43
40	4.32
50	6.8
60	9.7
70	13.2
80	17.3
90	21.9
100	27.0
110	33
120	39
130	46
140	53
150	61

Pipe braces at 15-ft OC attached to concrete deadmen are shown in Figure 11.3.

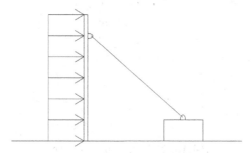

FIGURE 11.3   Wind distributed load.

Step 1: Determine the force from the wind as a 12″ strip load. From the chart, use the loading that corresponds to an 80-mph wind force. Table 11.1 indicates that an 80-mph wind produces 17.3 psf force on the wall panel. If the panel was higher than 30 ft, an additional 5 psf would be added because wind forces are greater the higher the obstruction. This will be discussed more in Example 11.2. The load on a 12″ strip is as shown here.

$$R \, (12'' \text{ strip}) = 17.3 \text{ psf} \times 20 \text{ ft} = 346 \text{ per linear foot of wall}$$

Step 2: Distribute the force to one brace. Since each brace supports 15 linear feet of wall, the 12″ strip load has to be multiplied by the 15-ft brace spacing.

R (one brace) $= 346$ plf of wall $\times$ 15 ft of wall $= 5190$ lb per 15 ft of wall. Figure 11.4 illustrates this FBD.

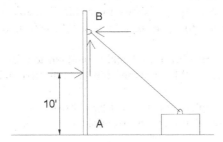

FIGURE 11.4   FBD of brace connection to back of form.

Step 3: Determine the force at the brace connection to the back of the wall panel (location B). The resultant force of this 20-ft strip load is 10 ft from the bottom because the resultant location for a rectangular load is one-half the distance of the rectangle. Apply this 5190 lb force 10 ft from the base of the wall and determine the reaction at the pipe brace connection toward the top of the wall. The brace is attached 3 ft from the top of the wall (17 ft from ground), the brace is sloped at 2V : 3H, and the deadman is 3 ft tall. The base of the panel (ground) is location A.

The horizontal component is F_{HB}, the vertical component is F_{VB}, and the axial force through the pipe brace (2 : 3 slope) is F. Summing the moments about the base of the panel will determine one unknown: the horizontal force at B. Since F_{VA} is in line with the connection point, there is no moment created by this force. Also the base of the panel is in line with point A as well, so again there is no moment generated from its reaction.

$$\Sigma MA = 0$$

$$5190 \text{ lb} \times 10' = F_{HB} \times 17', \quad F_{HB} = 3053 \text{ lb}$$

Figure 11.5 shows the F_{HB} at location B.

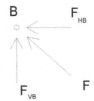

FIGURE 11.5   FBD at location B.

Using the 2V : 3H slope determine F_{VB}:

$$\frac{2}{3} = \frac{F_{VB}}{3053} \qquad F_{VB} = 2035 \text{ lb}$$

The axial force (F), therefore, is

$$F = \sqrt{(3053)^2 + (2035)^2} = 3669 \text{ lb}$$

The horizontal and vertical forces can be used to design the connection at the back of the panel. This will not be done in this example; but on a real design, the connection would have to resist these two forces for shear and tension on all members involved with the connection. The axial force will be used in the next step to design the pipe brace.

Step 4: Design the pipe brace for buckling and tension. Buckling is typically the weak link, so the brace will be designed for buckling first and then tension can be checked.

As mentioned in Chapter 2, the following is the calculation for allowable buckling stress. The factor of safety used in this example is 1.5 : 1.0.

$$F_{bs} = \frac{\pi^2 E}{(kL_e/r)^2 \times \text{FOS}} = \frac{P}{A}$$

where F_{bs} = allowable buckling stress (psi)
$\quad E$ = modulus of elasticity for steel, usually 29,000,000 (psi)
$\quad k$ = connection type multiplier, depending on fixed, hinged, etc., the connection to the back of the panel and deadman is hinged (1.0)
$\quad L_e$ = effective length of column, if there are supports, then the longest distance between supports (in), this pipe brace is not supported in the center
$\quad r$ = weakest radius of gyration (in). For pipe, there is only one radius of gyration due to its circular shape.

Refer to the pipe chart in Appendix 2 for pipe properties up to 12 in in diameter.

P/A is normal stress and is set equal to allowable buckling stress in order to determine $(P)_{\text{all}}$ or $(A)_{\text{rea}}$ allowed.

The trial-and-error method can be used. First determine the length of the pipe with the dimensions shown in Figure 11.6. Using the 2V : 3H slope and a vertical distance of 14 ft (20′–3′top–3′ btm), calculate the horizontal distance $\frac{2}{14}{}' = 3/H = 21$ ft.

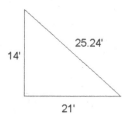

FIGURE 11.6 Pipe brace slope.

Now using the Pythagorean theorem, figure the pipe length on the diagonal:

$$\sqrt{(14')^2 + (21')^2} = 25.24 \text{ ft}$$

A $3\frac{1}{2}''$ standard pipe will be tested first. Look up the cross-sectional area (A) and the radius of gyration (r) for this pipe in Appendix 2.

$$A = 2.68 \text{ in}^2$$

$$r = 1.34 \text{ in}$$

Calculate F_{bs}

$$F_{bs} = \frac{\pi^2(29{,}000{,}000)}{\{[1.0(25.24 \times 12/\text{ft})/1.34]^2\} \times 1.5} = 3735 \text{ psi}$$

Set $F_{bs} = P_{all}/A$ and solve for $P_{all} = 3735 \text{ psi} \times 2.68 \text{ in}^2 = 10{,}009 \text{ lb}$

This 25.24-ft-long, $3\frac{1}{2}''$ standard pipe can support 10,009 lb before failing in buckling. If buckling was not an issue, then $P_{all} = 27{,}000 \text{ psi} \times 2.68 \text{ in}^2 = 72{,}260 \text{ lb}$ tensile strength. This is a 62,251-lb reduction because of buckling.

The required load in the pipe brace was only 3669 lb. There is already a factor of safety of 1.5 built into our allowable load of 10,009 lb; so this means if we use this pipe, the actual FOS is $10{,}009/(3735/1.5) = 4.02$. This is very high, so a smaller pipe should be considered.

Try a $3''$ standard pipe. Look up the cross-sectional area (A) and the radius of gyration (r) for this pipe:

$$A = 2.23 \text{ in}^2$$

$$r = 1.16 \text{ in}$$

Calculate F_{bs}:

$$F_{bs} = \frac{\pi^2(29{,}000{,}000)}{[(1.0(25.24 \times 12''/\text{ft})/1.16'')^2] \times 1.5} = 2799 \text{ psi}$$

Set $F_{bs} = P_{all}/A$ and solve for $P_{all} = 2799 \text{ psi} \times 2.23 \text{ in}^2 = 6242 \text{ lb} > 3{,}669 \text{ lb}$ (OK).

A $2\frac{1}{2}''$ pipe was tried by the author and could only support 3171 lb, so the $3''$ pipe is good with a little more FOS built in.

Check the pipe in tension: $3669 \text{ lb}/2.23 \text{ in}^2 = 1645 \text{ psi} < 27{,}000 \text{ psi}$ (OK).

Step 5: Transfer the horizontal and vertical pipe brace loads down the deadman anchor block and determine the minimum weight of the deadman.

In physics, friction was taught. At that time, the future construction management student probably never dreamed that this would come in handy in his field of study. Now is the chance to refresh one's memory of physics on how friction works. Figure 11.7 shows the pipe brace forces being transferred to the deadman anchor block. The same diagonal, vertical, and horizontal forces still apply.

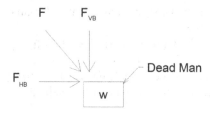

FIGURE 11.7 FBD at deadman anchor block.

Contrary to what most people believe or what common sense may appear to say, friction is not based on the surface area at the bottom of the deadman. Friction gets its resistance from the normal force and the coefficient of friction value associated with the two surfaces involved, the bottom of the concrete block and the ground. Depending on the construction site, the ground can be compacted soil, uncompacted soil, compacted base rock, asphalt, concrete, and the like. When the surface is unknown, always use a conservative coefficient, the lowest value. Table 11.2 shows some estimated friction values used in construction. The ranges used typically mean the difference between wet and dry surfaces. The wet surface would always be the lower coefficient value.

TABLE 11.2 Coefficient of Friction Values for Standard Materials in Construction

Surface 1	Surface 2	Coefficient (μ)
Concrete	Steel	0.35–0.45
Concrete	Loose soil	0.25–0.35
Concrete	Concrete (rough)	0.50–0.65
Concrete	Clay	0.20–0.40
Concrete	Sand	0.40–0.60
Concrete	Gravel	0.50–0.60
Concrete	Rock	0.50–0.70

The deadman size needs to be checked for the pipe in tension and in compression. When the pipe is in compression (which is how the pipe was sized), the deadman is being pushed down and to the right. When the pipe is in tension, the deadman is being pulled and lifted up to the left. It does not matter that the pipe was OK in tension. The fact of the matter is if the wind blows right to left, the pipe brace will still pull and lift the deadman.

Physics taught us that friction force is equal to the normal force (N) times the coefficient of friction (μ) mentioned above and shown in Figure 11.8. This force is a horizontal direction (x) force and the normal force is a vertical direction (y) force. The forces coming from the pipe brace can be converted back into an x and y force so that all the forces are either horizontal or vertical.

F_{HB} = horizontal brace component ($-x$ when pulling left, $+x$ when pushing to right)

F_{VB} = vertical brace component ($-y$ when downward, $+y$ when upward)

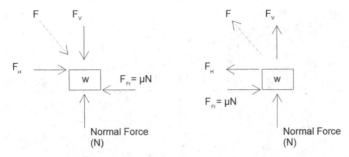

FIGURE 11.8 FBD push and pull friction forces.

W = weight of deadman vertically downward $(-y$, downward only)
N = normal force vertically upward $(+y$ upward only)
F = horizontal force $(N\mu)$, $(-x$ when resisting left, $+x$ when resisting right)

Draw two free-body diagrams, one representing the downward force of the pipe brace on the deadman and the other representing the upward force of the pipe brace on the deadman. From these free-body diagrams, total the forces in both the x and y direction in order to solve for the unknowns.

Using Table 11.2, a coefficient of friction of 0.35 is estimated since the project conditions will be different for each wall location. Sometimes the deadman will be on compacted aggregate base and sometimes on soil. A lower number would be more conservative and require a heavier deadman. There are five variables listed above. Friction is based on the normal force (unknown) and the coefficient (known). This means there are really only four variables; two are known (from the pipe brace), and two are unknown. Figure 11.9 shows anchor blocks being used as a support system for bridge falsework.

Do we add the forces in the x or y first? Looking at the FBDs, it is apparent that, besides the known forces, there are three forces vertically in the y and only two forces horizontally in the x. Therefore, it would make sense to add the forces where there are fewer unknowns and then use this information to figure out the remaining forces. From the pipe brace there is $F_{HA} = 3053$ lb and $F_{VA} = 2035$ lb. A factor of safety also needs to be considered so a $1.5 : 1.0$ will be included in the following illustration.

$\Sigma Fx = 0$ and first solve for the normal force (N). $F_{HA} \times FOS - \mu N = 0$

$(+3053 \text{ lb} \times 1.5) - 0.35N = 0 \qquad N = 13{,}084 \text{ lb}$

Now we have solved one of the two unknowns $(N = 13{,}084 \text{ lb})$ and it already includes an FOS. The forces can now be added in the y direction with a value for N.
$\Sigma F_y = 0$, using the new value for N. Remember, N has the FOS built in so do not apply another FOS.

$-F_{VA} - W + N = 0$

$- 2035 \text{ lb} - W + 13{,}084 \text{ lb} = 0 \qquad W = 11{,}049 \text{ lb}$

FIGURE 11.9 Falsework bent supported by pipe braces.

The deadman weight has been determined when the pipe is pushing downward, wind blowing left to right. Now, reverse the direction of the pipe brace force, wind blowing right to left and the pipe brace lifting and pulling in this same direction. The normal force will be the same because, even though the direction reversed, they all reversed; therefore, the positive and negative signs reverse, but the magnitude of the normal force is the same value.

$\Sigma Fx = 0$ and first solve for the normal force (N). $F_{HA} \times \text{FOS} - {}^-N = 0$

$(+3053 \text{ lb} \times 1.5) - 0.35N = 0 \qquad N = 13,084 \text{ lb}$

The forces can now be added in the y direction with a value for N.

$\Sigma Fy = 0$, remember, N has the FOS built in so do not apply another FOS.

$+ F_{VA} - W + N = 0$

$+ 2035 \text{ lb} - W + 13,084 \text{ lb} = 0 \qquad W = 15,119 \text{ lb}$

The deadman in this case is much larger (Figure 11.10). You can see by the sign change of the F_{VA}, the 2035-lb value was now added to the normal force instead of subtracted. This increases the deadman requirement by 2×2035 lb.

How big does a concrete anchor block have to be to weigh 15,119 lb? Concrete weighs approximately 150 pcf. With this information, we can determine the volume required for a deadman to weigh 15,119 lb.

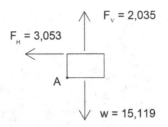

FIGURE 11.10 Tipping analysis.

15,119 lb $= (b \times w \times h)$ 150 pcf. If the dimensions of the deadman were all the same, say c, we could take the cube root of the volume:

$$c = \sqrt[3]{\frac{15,119}{150}} = 4.65 \text{ ft}$$

The volume of this deadman needs to be 100.8 ft^3 if it is made of 150 pcf concrete. What if the base of the deadman was set at 5 ft $\times$ 5 ft and we needed to know the height (h)?

$$h = \frac{100.8 \text{ ft}^3}{5 \text{ ft} \times 5 \text{ ft}} = 4.03 \text{ ft high}$$

Here are two deadman sizes that would work. However, will they tip over if the pipe brace is connected to the top? Let's try the first deadman since it is taller and more susceptible to tipping.

The upward forces will be used since that is how the deadman was sized. There is a deadman weight acting downward of 15,119 lb and in the middle of the deadman ($\frac{1}{2}$ of 4.65 ft). There is F_{VA} acting upward with 2035 lb in the middle of the deadman ($\frac{1}{2}$ of 4.65 ft). Finally, there is F_{HA} acting to the left with 3053 lb at the top of the deadman, which is 4.65 ft high. If the moments are added about the bottom left corner of the deadman and the weight (W) is left unknown, the stability can be determined. Clockwise will be negative and counterclockwise will be positive.

$$\Sigma MA = 0$$

$$(+3053 \times 4.65) - (W \times 2.33') + (2035 \times 2.33) = 0$$

$$W = 8128 \text{ lb} < 15,119 \text{ lb} \quad \text{(OK)}$$

The minimum weight of the deadman so that it does not tip over is 8128 lb. The actual deadman weight based on sliding has to be 15,119 lb. Therefore, the deadman will not tip over. Figure 11.11 shows pier forms guyed with a wire rope cable inside a highway median. Sometimes the real estate to place anchor blocks is limited and K-rail is used. However, the weight and geometry of the K-rail must be analyzed in the same way the anchor blocks were designed.

The attachment of the pipe brace to the top of the anchor blocks should be designed. The horizontal force can be used to check the shear of the connection. The vertical, uplift force can be used for the connection in tension. Both of these must be satisfied. This analysis will be performed in Example 11.2.

FIGURE 11.11   Pier forms guyed with wire rope.

11.2.1 Life Application of Friction Forces

For the student who likes real-life examples to help understand certain concepts, an illustration from the gym will be used. This example shows a trainer in Figures 11.12 and 11.13 exercising by pushing and pulling a weight inside a metal wheelbarrow tub—without the wheels, of course.

FIGURE 11.12   Pushing the sled.

FIGURE 11.13   Pulling the sled.

In the gym, pushing or pulling a steel wheelbarrow tub full of weight on rubber mats can give a person an appreciation for friction forces. In the photos, the trainer is shown pushing and pulling the metal tub in two different directions. In Figure 11.12, the trainer is pushing the tub. Notice the angle of the trainer's arms. Like Example 11.1, the angle of the pipe brace can greatly affect the resistance of the deadman as the wheelbarrow will with the trainer. The rubber mats are going to have a different coefficient of friction with steel then the deadman had with the soil or the rock.

In Figure 11.13, the trainer is pulling the tub. This is the same as the pipe brace pulling and lifting the deadman. In Example 11.1, the pulling and lifting of the deadman was what determined the minimum weight of the deadman. So, it would make sense that it is easier for the trainer to pull the tub than push the tub if a correlation is made between the two examples. The coefficient of friction between steel and rubber is about 0.60–0.70. Therefore, the same forces in both examples would result in greater forces from the pipe brace or the trainer to move the wheelbarrow tub (deadman).

11.3   REBAR GUYING ON HIGHWAY PROJECTS

As mentioned at the beginning of this chapter, guying and bracing rebar cages on highway projects is risky and dangerous. Recently, in 2013, the Construction Institute of the American Society of Civil Engineers sponsored a "Rebar Cage Construction and Safety—Best Practices" study. The study is one of the first to address the specific concern of rebar cage stability during construction. During the writing of this book, the author toured two projects in northern California that had many challenging rebar guying conditions.

The problem and concern with guying tall, unstable rebar cages has always been an issue; however, in light of accidents that have occurred in the past 15–20 years, measures have been taken by state and federal agencies, owners, engineers, and contractors to eliminate accidents and even near-misses. The following text and example is modeled from an actual project in northern California with state jurisdiction and oversight. The design calculation includes current factors of safety used by engineers and state and federal agencies. The designs on this and similar projects are reviewed and approved by at least two separate engineers, either employed by the contractor and/or the owner/engineer. If this strict requirement of reviews is not specified by the contractor, then they should be at least specified by the owner. Hopefully, they are standard practice for both parties.

Example 11.2 The guying of a 40-ft-tall × 10-ft-wide rebar cage and pier form is shown in Figure 11.14.

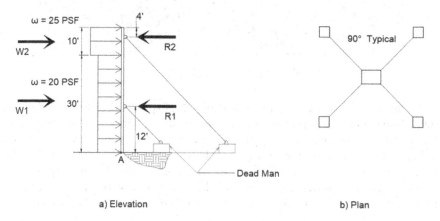

a) Elevation   b) Plan

FIGURE 11.14 Example guying sketch.

Step 1: Determine all the wind forces and the reactions at the cable locations.

In California and other state and federal projects, the forces on vertical obstructions such as bridge piers are dictated by the data shown in Table 11.3 and is based on the height of the structure. As the table shows, the minimum force used in determining the guying and bracing system is 20 psf at 0–30 ft in height above grade. As the structure gets taller, the design force increases, as shown in Table 11.3. Now it would make sense that if the construction was taking place in an area that had more than

TABLE 11.3 Wind Pressures Based on Height above Grade

Height Zone (height above grade in feet)	Wind Pressure (psf)
0–30	20
31–50	25
51–100	30
Over 100	35

Source: CalTrans Falsework Manual.

normal wind forces, the pressures used would represent the local conditions and the values would scale up from the height above grade.

In this example, the height of the column is 40 ft. One interpretation would be to apply 25 psf from 0–40 ft as a uniform load. The intent of Table 11.3, however, is to apply the 0–30 ft pressure (20 psf) as a rectangular load from the bottom of the form to the 30-ft elevation and then 25 psf across the top 10 ft. Once the pressure is determined, the location of the guy cables should be decided. Sometimes, in order to decide on one level of cables versus two or three levels of cables, the designer should run some simple resultant calculations in order to see if a smaller number of guys will be too much for one level. For this example, it has been decided to place a set of four guys 12 ft from the ground and place a set of four guys 36 ft from the ground. The guys will be placed at a 45° angle. With this information, the pressure diagram will look like Figure 11.14. The widest form face is 10 ft.

Calculate W_1 and W_2 if the widest form face is 10 ft.

$$W_1 = 20 \text{ psf} \times 30' \times 10' = 6000 \text{ lb}$$
$$W_2 = 25 \text{ psf} \times 10' \times 10' = 2500 \text{ lb}$$

Point A is located at the bottom of the form at ground level, so there is a point to sum moments. Because two levels of cables were selected, there are two unknowns. This is referred to as an indeterminate loading condition and sum algebra will be used to determine R_1 and R_2. Below this calculation, a computer program will be used to check the resultant values:

$$\Sigma MA = 0$$
$$(W_1 \times 15') + (W_2 \times 35') - (R_1 \times 12') - (R_2 \times 36') = 0$$
$$\Sigma Fx = 0$$
$$W_1 + W_2 = R_1 + R_2, \ W_1 + W_2 = 6000 \text{ lb} + 2500 \text{ lb} = 8500 \text{ lb}$$

therefore $R_2 = 8500 \text{ lb} - R_1$.

Now go back to summing the moments about point A and plug R_2 in as $8500 - R_1$.
$(6000 \times 15') + (2500 \times 35') = 12' R_1 + 36' (8500 - R_1)$, with algebra, solve for $R1$.

$90,000 + 87,500 = 12' R_1 + 306,000 - 36' R_1, 90,000 + 87,500 - 306,000 = 12' R_1 - 36' R_1$

$R_1 = 128,500/24' = 5354 \text{ lb}$

Finally, $R_2 = 8500 \text{ lb} - 5354 \text{ lb} = 3,146 \text{ lb}$

At this point, W_1, W_2, R_1, and R_2 have been solved as values in pounds of force or reactions.

The computer-generated force diagram in Figure 11.15 confirms the values for R_1 and R_2 at 3.15 k and 5.35 k, respectively.

The arrows in Figure 11.15 represent the W forces from the wind, and the supports indicate the reactions where the cables are connected. The cables are connected at 45° angles, so these horizontal forces will be converted to vertical and diagonal forces in order to size the wire rope guy cables and the weight of the anchor blocks.

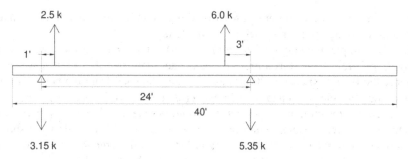

FIGURE 11.15   Computer force diagram.

Step 2: Determine the minimum requirement for the wire rope size of the lower and upper guy cables. Draw FBD similar to Figure 11.16.

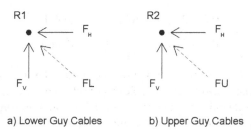

a) Lower Guy Cables   b) Upper Guy Cables

FIGURE 11.16   FBD of upper and lower guy cables.

Lower Guy Cables

$$F_L = F_H \text{ or } F_V \times 1.414 \text{ (45}°\text{ angle)} = 5354 \text{ lb} \times 1.414 = 7571 \text{ lb}$$

Upper Guy Cables

$$F_U = F_H \text{ or } F_V \times 1.414 \text{ (45}°\text{ angle)} = 3146 \text{ lb} \times 1.414 = 4448 \text{ lb}$$

F_L and F_U represent the lower and upper cable forces. Depending on the FOS used, this will determine the SWL of the cable and, ultimately, the size of the cable. Table 11.4 shows wire rope breaking strength and the SWL depending on the FOS chosen by the engineer. Earlier in this book, it was implied that a 5.0 or 6.0 : 1.0 FOS should be used on all wire rope. In this case, it is up to the discretion of the design engineer and the governing agency specifications. The wire rope charts in these tables have been developed from similar IWRC charts, and averages were used to represent several different manufacturers of wire rope capacities.

Wire rope clips are used to join wire rope back to itself when creating a loop at the end connection point. The wire rope industry has determined that when these clips are used, the capacity of the wire rope can be compromised. For that reason, the minimum breaking strength and allowable loads are reduced to 80 or 90% of the wire rope capacity. Table 11.5 lists the same information except includes the reduction of

TABLE 11.4 Estimated Allowable Wire Rope Capacities[a]

Minimum Breaking Force and Allowable Loads (tons) for 6 × 19 IWRC-XIPS Wire Rope

Wire Rope Diameter (in)	Breaking Strength (tons)	SWL with 2.0 FOS (tons)	SWL with 3.0 FOS (tons)	SWL with 4.0 FOS (tons)	SWL with 5.0 FOS (tons)	SWL with 6.0 FOS (tons)
$3/8$	7.6	3.8	2.5	1.9	1.5	1.3
$1/2$	13.3	6.7	4.4	3.3	2.7	2.2
$5/8$	20.6	10.3	6.9	5.2	4.1	3.4
$3/4$	29.4	14.7	9.8	7.4	5.9	4.9
$7/8$	39.8	19.9	13.3	10.0	8.0	6.6
1	51.7	25.9	17.2	12.9	10.3	8.6

[a]SWL = safe working load, IWRC

TABLE 11.5 Estimated Allowable Wire Rope Capacities[a]

Minimum Breaking Force and Allowable Loads for 6 × 19 IWRC-XIPS Wire Rope with Wire Rope Reduction of 80% < 1″ and 90% 1″ and Greater

Wire Rope Diameter (in)	Breaking Strength (tons)	SWL w/2.0 FOS & 80–90%	SWL w/3.0 FOS & 80–90%	SWL w/4.0 FOS & 80–90%	SWL w/5.0 FOS & 80–90%	SWL w/6.0 FOS & 80–90%
$3/8$–80%	7.6	3.0	2.0	1.5	1.2	1.0
$1/2$–80%	13.3	5.3	3.5	2.7	2.1	1.8
$5/8$–80%	20.6	8.2	5.5	4.1	3.3	2.7
$3/4$–80%	29.4	11.8	7.8	5.9	4.7	3.9
$7/8$–80%	39.8	15.9	10.6	8.0	6.4	5.3
1–90%	51.7	23.3	15.5	11.6	9.3	7.8

[a]SWL = safe working load, IWRC.

wire rope capacity to 80 or 90% when these clips are used. Figures 11.17 and 11.18 show a wire rope clip and how they are used to make connections, respectively.

Using an FOS of 4:1, the wire rope size required by Table 11.5 would be:

Lower 7571 lb/2000 lb/T = 3.8T, use 5/8″ wire rope (SWL = 5.2T > 3.8T, but above $\frac{1}{2}$″ 3.3T).

Upper 4448 lb/2000 lb/T = 2.2T, use $\frac{1}{2}$″ wire rope (SWL = 3.3T > 2.2T, but above $\frac{3}{8}$″ 1.9T).

Using a FOS of 4 :1, the wire rope size required by Table 11.6 would be as follows per level:

Lower 7571 lb/2000 lb/T = 3.8T, use $\frac{5}{8}$″ wire rope (SWL = 4.1T > 3.8T, but above $\frac{1}{2}$″ 2.7T).

FIGURE 11.17 Wire rope clip.

FIGURE 11.18 Wire rope connection with clips and thimbles.

FIGURE 11.19 Pier guying system.

Upper 4448 lb/2000 lb/$T = 2.2T$, use $\frac{1}{2}''$ wire rope (SWL $= 2.7T > 2.2T$, but above $\frac{3}{8}''$ 1.5T).

Figure 11.19 is a photo of a pier form being guyed with wire rope and deadmen anchors.

The wire rope selection shows that the reduction factor for the use of wire rope clips did not affect the size required with a 4 : 1 FOS. Also, the 4 : 1 FOS was selected for this example but does not necessarily mean that it is the standard. The designer must check with local and in-house requirements for the proper FOS to use in his situation.

Table 11.5 uses 80 % wire rope clip reduction for rope $< 1''$ and reduction to 90% capacity for rope $1''$ and greater.

There are a few terms used with wire rope that need to be explained. The following list describes these in no particular order.

IWRC: International Wire Rope Corporation sets the standards for wire rope manufacturers.

XIPS: Extra improved plowed steel is the wire type used in the more widely used wire rope in construction.

6 × 19: Wire rope designation gives the number of strands in a single wire rope diameter and the number of individual wires within one strand. Therefore, 6 × 19 means there are 6 strands in the rope and each strand contains 19 wires. 6 × 19 is a popular combination for construction use. Figure 11.20 shows the anatomy of a 6 × 19 wire rope's cross section.

SWL: Safe working load is the allowable load of anything after a FOS has been applied.

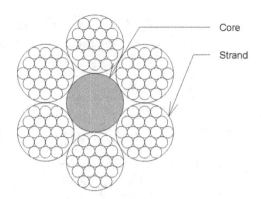

FIGURE 11.20   Wire rope anatomy.

Lay: A lay is the linear distance in inches along the length of the rope that it takes for one strand to make a complete 360° revolution. During the manufacturing process, wire rope strands are twisted. This improves the strength, allows some stretch under load, and helps keep the strands together.

Rule of Eighths for Sizing Wire Rope with a 6:1 FOS   A "trick" or quick calculation that was developed years ago in the field of construction enables one to quickly determine the capacity of a certain diameter wire rope. The rule says:

1. Convert the wire rope size fraction to eighths if it is not already in eighths (e.g., $\frac{1}{4} = \frac{2}{8}$, $\frac{1}{2} = \frac{4}{8}$, etc.).
2. Square the numerator.
3. Divide the new numerator by the denominator 8.

This is the wire rope capacity in tons with close to a 6 : 1 FOS. Table 11.6 shows these values from $\frac{1}{4}''$ diameter to $1''$ diameter.

TABLE 11.6   Rule of Eighths

Wire Rope Size	Size in 8ths	Square the Numerator	SWL (tons)
$\frac{1}{4}$	$\frac{2}{8}$	$\frac{4}{8}$	0.5
$\frac{3}{8}$	$\frac{3}{8}$	$\frac{9}{8}$	1.1
$\frac{1}{2}$	$\frac{4}{8}$	$\frac{16}{8}$	2.0
$\frac{5}{8}$	$\frac{5}{8}$	$\frac{25}{8}$	3.1
$\frac{3}{4}$	$\frac{6}{8}$	$\frac{36}{8}$	4.5
$\frac{7}{8}$	$\frac{7}{8}$	$\frac{49}{8}$	6.1
1	$\frac{8}{8}$	$\frac{64}{8}$	8.0

TABLE 11.7 Wire Rope Termination Specifications

Wire Rope/ Clip Size	Minimum Number of Clips	Turn Back (in)	Torque (ft-lb)
$\frac{1}{4}$	2	4.75	15
$\frac{3}{8}$	2	6.5	45
$\frac{1}{2}$	3	11.5	65
$\frac{5}{8}$	3	12	95
$\frac{3}{4}$	4	18	130
$\frac{7}{8}$	4	19	225
1	5	26	225

Source: Crosby Corporation, Wire Rope Clip Manufacturer.

This rule is very conservative and always produces a value that slightly exceeds the 6 : 1 FOS. Compare these SWL values to the first wire rope chart's 6 : 1 FOS column. The values should be close.

Properly Installing Wire Rope Clips The proper installation of wire rope clips can be found in any manufacturer's published data. Table 11.7 has been provided here to explain the different specifications one needs to follow to make a proper wire rope connection that will stay within 80% of the wire rope's SWL capacity. For actual specifications of wire rope accessories, the design engineer should refer to the product data provided by his supplier.

The wire rope size usually, if not always, matches the clip size. The minimum number of clips is how many clips are required to match the strength of the rope itself and increases as the rope size increases. In order to fit the number of clips required, the next column indicates the amount of rope "turn back" that is necessary so that the clips are spaced properly and are not too close or too far from one another. Finally, since each wire rope clip has two nuts that attach to one U bolt, when tightened these nuts have a required torque value in order to match the capacity of the wire rope clip. In addition to this information, the user should read the manufacturer's installation requirements, such as the proper orientation of the clip and the use of accessories like thimbles (cable softeners).

Step 2 is complete. The next step is to determine the anchor block requirement.

Step 3: Determining the size and number of anchor blocks.

When starting this step, it should be pointed out that most companies already have deadmen in their inventory that meet a particular size. For instance, one company interviewed for this chapter indicated that all their deadmen weighed 9600 lb. At 150 pef concrete weight, this is the equivalent of 64 ft^3 ($4' \times 4' \times 4'$ or equivalent). Figure 11.21 shows an example of how two deadmen were doubled up for twice the sliding capacity. The wire rope was also run between the two so that tipping was minimized.

FIGURE 11.21 Deadmen for a guying system.

This same 9600-lb deadman will be used throughout this step of the example. The minimum weight of the anchor blocks will be determined, and then the quantity per wire rope connection will be known. This method is realistic and often uses a lot more weight than required because the number of blocks is usually rounded *up* to the nearest block. On occasion, the engineer will round *down* to the lower number of deadmen if it stays within a reasonable FOS. For instance, if an FOS of 2 was desired and using only one deadman produced an FOS of 1.9, the engineer could possibly accept this condition, maybe knowing that there was redundancy somewhere else in the design.

The forces that were determined earlier are going to be used in the design of the anchor blocks. For the required weight, the horizontal and vertical forces will be used rather than the axial cable force. In addition, the coefficient of friction is required. This example will use a value of 0.40 for μ. In Example 11.1, it was learned that the deadman weight was greatest when the pipe brace was pulling and lifting. Therefore, it would be an additional and unnecessary exercise to calculate the required weight while pushing the deadman. More importantly, since this example is using cables, it would also be impossible to push the anchor considering cable cannot act in compression. With that said, using the F_H and F_V forces for each level anchor system, determine the minimum number of deadmen required.

Upper Level Deadman $\Sigma Fx = 0$ and first solve for the normal force (N). $F_H \times$ FOS $- {}^-N = 0$

$$(+3146 \text{ lb} \times 1.5) - 0.4N = 0 \qquad N = 11{,}798 \text{ lb}$$

The forces can now be totaled in the y direction with a value for N.

$\Sigma F_y = 0$, using the new value for N. Remember, N has the FOS built in so do not apply another FOS.

$+F_{VA} - W + N = 0.$

$+3146$ lb $- W + 11,798$ lb $= 0 \qquad W = 14,944$ lb

14,944 lb/9600 lb/deadman $= 1.6$ deadmen, use 2 ea $-$ 9600 lb deadman

Lower Level Deadman

$(+5354$ lb $\times 1.5) - 0.4N = 0 \qquad N = 20,078$ lb

$\Sigma F_y = 0$, using the new value for N.

$+5354$ lb $- W + 20,078$ lb $= 0 \qquad W = 25,432$ lb

25,432 lb/9600 lb/deadman $= 2.6$ deadmen, use 3 ea $-$ 9600 lb deadman

If two deadmen were used for the lower level guy cables, the FOS would decrease. In order to know by how much, one would have to recalculate the required weight without an FOS and then divide into 9600 lb $\times 2 = 19,200$ lb and this would be the new FOS.

$(+5,354$ lb$) - 0.4N = 0 \qquad N = 13,385$ lb

$+5,354$ lb $- W + 13.385$ lb $= 0 \qquad W = 18,739$ lb

FOS $= 19,200$ lb/18,739 lb $= 1.02$, this is almost the equivalent of not having an FOS at all, so using three deadmen in this case is necessary.

11.4 ALTERNATE ANCHOR METHOD

When the solution is not acceptable, as in the case of the number of deadmen needed, as always in temporary structure design, the designer could go back into the problem and rearrange some assumptions. Such as in the previous example, the guy cables could be relocated vertically and the angle could even be changed, in hopes of lowering the forces and maybe using only two deadmen per guy level.

Another method of anchoring guy cables to the ground level could be accomplished by using driven or drilled pile encased in concrete. Figure 11.22 is an example of a set of guying cables attached to a cast-in-drilled-hole (CIDH) anchor. Rebar hooks have been placed in the top of the anchor in order to have a place for the screw pin shackles to connect. As a student of temporary structure design, one should consider what would go into this design. What determines this pile's capacity?

Length of pile
Pile diameter
Skin friction between pile and soil

FIGURE 11.22   Drilled pile anchor.

Rebar size

Rebar embedment into CIDH

Strength of concrete

Angle of wire rope guying

Number of guys per CIDH

Most, if not all, of these make up the anchor capacity.

Anchoring guy wires to a concrete-encased pile has some advantages and disadvantages. The advantages are:

They take up less room.

They can be designed with more capacity (compared to more than two deadmen.

The disadvantages are:

They are more costly to install (an estimated $2500–$6000).

They also may have to be removed.

If they are not removed, the beams are liquidated.

Another item that bracing can be attached to is the shoring used for a cofferdam surrounding the foundation excavation if used. This could include sheet pile and soldier pile. The design would have to be checked so that these piles can support the bracing load as well as act as the support of excavation at the same time. A combined stress analysis would also have to be considered.

Anchoring to Concrete Many times, the vertical elements being anchored or guyed are part of a larger foundation slab on grade such as in commercial and treatment plant work. In this case the bracing or guying is attached to a foundation using plates, bolts, and nuts. It is assumed that the foundation is more than capable of supporting the forces and the weak link is the connection hardware. This connection could also represent how the pipe or cable is attached to the top of a concrete deadman as well. Figure 11.23 would be an example of the two main forces being considered at this connection point. Therefore, this calculation should be done in either case.

Table 11.8 has been included for drill-in anchor bolts in concrete, taking the average capacities of similar anchors available in the market. It is very important to know the f_c' strength of the concrete at the time of anchor use. If the concrete strength has to be estimated, use the more conservative value. Additional concrete cylinders can be cast if the concrete strength during anchor installation and use is critical to the design. The engineer of record would want to know this value before loading the system to make sure its value is within the design criteria.

Since the drill-in anchor capacities are shown as "ultimate values," the designer must apply a FOS to the values shown. In the case of these anchors, a 2.0:1.0 FOS is sufficient. With this information, and the forces on the top of the deadman

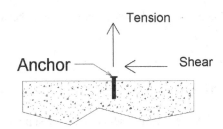

FIGURE 11.23 Anchoring system to concrete.

TABLE 11.8 Tension and Shear Values for Drilled-in Concrete Anchors[a]

Ultimate Tension and Shear Values for Drill-in Anchor Dolts (lb) in Concrete

Anchor Dia. (in)	Required Torque (ft-lb)	Embedment depth (in)	$F_c' = 2000$ psi		$F_c' = 4000$ psi	
			Tension	Shear	Tension	Shear
$\frac{1}{4}$	8	$2\frac{1}{8}$	2,260	1,680	3,300	1,680
$\frac{3}{8}$	25	4	4,800	4,000	5,940	4,140
$\frac{1}{2}$	55	6	5,340	7,240	9,640	7,240
$\frac{5}{8}$	90	$7\frac{1}{2}$	7,060	9,600	15,020	11,900
$\frac{3}{4}$	175	$6\frac{5}{8}$	10,980	20,320	17,700	23,740
$\frac{7}{8}$	250	8	14,660	20,880	20,940	28,800
1	300	$9\frac{1}{2}$	18,700	28,680	26,540	37,940

[a]Values in this table are industry averages.

in Examples 11.1 and 11.2, the designer should be able to size the anchors that will sufficiently support the connection of the wire rope or pipe brace to the top of the deadman.

For each example, determine the bolts required to attach either the pipe brace or wire rope to a concrete slab or deadman that consists of 4000 psi concrete.

Example 11.3   Pipe Brace

$$F_V = 2035 \text{ lb} \times 2.0 \text{ FOS} = 4070 \text{ lb} \qquad \text{(tension)}$$

$$F_H = 3053 \text{ lb} \times 2.0 \text{ FOS} = 6106 \text{ lb} \qquad \text{(shear)}$$

From Table 11.8, the proper anchor is selected that can support the shear and tension force at the same time. The tension force would require only a $\frac{3}{8}''$ anchor (3300 lb < 4070 lb < 5940 lb). For the shear, since it is governing, the $\frac{1}{2}''$ anchor would be required (4140 lb < 6106 lb < 7420 lb).

Conclusion: Use $\frac{1}{2}''$ anchor governed by *shear*.

Example 11.4   Wire Rope

The wire rope vertical and horizontal forces were the same because the cables were placed at a 45° angle.

Lower Brace

$$F_V = F_H = 5354 \text{ lb} \times 2.0 \text{ FOS} = 10{,}708 \text{ lb} \qquad \text{(tension and shear)}$$

Again, from Table 11.8, select the anchor that can support both the shear and tension force from the cable load. The tension and shear force would both require the same anchor in this case, which is a $\frac{5}{8}''$ anchor.

Conclusion: Use $\frac{5}{8}''$ anchor governed by *shear* and *tension*.

Upper Brace

$$F_V = F_H = 3146 \text{ lb} \times 2.0 \text{ FOS} = 6292 \text{ lb} \qquad \text{(tension and shear)}$$

Finally, from Table 11.8, select the anchor that can support both the shear and tension force from the cable load. The tension and shear force would both require the same anchor in this case, which is a $\frac{1}{2}''$ anchor.

Conclusion: Use $\frac{1}{2}''$ anchor governed by *shear* and *tension*.

In many cases, the engineer or contractor may elect to use $\frac{5}{8}''$ anchors everywhere, so there is no confusion of where the $\frac{1}{2}''$ is used and where the $\frac{5}{8}''$ is used.

Internal Bracing   Internal bracing is used to resist external lateral forces without extending guying or bracing beyond the footprint of the support structure itself. Internal braces are common with falsework frames (bents) as shown in Chapters 10

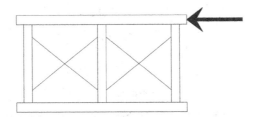

FIGURE 11.24   Falsework bent with 2% lateral force.

and this chapter. The bracing can be 2× wood members nailed or bolted to timber posts or rebar welded to steel pipe. Regardless of the type, the bracing should be designed to resist the external forces applied to the system. The external forces can come from:

Wind

Earthquake

Impact from construction equipment or third-party (public) vehicular traffic

Wind on falsework would typically be low due to the small amount of surface area being contacted by the wind. However, sometimes a conservative approach is taken that applies the wind force to the side of the falsework as if it were a solid mass from the ground to the top deck of the bridges. This was discussed in Chapter 10. Figure 11.24 illustrates a falsework bent with transverse bracing.

Earthquake forces have become better understood in the last several decades and seem to dominate (at least on the West Coast) the cause of lateral loads to our temporary, as well as permanent, structures. A common force calculation takes the dead load of the structure within the tributary area of the support being analyzed and induces 2% as a lateral force at the upper portion of the structure. The sketch in Figure 11.24 shows how this force would be applied to a single-support system.

Impact forces from equipment or traffic can be estimated with a few assumptions and physics. However, this may not govern the worst-case scenario for the design. Also, other precautions are usually taken by the engineer and the contractor to avoid allowing this to control the lateral bracing. For instance, on state highway work, additional struts are added to falsework bents, and loads/forces are *doubled* when these supports are near traffic. Sometimes this can double the number of falsework posts required.

Sample Calculation   Bridge superstructure weight, 10,000 plf of bridge

Weight associated with the heaviest loaded support (34′), 10,000 lb/34 ft span = 340,000 lb

2% of this weight as a lateral force, 340,000 lb × 2% = 6800 lb

If this 6800-lb force is applied to the top of the frame shown in Figure 11.23, how big would the cross bracing need to be?

Apply the 6800-lb force to either side of the frame. Then determine the slope of the diagonal braces. In this case, the space between the posts is 16 ft and the distance between the top and bottom connections is 14 ft. Figure 11.25 illustrates this lateral force and the dimensions of the bent.

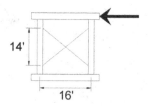

FIGURE 11.25 Typical bent with internal bracing.

A force triangle can be created with the geometry of the height and width of the bracing from connection to connection. The $14'H \times 16'W$ slope is used to develop F_V, F_H and, ultimately, F (diagonal) force. By proportioning the known force (F_H of 6800 lb) with the known force diagram dimensions, the F_V and F forces can be determined.

$$\frac{6800 \text{ lb}}{F_V} = \frac{16'}{14'}$$

$$F_V = 5950$$

$$F = \sqrt{(6800)^2 + (5950)^2} = 9036 \text{ lb}$$

Figure 11.26 shows the force diagram for this example as well as the vertical, horizontal, and diagonal dimensions.

The 9036-lb force is in *tension* as shown with the forces acting in the direction shown. Seismic forces can work in any horizontal direction in the way a wind load could act. If the 6800-lb horizontal force is reserved and acts from right to left, the vertical and diagonal force will be the same because the force triangle has not changed. The only difference is that the opposite diagonal is in compression.

Wire Rope System   If wire rope was used to brace this system, the wire rope would have to be selected to match the 9038-lb force with a factor of safety of 2 or 3:1. Using Table 11.6 with a 3:1 FOS and 80% reduction for wire rope clips, the wire rope size can be determined:

$$9036 \text{ lb}/2000 \text{ lb/ton} = 4.51 \text{ tons}$$

$$\text{Use } \frac{5}{8}'' \text{ wire rope.}$$

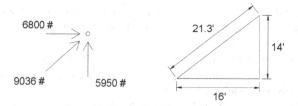

FIGURE 11.26 Force diagram and dimensions.

FIGURE 11.27 Guying frame.

Internal bracing can take the form of a four-sided frame, as in Figure 11.27, which was taken on a project over a northern California river where there is no place to attach the guy externally and the ground was too far below the work surface. The frame was designed so that the rebar cages and formwork could be guyed to the frame, which would transfer these forces through its components to something more substantial below, such as a work trestle.

To be most economical, the frame should be designed to support all scenarios of guying throughout the sequence of construction of one pier and, preferably, the reuse on other piers that may differ slightly in geometry but require the same support. In other words, it should be designed for as many reuses as feasible.

Longitudinal Bracing Longitudinal bracing connects one falsework bent to an adjacent falsework bent in the longitudinal direction of the bridge or structure. These braces should withstand the same 2% lateral force as the transverse bracing and would usually require that the stringers also be connected to the top cap so the forces are transferred properly. Welding is always acceptable, but the preferred method would be 2000-lb capacity beam clamps that attach the bottom flange of the stringer to the top flange of the cap beam.

Wire rope attachments to top and bottom cap beams would usually come in the form of shackles through a hole in the beam flange, wire rope, and wire rope clips and with beam clamps at the upper stringer locations as shown in Figures 11.28 and 11.29.

FIGURE 11.28 Longitudinal bracing connected to bottom cap.

FIGURE 11.29 Beam clamp attaching a stringer to a top cap.

CASE STUDY: REBAR BRACING FOR BRIDGE PIER

Project Overview

Highway 162 in Oroville, California, had an old bridge section spanning the Feather River that was built in the 1920s. In the latter part of the 20th century, engineers were concerned that the center pier (in the middle of the Feather River) was compromised and may not survive many more high-water experiences. CalTrans decided to build a new bridge to replace the existing antiquated bridge. In the fall of 1996 the new project was put out to bid and awarded. In January of 1997, the Feather River came to within a few feet of cresting and flooding Highway 70. The contractor and state agency grew more and more concerned about potential risks in the next 3 years of construction.

As mentioned previously, the bridge piers were extremely large and awkward. The base of the pier was 35 ft × 6.5 ft and the top (45 ft above water) measured 70 ft × 6.5 ft. The pier reinforcing steel was prefabricated on site in slightly less than half-sections. The hoisting of less than half the cage was done by two cranes: Manitowoc 4000 and 3900W. The lift was designed so that the larger crane would have the whole load eventually. The cage was very flexible and internal bracing was added. This was all done by experienced iron workers who had designed and built similar lifts on past projects. This was years before state authorities began putting specifications together to reinforce cages such as the one built for this project (Figure 11.CS1).

FIGURE 11.CS1   Hoisting of rebar cage.

(*continued*)

CASE STUDY: REBAR BRACING FOR BRIDGE PIER
(Continued)

Pipe braces were engineered and placed on the cage prior to hoisting into place. These braces were positioned on the cage so that they could attach to the top of the sheet pile cofferdam and be at an acceptable angle that would not overstress the pipe braces. Since the braces were pipe, they acted as push-pull bracing and could be placed on one side. The bottom of the cage rested on the pier footing (Figure 11.CS2).

FIGURE 11.CS2   Installing braces.

FIGURE 11.CS3   Placing of intermediate hoop reinforcing.

The remaining rebar had to be placed by hand with crane assistance to attach the two halves in the center. The forms were placed once the rebar was completed. About one-third of the pier's top 70-ft dimension was tucked underneath the existing bridge (Figure 11.CS3).

There were three piers in total. The same process was followed for all three. The production improved slightly, but the methods did not change from the first cage to the third cage.

11.5 CONCLUSION

This chapter has covered different types of bracing and guying methods and, more importantly the dangers when not accounted for. The contractor should seek professional help if they lack in-house expertise. The student of temporary structures should be aware of the system selections and complexity of each before pursuing solutions. Factors of safety (FOS) are extremely important to the success of any system.

CHAPTER 12

TRESTLES AND EQUIPMENT BRIDGES

Trestles or equipment bridges help contractors access their work in conditions that are not conducive to standard methods. Trestles are used in construction to access work that is not the typical flat land access, or it may be flat but environmentally sensitive or over water. For these and other reasons, it is necessary for contractors to design trestles and work bridges in order to perform the work. Access over water or rough terrain is fairly common in the heavy civil arena in order to get the resources close to the work and reduce the need for floating equipment, such as derricks and barges, or even "high-line" construction methods. High-lines are wire rope and pulley systems that span canyons, valleys, gorges, and even steep terrain in order to suspend block and tackle that hoists materials to these remote areas.

Locations, such as these, can greatly affect access at projects such as the Hoover Dam, where the topography dictates a method other than a work trestle. Figure 12.1 shows a Manitowoc 3900 W working from a trestle on the Feather River in northern California.

12.1 BASIC COMPOSITION OF A STANDARD TRESTLE

A trestle or work platform can be designed in all shapes and sizes. However, in the author's experience, trestles fall within the same design format the majority of the cases. This format consists of driven pile, framework from wide-flange beams, and small miscellaneous metal sections and wood decking. This format will be described in the following section, and the examples of this chapter will, for the most part, maintain this philosophy.

FIGURE 12.1   Driving sheet pile from work trestle.

12.1.1   Foundation—Pipe, H Pile, and Wide-Flange and Composite Piles

Trestle foundations are rarely, if ever, founded on spread footings. More often, a trestle is supporting the weight of a 150,000- to 300,000-lb crane and requires substantial support, usually driven pile. Also, when over water, a trestle has to be on piles in order to elevate above the high-water mark. So for several reasons, most trestles will be always founded on driven or drilled pile. Finally, if we consider that these piles are typically removed after the trestle life is over, then one would conclude that driven pile (and usually steel pipe or beam) is more easily removed.

Piles for trestles have to withstand several forces, some of which are magnified depending on how high the trestle is above grade or dredge line. Typical forces include:

Bearing and Friction: The pure dead load of the anticipated loads, which is not only the crane itself but the crane's weight while lifting its maximum load

Bending/Lateral Stability: Due to the dynamics of the crane or other equipment moving, starting, and stopping

Bending/Lateral Stability: Due to water movement, which could come from wave action or flowing water such as in a river

Bending/Lateral Stability: Due to impact by marine equipment or vessels

12.1.2   Cap Beams—Wide-Flange Beams with Stiffeners

Cap beams are the lateral members above the pile that support the stringers or girders. The cap beams transfer the stringer/girder loads into the pile. Cap beams can be

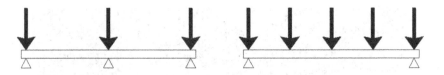

a) Load Directed Through Web   b) Cap Beam in Bending Condition

FIGURE 12.2   Cap beam loading conditions.

designed so that they only transfer shear and reaction forces, or they can be placed in a bending condition as well. Figure 12.2 illustrates these two scenarios.

Web yielding, as we learned in Chapter 7, occurs when a large load is transferred from one member to another, in which case the supporting member's web is suscepti- ble to buckling (yielding). The probability of this occurring is higher when the stringer (arrow) lines up with the pile (supports). In this case, the load is so high and the web of the beam is so tall and/or thin that it buckles like a column. Stiffeners or gusset plates are used to combat web yielding at high shear concentrations. High shear concentra- tions occur at reaction points such as pile to cap beam location or stringer/girder to cap beams. Typically, it does not take a very large gusset to resist the web yielding; it just takes something added to the web to act like a small column, such as $\frac{1}{2}''$–$\frac{3}{4}''$ thick flat plate welded against the beams web between the top and bottom flanges. In the case of falsework, contractors sometimes use small pieces of 4 × 4's and wedge them between the top and bottom flanges of the beam. Figure 12.3 shows a large stringer supported by a stout cap beam. Figure 12.4 is a sketch of how web yielding would affect a beam.

FIGURE 12.3   Stringer supported by cap beam.

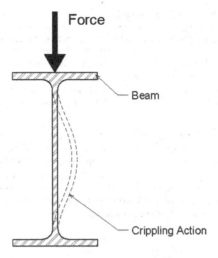

FIGURE 12.4 Web yielding.

12.1.3 Stringers/Girders—Wide-Flange Beams Braced Together

Stringers or girders span from one bent cap to the next bent cap similar to stringers for bridge falsework. One difference between the stringers for trestles and stringers for falsework is the lateral bracing between the stringers. Typically, stringers and girders are designed for bending stress requirements and are then checked to see how much bracing is needed to resist lateral buckling (which will be discussed next). Since there are so many more dynamics and chances of impacts with trestles, the lateral bracing tends to be steel angle or channel welded to the inside web or flange of the beams. With falsework, as shown in Chapter 10, lateral bracing can be achieved using only 4 × 4's and steel banding since there is less chances of dynamics.

Depending on the depth of the stringers, they may also require stiffeners directly over the cap beams similar to the stiffeners resisting web yielding. Stringers are often put together as "teams" prior to placement. Teams are two beams attached together by their lateral bracing (which is already required) so that they can be placed as a unit, thus requiring less handling of pieces at the job site. Teams are usually put together either off-site or on-site in a fabrication yard to maximize the efficiency of the welders.

12.1.4 Lateral Bracing

Once a beam is sized for bending stress, it should always be checked to see if bracing is required. This problem can be solved in two ways:

1. Compute the critical length (L_b) that the beam can span without bracing, and place braces so that this length is not exceeded.
2. Compute the allowable stress that a beam can endure while complying with bending and lateral bracing requirements, and size the beam for both.

The second method will always result in a larger beam, but with no, or little, bracing required. The first method is usually more common because the difference in

TABLE 12.1   Lateral Bracing Cost Comparison[a]

Cost Comparison between Larger Beams and Bracing Smaller Beams	
Larger Stringers without Bracing	Smaller Stringers with Braces at $\frac{1}{3}$ Points
There are five stringers required	There are five stringers required
They are W36 × 230 without braces	They are W30 × 173 but they require 4 braces
They are 40 ft long	They are 40 ft long
Steel weight = 230 × 40 × 5 ea = 46,000 lb	Steel weight = 173 × 40 × 5 = 34,600 lb
At a steel price of $0.60 per lb = $27,600	At a steel price of $0.60 per lb = $20,760
	Add 4 braces at $500 ea (labor and material)
	Total cost = $22,760

[a]The extra steel required to avoid bracing cost the project an additional $4840. The cost of the extra steel would become more economical if 14 or more braces were needed.

cost between adding some bracing or increasing steel weight so lateral bracing is unnecessary is usually very significant. Table 12.1 compares these options.

Lateral Bracing Calculations to Avoid Flange Buckling   According to AISC, 9th edition formula (F1-8), the following length terminology is used:

L_c is the maximum unbraced length for $0.66F_y$

L_u is the actual unbraced length for $0.6F_y$

L_b is the actual unbraced length

$$L_b = \frac{20,000}{F_y(d/A_f)}$$

where L_b = critical length of beam span at which the beam is subject to lateral buckling

F_y = yield stress of steel, i.e., A36 or 36 ksi

d = depth of beam from the beam charts in Appendix 1

A_f = area of flange from the beam charts in Appendix 1

20,000 ksi = constant

In this instance, A_f is the area of the compression flange in the chosen beam. The larger this area, the more resistance the beam has to flange buckling. Since A_f is divided into d (depth of beam), the d/A_f value from the beam charts indicates what L_u value to expect. As the beams get larger, the d/A_f becomes smaller, thus increasing the L_b in the formula $L_b = 20,000/F_y\,(d/A_f)$. The units of d/A_f are a value over inches (in/in²).

If $L_u < L_c$, $F_{all} = 0.60F_y = 0.60(36\text{ ksi}) = 21.6$ ksi beam span is fine for bending only

If $L_u > L_b$, $F_{all} = 12,000$ ksi$/L_u(d/A_f) =$ new allowable stress in ksi, which is less than $(0.60F_y)$, where $L_u = $ is the actual span of beam and 12,000 ksi is a constant.

d/A_f has its own column in Appendix 1 and 3.

Example 12.1   Lateral Bracing

An A36, W24 × 146 spans 40 ft. Determine whether this beam needs lateral bracing; and, if so, determine the lightest beam that would replace this beam and would not need lateral bracing.

Step 1: Calculate the critical length of the beam and compare it to 40 ft.
$L_b = 20,000$ ksi$/[(36$ ksi$)(1.76)] = 315.7$ in$/12''/$ft $= 26.3$ ft $<$
40 ft. This beam needs one brace.

Step 2: Determine how many braces would be needed.
To determine how many braces are needed, take L_b and determine if it is greater or less than $\frac{1}{2}$ the beam span. If it is greater, then one brace is required with this beam. If not greater, then more than one is required. If L_b is less than $\frac{1}{2}$ the span, but greater than $\frac{1}{3}$ span, then two braces are required. If L_b is less than $\frac{1}{3}$ span, but greater than $\frac{1}{4}$ span, than three braces are required. Table 12.2 compares L_b with L.

TABLE 12.2   Critical Length vs. Actual Length

Critical Length	Braces
$L_b > L$	No braces
$\frac{1}{2}L < L_b < L$	One brace
$\frac{1}{3}L < L_b < \frac{1}{2}L$	Two braces
$\frac{1}{4}L < L_b < \frac{1}{3}L$	Three braces
$L_b < \frac{1}{4}L$	Change beam size (recommended)

Figure 12.5 illustrates how these beams would be braced.

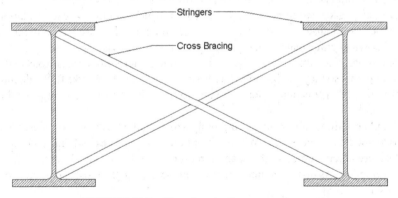

FIGURE 12.5   Cross bracing between stringers.

This W24 × 146 needs one brace because 26.3 ft is less than 40 ft but greater than 20 ft ($\frac{1}{2}L$).

Step 3: If bracing is not desired, determine the lightest beam that works for bending and flange buckling by calculating a new F_{all}. For bending only, F_{all} is 21,600 psi (or $0.60F_y$). In order to omit bracing, the new F_{all} must be used to determine the maximum allowable moment.

$F_{all} = 12,000$ ksi$/[(40' \times 12''/\text{ft})(1.76)] = 14.17$ ksi. Use this allowable stress to determine the maximum bending moment for this beam with this span.

For $L_u > L_c$; $M_{max} = S_x \times F_{all}$ (use S_x for the W24 × 146); $M_{max} = 371$ in^3 × 14.17 ksi = 5256 in-k.

For $L_u \leq L_c$; $M_{max} = S_x \times F_{all}$ (use S_x for the W24 × 146); $M_{max} = 371$ in^3 × 21.60 ksi = 8014 in-k.

So what does this mean? This means that if the L_u (40-ft span) value is greater than L_c, then in order to use the W24 × 146 beam, a maximum moment of 5256 in-k cannot be exceeded. If L_u is less than L_c, then no bracing is needed and the W24 × 146 is fine as long as it does not exceed a maximum moment of 8014 in-k. For this to occur, the span or the load or a combination of both, would have to be reduced.

12.1.5 Decking—Timber or Precast Concrete Panels

The decking is the component of the trestle that the equipment makes contact with and transfers the equipment loads to the stringers. Because all the equipment and construction materials are over the top of the deck, means should be taken to avoid having materials, soil, lubricants, fuels, and the like from getting between the decking components. It is common for the contractor, when over water or traffic, to have to seal the gaps between the voids in the decking. Sometimes plastic sheeting or filter fabric is used to keep contaminants from falling below the trestle. In extremely environmentally sensitive areas, sealant between the decking members could be necessary to make leaks impossible to pass.

The most common material used for trestle decking is timber and usually 12 × 12 × 30 ft timbers are combined into five-timber mats. The alternative would be to individually place and remove single 12 × 12's, which would cost extra in labor and equipment and cause excessive damage to the timbers. When used as combined mats (also known as crane mats), one rigging moves five timbers at once, so they are moved around more efficiently. The timbers are bolted together with through-rods and have picking points so they can be moved with a crane and not just a forklift. While moving mats with a forklift is okay, more damage occurs from the forks and reduces the life of the mats.

Another option, though not as common, is to deck the trestle with precast concrete slabs as shown in Figure 12.6. While this option may be more costly in up-front costs, if slabs are taken care of so they can be reused a number of times, it may become a feasible option. As in most temporary structure decisions, the most economical option usually is the best.

Impact Factor   This text uses an impact factor multiplier to combat the unknowns of equipment dynamics when starting and stopping quickly. Traditionally, instead of trying to calculate the actual forces created by these unfortunate events, the dead load is increased by at least 30% to account for additional forces. This multiplier is calculated as follows.

$$\text{Impact factor, IF} = \frac{50}{125 + L}$$

where L is the length of the span and 50 and 125 are constants.

FIGURE 12.6 Precast concrete trestle deck.

The impact factor is dependent on the span of the beam involved: therefore, the longer the beams, the lower the factor. The impact factor used is never greater than 30%: therefore, if the IF calculates to anything greater than 0.30, then the impact factor would be 1.30 minimum. However, if the IF is less than 0.30, then the actual number would be used. The following examples should give the reader a good understanding.

Example 12.2 When a beam span is 35 ft long, the IF = 50/125 + 35 = 0.3125, use 1.30 as the multiplier.

Example 12.3 When a beam span is 50 ft long, the IF = 50/125 + 50 = 0.2857, use 1.286 as the multiplier.

The multiplier is used to increase the shear and bending stresses on the beams in question. This is an additional factor of safety, but for impact and dynamics. The standard FOS (e.g., $F_b = 0.60F_y$) will still apply to the steel and wood members. This is not considered redundancy, these are two different factors.

Some equipment bridges that will support only slow-moving equipment, such as crawler cranes, may not have an impact factor, or may have a slight impact factor such as 5–10%. This is discussed among engineers quite often. Redundancy in design is common and the question of whether to include impact factors or how much of an impact factor to include will continue to be discussed. Most often, the engineer of record and the check engineer come up with an acceptable solution that pleases both.

In Example 12.4, a true impact factor will be used for fast-moving vehicles. In the other examples, 5–10% will be used to show the variances.

12.1.6 Environmental Concerns

When access is difficult on a project, it is often because the project site is on or adjacent to a body of water, wetlands, steep terrain, and the like. Such areas are commonly environmentally sensitive areas as well. They may contain plant, fish, and wildlife species; noise and/or vibration restrictions; or any of a number of other conditions. In the Carquinez Straights during the Benicia Bridge construction in the early part of this century, the construction operations included pile driving of cofferdam sheet piling and large-diameter permanent pile. The impact of these elements, whether permanent or temporary to the project, had an effect on the marine wildlife. The owner and contractor had to mitigate any harm to the affected species. A bubble curtain was employed to dampen the vibration and sound waves through the water that could cause harm to the marine environment.

When piles are driven into a riverbed, as shown in Figure 12.7, it is common to have silt from the river bottom contaminate the water. When these areas contain fish habitat, the fish and game or water quality control permits typically require the control of this silt release, which can be accomplished by using some sort of silt curtain that contains the silt in one area, as well as mitigation for any silt contamination that does occur. Some permits that may determine the means and methods of the contractor are:

DWR, Department of Water Resources

FEMA, Federal Emergency Management Agency

FIGURE 12.7 Work trestle over the Sacramento River.

WQCB, Water Quality Control Board

DFG, Department of Fish and Game

As mentioned earlier in the chapter, the most common trestle uses timber mats on steel wide-flange stringers, over steel wide-flange cap beams, supported by pile. The Example 12.4 will follow this design.

CASE STUDY: ANTLERS BRIDGE

Project Overview

The Antlers Bridge project replaced a 72-year-old, 1328-ft-long steel deck truss bridge with a 1970-ft-long reinforced concrete cast-in-place segmental bridge. The bridge, which spans the Sacramento River arm of Shasta Lake in northern California, is a vital part of the Bureau of Reclamation's Central Valley Water Project in California. It provides water for agriculture, power generation, flood control, recreation, and environmental needs. The water can be 120 ft deep at the project location, and it is common for the water depth to fluctuate 100 ft in a single season. The many environmental, design, and construction challenges on the project included:

- Major interstate route
- Deep water design and construction
- Hazardous waste consideration
- Difficult access
- Limited construction staging areas
- Storm water pollution prevention
- Variable water depth consideration
- Environmental construction windows
- Pier height
- Challenging bridge demolition demands
- Weather extremes: high rainfall, freeze, and snow area

Access to Work

One of the more difficult and expensive challenges to overcome was lake access. In general, the seasonal drawdown of the lake is predictable. The Bureau of Reclamation is required to draw the lake down to an elevation 46' below full pool every year due to flood control requirements. Agricultural drawdown is also predictable. It is governed by irrigation and water supply contracts between the Bureau of Reclamation and its clients, and generally results in a drawdown of approximately 40'–50'. Depending on the lake level at the beginning of the season, one can predict that the lake will be, at a minimum, 50' lower at the end of the season. The largest unknown is lake level

(continued)

CASE STUDY: ANTLERS BRIDGE (*Continued*)

recharge because the weather in the area varies considerably. In reviewing historical water cycles, the low-water period can be as long as 2 months in wet years and 5 months in dry years (Figure 12.CS1).

FIGURE 12.CS1   Trestle on north side of lake.

The facts pointed to many options, each having a certain degree of risk.

Option 1

If lake levels were to stay low for two consecutive years, the trestle could be built 46′ lower. With a good deal of certainty, the trestle would be exposed for 5–6 months, and with an aggressive schedule the foundations for the main piers could be completed.

A crane-mounted barge was going to be necessary on the project anyway for delivery of material from one side of the lake to the other. The barge crane could be used during the high-water periods (Figure.12.CS2).

Option 2

Build trestle to full pool elevation and guarantee access all year.

The contractor chose option 1. The water levels in the previous 2 years were very low with the highest elevation not even reaching the flood control elevation of 46′ below full pool. Figure 12.CS3 shows the lake level rising in the first

FIGURE 12.CS2   Trestle under construction—year 1.

FIGURE 12.CS3   Lake level rising.

construction season. If water levels were going to continue to stay this low, 12-month access was available. Another reason is the cost of constructing the trestle to full pool elevation was more than double due to additional material and bracing requirements.

(continued)

CASE STUDY: ANTLERS BRIDGE (*Continued*)

Unfortunately, the weather did not cooperate. Within a month of starting trestle construction (December 2009), the weather changed and the lake level increased and inundated the work. Within 2 months the lake level increased to above the design trestle deck height and work stopped. That year the lake level increased 127′ to full pool and did not recede to flood control elevation until September 2010. It again increased above the flood control level by December 2010, allowing only 3 months of access to the uncompleted trestle. As a result, the contractor's only option was to redesign and raise the trestle to full pool elevation.

Example 12.4 Trestle Design Example

Design an access trestle for concrete and earth-moving trucks given the following criteria:

Project Features: Two-span bridge with two abutments and one center bent. Use two stringers, 11 ft apart and a double bent cap. For other criteria use, the drawings provided in Figures 12.8–12.15.

The truck has two axles and single tires front and rear. The rear axle load is 50 k and the front axle load is 40 k. These loads are split in half for left and right wheels. Therefore, front-left = 20 k, front-right = 20 k, rear-left = 25 k, and rear–right = 25 k. See the sketch in Figure 12.8 for clarity.

Design Criteria

Use timber decking and all steel members below the decking. Use A36 steel with $F_b = 0.60F_y$ and an additional impact factor (IF). For decking, use F_b allowable = 1800 psi and F_v allowable of 160 psi. Figure 12.8 includes additional information for design.

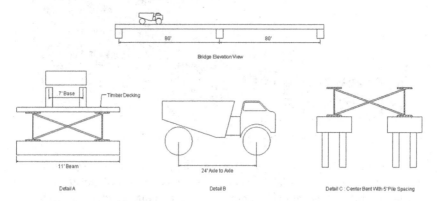

FIGURE 12.8 Example 12.4 Truck bridge details.

Step 1: Position the truck in detail A in Figure 12.8 so that the wheel locations of the heaviest wheels (50-k axle, 25 k each) generate the maximum shear force in the timber decking. This location, by rule of thumb, is when one of the wheels is 12 in from the stringer below. Technically, the wheel can be moved to within 2–3 in and a higher shear value can be achieved; however, 12 in is used to simplify the distances to 1 ft increments. Figure 12.9 shows the 12-in distance from the centerline of the stringer to the centerline of the wheel. Sum the moments using the two 25-k loads and determine the reactions, which are the maximum and minimum shear loads. Since the decking is specified to be timber, the size of the timber can be determined by shear and then later checked for bending. Unlike steel, timber is susceptible to shear failure just as much as bending failure.

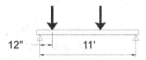

FIGURE 12.9   Wheel loading for maximum shear.

Shear:

ΣM about left side, $R_{right} = [(25 \text{ k} \times 1') + (25 \text{ k} \times 8')]/11' = 20.5$ k.

$$\Sigma F_v = 0, \; -25 \text{ k}-25 \text{ k} + R_{left} + 20.5 \text{ k} = 0, \; R_{left} = 29.5 \text{ k}.$$

The maximum shear load from this arrangement is 29.5 k at the left reaction (support stringer). Draw a shear diagram similar to Figure 12.10 to visualize this.

FIGURE 12.10   Shear diagram.

Step 2: Calculate the impact factor for this 11-ft span.
Impact factor $= 50/125 + 11' = 0.37 > 0.30$; therefore use 1.30 multiplier.
Step 3: Determine the maximum shear in one timber and size the decking based on shear. Now consider that, based on the size of one of these wheels, one wheel will make contact with at least two and maybe three timbers at one time, the tire pressure will also have an effect on this as well. Assume a wheel contacts three 12″ timbers at once, then calculate the "new" maximum shear on one single timber with the impact factor. Figure 12.11 is a representation of this loading.

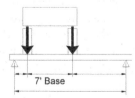

FIGURE 12.11   Wheel loading on timber.

$V_{max} = 29.5$ k $\times 1.30/(3$ timbers$) = 12.8$ k per one single timber.
Since $F_v = 1.5V/A$ for rectangular sections, then $A_{min} = 1.5V_{max}/160$ psi,
$A_{min} = 1.5\,(12.8$ k$)/0.160$ ksi $= 120$ in^2

As a square section, this would require a minimum of $10.95'' \times 10.95''$. The closest
timber that is larger than these dimensions is an S4S 12×12. Now check to see if this
12×12 can withstand the maximum bending moment. An S4S 12×12 measures
$11.25'' \times 11.25''$. Figure 12.12 shows a trestle with wood decking and hand railing.
This is a very common trestle arrangement between the decking and the stringers.

FIGURE 12.12   Trestle with wood decking.

Step 4: Assuming a 12×12, check bending stress. The section modulus of an S4S
12×12 is 237 in^3. The allowable bending stress for the timber decking was given at
1800 psi. Now, position the truck so that the wheels generate the maximum possible
bending moment. The following rule explains how this is done.

Rule: To generate the maximum bending moment in a beam with two forces either
equal or not, calculate the *center of gravity* between the two forces (wheels). Then
position the *centerline of the beam* span in between the *heaviest force/wheel* load
(either one if they are equal) and the *resultant* (center of gravity) of the forces/wheels.

Using this rule, position the truck accordingly. The center of gravity between the
left and the right wheel is in the middle because both wheels are 25-k loads. Convert
the 25-k wheel load into a single timber force considering the impact factor and the
three timbers it makes contact with. Call these forces V_1 and V_2.

$$V_1 = V_2 = \frac{25 \text{ k} \times 1.30}{3 \text{ timbers}} = 10.8 \text{ k}$$

Now place the vehicle in a location on the 11-ft span so the centerline of the beam is between the 10.8-k wheel on either side and the resultant location, which is 5.5 ft from the left or 5.5 ft from the right as shown in Figure 12.13.

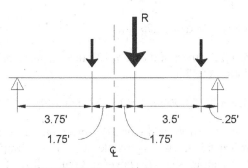

FIGURE 12.13 FBD of truck position.

Left support to left wheel = 3.75 ft
Left wheel to centerline of span = 1.75 ft
Centerline of span to resultant = 1.75 ft
Resultant to right wheel = 3.5 ft
Right wheel to right support = 0.25 ft

All these distances should add up to the 11-ft timber span. If they do, calculate the maximum moment generated by this configuration. First, redraw the free-body diagram without the unneeded dimensions and remove the resultant arrow and beam centerline as in Figure 12.14.

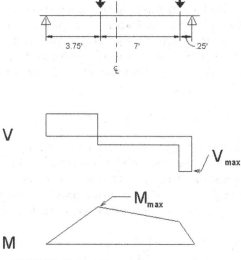

FIGURE 12.14 Shear and moment diagram.

$$\Sigma M = 0, \; R_{\text{right}} = [(10.8 \text{ k} \times 3.75') + (10.8 \text{ k} \times 10.75')]/11' = 14.2 \text{ k}$$

$$\Sigma F_v = 0, \; R_{\text{left}} = -10.8 \text{ k} - 10.8 \text{ k} + 14.2 \text{ k} = 7.4 \text{ k}$$

$$V_{\text{max}} = 14.2 \text{ k}$$

$$M_{\text{max}} = 7.4 \text{ k} \times 3.75 \text{ ft} = 27.75 \text{ ft-k}$$

Solve for the bending stress in the 12×12 timber using a moment of 27.75 ft-k and a section modulus of 237 in^3.

$$F_b = (27.75 \text{ ft-k} \times 12''/\text{ft})/237 \text{ in}^3 = 1.405 \text{ ksi or } 1405 \text{ psi} < 1800 \text{ psi} \quad \text{(OK)}$$

Step 5: Determine the front and rear axle forces on one stringer using impact factors and with the truck in the worst possible case for the stringer in order to design the main stringers spanning 80 ft, as shown in Figure 12.15. Since the beams are simply supported single spans, the design of one will represent both of them. Start by calculating a new impact factor for the 80-ft span.

$$\text{IF} = 50/125 + 80 = 0.24$$

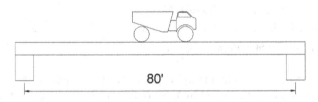

FIGURE 12.15   Truck positioned on 80-ft span.

Since this value is less than 0.30, use 1.24 as the multiplier. Now multiply this value times the maximum wheel load on the girder in the worst-case scenario. This worst-case scenario was determined in the timber design portion of this example, which used a 29.5-k load in order to generate the highest shear. In other words, the stringer experiences the heaviest load when the truck drifts to the left or right and gets within one foot of the centerline of the stringer. Another thing to consider is that if we use 29.5 k instead of 25 k, we should use a higher load that the other 40-k front axle load. First, calculate the heavier load with the 1.24 impact multiplier, and then figure out what the front axle load will be.

$$\text{Rear} = 29.5 \text{ k} \times 1.24 = 36.6 \text{ k}$$

Now for the front axle single wheel load on the stringer, use a proportioning technique from the rear to the front. We know that the original rear load for a single wheel was 25 k and the front was 20 k. Proportion these figures to achieve the missing front axle load, considering worst-case positioning of the vehicle and the impact factor 36.6 k is to 25 k as (front axle force, F_f) is to 20 k; front axle force $= 36.6 \text{ k} \times 20 \text{ k}/25 \text{ k} = 29.3 \text{ k}$

$$\frac{36.6 \text{ k}}{25 \text{ k}} = \frac{F_f}{20 \text{ k}}$$

Now both forces from either the left or right side of the truck are known to be on one single stringer, using the worst-case position and the impact factor.

Step 6: Stringer design. Using the forces from step 5, size the stringer for bending stress (ignoring shear) and then check for flange buckling. The rule that was used previously will be used again. The rule has been restated below with customization for the stringer design and two unequal loads, front and rear.

Rule: To generate the maximum bending moment in a beam with two unequal forces, calculate the *center of gravity* between the two forces (front and rear). Then position the *centerline of the beam* span in between the *heaviest force/wheel* load (rear in most cases) and the *resultant* (center of gravity) of the two forces/wheels.

Determining center of gravity for two unequal forces requires that the moments be added in order to find the location of the center of gravity. The forces involved are the rear 36.6-k force and the front 29.3-k force. The total resultant force from these two are $36.6 + 29.3 = 65.9$ k. The axle spacing of this truck was given at 24 ft wide. Therefore, the center of gravity will be somewhere between the middle and the 36.6-k load. Choose one side and add the moments, using the wheel loads and the resultant force of 65.9 k.

$$\Sigma M \text{ (rear wheel)} = 0$$

$(65.9 \text{ k} \times X') = (29.3 \text{ k} \times 24')$, where X' is the distance from the rear wheel and the resultant:

$$X' = \frac{29.3 \text{ k} \times 24'}{65.9 \text{ k}} = 10.7 \text{ ft}$$

Figure 12.16 indicates the dimensions between the resultant and the front and rear axles.

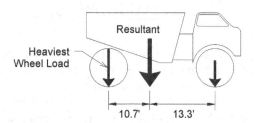

FIGURE 12.16   Resultant location.

Now follow the rule for maximum moment placement and set the centerline of the 80-ft-span between the 36.6-k load and the resultant. Divide the 10.7 dimension in half and place the centerline 5.35 ft from the 36.6-k load.

Redraw the free-body diagram with only the information needed and make sure all three dimensions add up to 80 ft. See Figure 12.17 for the proper placement of the truck on the span.

$$\Sigma M = 0, \ R_{\text{right}} = [(36.6 \text{ k} \times 34.7') + (29.3 \text{ k} \times (34.7' + 24')]/80'$$
$$= 37.4 \text{ k}$$
$$\Sigma F_v = 0, \ R_{\text{left}} = -36.6 \text{ k} - 29.3 \text{ k} + 37.4 \text{ k} = 28.5 \text{ k}$$

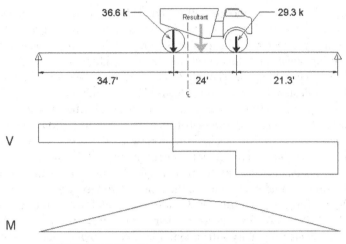

FIGURE 12.17   Placement of truck on 80-ft span.

$$V_{max} = 37.4 \text{ k}$$

$$M_{max} = 28.5 \text{ k} \times 34.7 \text{ ft} = 989 \text{ ft-k}$$

Using $F_b = 21.6$ ksi for allowable bending stress and 989 ft-k for maximum moment, determine the three lightest W shape beams that will work. Then check all three for flange buckling and determine the bracing needed.

$$S_{x(min)} = \frac{(989 \text{ ft-k})(12''/\text{ft})}{21.6 \text{ ksi}} = 550 \text{ in}^3$$

Looking at Table 12.3, all three of these beams were selected on the basis of bending requirements. Two of the three need four braces according to L_c requirement, and the other, five braces. The braces are recommended at each end and intermediates are required per the L_c calculation. Based on its need for an additional brace, it might seem logical to rule out the W30 × 191. The other two beams are very different in weight. In this case, using the lighter of the two beams (W27) would seem to make sense because it satisfies bending requirements and needs the same number of braces to install. With that said, we could have considered the W30x191 as good as choice 2 (W24 × 229) because the difference in weight between the two at $0.60/lb could easily pay for the extra brace. A simple analysis, using the cost of $1000 to furnish, install, and remove a single brace, tells the whole story.

TABLE 12.3   Flange Buckling Check for BEST beam

Beam	S_x (550 in³ min)	d/A_f	L_c (ft)	Braces, Conclusion
W27 × 194	556	1.49	28.3	80/28 = 3, 4 braces
W30 × 191	598	1.72	24.5	80/24 = 4, 5 braces
W24 × 229	588	1.15	36.6	80/36 = 3, 4 braces

W27 × 194, 80 ft × 194 = 15,520 lb × $0.60 = $9312 plus four braces @

$$\$1000/\text{ea} = \$13,312$$

W30 × 191, 80 ft × 191 = 15,280 lb × $0.60 = $9168 plus five braces @

$$\$1000/\text{ea} = \$14,168$$

W24 × 229, 80 ft × 229 = 18,320 lb × $0.60 = $10,992 plus four braces @

$$\$1000/\text{ea} = \$14,992$$

As suspected, based on this cost analysis, the W27 is indeed the most economical choice, the W30 was second best, and the W24 was third best.

Step 7: Design the center bent. Determine the cap beam sizes and the pile load requirement. Ignore the abutments, as it is assumed they are on mudsills directly on the ground surface. The pile sets at each of the individual bent caps are spaced at 5-ft centers. The force arrow is the stringer load bearing on the bent cap beam (see Figure 12.18). The first question is: What is the largest load one bent cap will support? This can be answered by going back to the stringer design and positioning the load close to the cap beams.

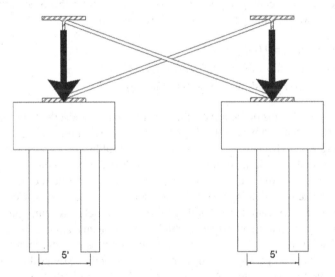

FIGURE 12.18 Stringers loading cap beams.

It can be assumed that only one truck is on the bridge at any given time or it can be assumed that two could mistakenly be on the bridge, one per span, at the same time. The latter, more conservative approach is usually always taken and will be in this example. The reactions from step 6 indicate that 37.4 k was the reaction for one stringer, so if there are two stringers, then the total load on the bent cap is twice the stringer reactions.

$$\text{Bent cap load} = 37.4 \text{ k} \times 2 = 74.8 \text{ k}$$

The maximum moment on the beam then is $M_{max} = PL/4$; $M = (74.8)(5 \text{ ft})/4 = 94$ ft-k.

$$V_{max} = 74.8 \text{ k}/2 = 37.4 \text{ k}$$

Using the same allowable bending stress of 21.6 ksi, the minimum section modulus is

$$S_x = \frac{(94 \text{ ft-k})(12''/\text{ft})}{21.6 \text{ ksi}} = 52.2 \text{ in}^3$$

Select two HP shapes that have a minimum section modulus of 52.2 in³ and then check for web yielding. The two H-Piles meeting this criteria are listed in Table 12.4.

TABLE 12.4 HP Shape Selection

H-Pile	S_x (in³)	k	t_w
HP10 × 57	58.8	1.0625	0.565
HP12 × 53	66.8	1.125	0.435

Step 8: Web yielding.

The potential yielding occurs when a concentrated load or reaction force occurs and the beam's web cannot support the load and, therefore, becomes overstressed. The allowable web yielding stress for any beam according to AISC, 9th edition, is

F_{wy} (allowable) = $0.66F_y$.

A36 steel, F_{wy} (allowable) = 0.66(36 ksi) = 23.76 ksi.

Grade 50 steel, F_{wy} (allowable) = 0.66(50 ksi) = 33 ksi.

There are two beams in Table 12.4 that are adequate to resist the bending moment. The beams have been determined to have a high enough section modulus to withstand any bending stress. However, is either beam susceptible to web yielding? The best beam, because of its weight per linear foot, is the HP12 × 53, so this beam should be checked for web yielding. If it passes the check, then it will be the best beam. If it does not pass the web yielding test, then the HP10 × 57 should be checked. If neither beam passes, the designer can (1) add stiffeners or (2) select a beam that has a high S_x and better web yielding resistance, which usually means a thicker web.

Web yielding occurs at the area of the web thickness and the length developed from the width of contact (beam flange or pile width) and five times the K dimension. K is designated in the beam values (see Appendix 1) and is the distance in inches from the top of a beam to the point of tangent where the steel has a radius to the web thickness. The concentrated load is a force acting downward from a beam above and projects the force at an angle of approximately 22° (2.5 : 1). The beam's flange width dictates where this angle of projection begins. For a good understanding of this concept, see Figure 12.19.

Scenario 1: Downward Projection

$$F_{wy} = \frac{P}{[(b_f + (5K)) \times (t_w)]} \leq 0.66 \text{ ksi}$$

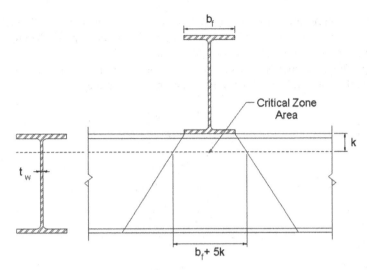

FIGURE 12.19 Web yielding downward projection.

where P = concentrated force or reaction
  b_f = flange width of upper beam
  t_w = web thickness of lower beam
  K = dimension from top of lower beam to the tangent of the web thickness (see beam values in Appendix 1)
  F_y = yield strength of steel
  0.66 = allowable stress multiplier

Scenario 2: Upward Projection (refer to Figure 12.20)

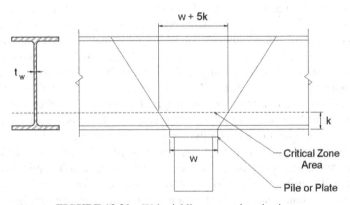

FIGURE 12.20 Web yielding upward projection.

The stress is a normal case as in P/A. The difference is that $A = [f_b + 5(K)] \times t_w$. This is the rectangular section of the yielding section at the web located distance K from the top of beam.

When the concentrated load is a reaction and comes from underneath the beam in question, project the force at a 2.5 : 1 angle upward. Instead of f_b for the upper beam, use the dimension of the plate and/or pile width.

Step 8 (continued): Now, using this same principle in scenario 2, calculate the stress on the pile cap beam from the concentrated load at the pile.

$$P = 37.4 \text{ k}$$

$$\text{Cap beam} = \text{HP}12 \times 53$$

$$t_w = 0.435 \text{ in}$$

$$K = 1.125 \text{ in}$$

$$\text{pile} = \text{HP}14 \times 89$$

$$b_f = 12.695 \text{ in}$$

$$d = 13.83 \text{ in}$$

Depending on the pile orientation, either dimension can represent the beginning of the projection.

Assuming the pile is turned so the f_b dimension is transverse to the cap beam, use the $f_b = 12.695$ in and *no plate is used.*

$$F_{wy} = \frac{37.4 \text{ k}}{[12.695'' + (5 \times 1.125'')] \times (0.435'')}$$

$$= 4.69 \text{ ksi} < 0.66F_y \text{ (or 23.76 ksi)} \qquad \text{(OK)}$$

Example 12.5   Crane Trestle

A crane trestle is needed for a Manitowoc 4000W over a river. The river is normally 8 ft deep (20 ft during flooding) and flows at a rate of 9 ft/s. The work can only be done in the low flow time, so 8 ft is the most the trestle will experience. The crane has to access a pier in the middle of the river approximately 150 ft from the normal water level (spring, summer, and fall) shore line. In the past, your company has built trestles for similar size cranes using spans between 28 and 34 ft long. At 30 ft, this would create five spans of trestle. This will be the goal since the available beams in the company's yard average 40 ft in length. Longer spans could be checked, but a four-span trestle would require 37.5-ft spans (150/4). This would not allow the stringers to have proper overlap on the cap beams, so five spans is the best option.

The maximum load the crane will lift during the life of this trestle is 39.9 k. Some other characteristics of this lift are as follows. The boom length required in order to reach the work during all phases of construction is 140 ft. All cranes have charts that indicate the maximum capacity of the crane in multiple cases of boom length, radius, and angle. Crawler cranes have two cases of maximum capacity, with the tracks extended or retracted. It is most common to operate cranes with the tracks extended, unless there are tight access restrictions. In this case, the trestle will be constructed to accommodate the crane with extended tracks and a walkway for workers. The crane radius (distance between the crane rotation and the load during maximum lift) with 140 ft of boom and lifting 39.9 k is approximately 65 ft. The track dimensions are 20.5 ft long (center of roller to center of roller) and 4 ft wide. The crane weighs 330,000 lb.

The total load of the crane while lifting 39.9 k is 369,900 lb (rounded to 370,000 lb or 370 k).

Operating over the Side

The crane will be operating over the side while servicing the pier construction. There-fore, this design analysis will be with the 39.9-k load over the side at a 65-ft operat-ing radius. When a crane is lifting a maximum lift at its maximum radius, 80–90% (tipping) of the total load shifts to the side of the lift. In this case, the more conserva-tive 90% will be used. A value as low as 80% can be used if the designer wants to be aggressive.

Use 90% tipping:

90% of 370 k = 333 k, 333 k/20.5 ft (track length) = 16.24 klf, 16.24 klf/4 ft (track width) = 4.1 ksf.

Operating over the Front

The gradient (triangular loading diagram) for the maximum load over the front of a crawler crane is developed by determining how much of the tracks overall length (20.5 ft) receives the load. One rule of thumb is that one-half of the track's length receives 100% of the load as a varying uniform load (triangular). Figure 12.21 shows what this loading diagram would look like along one-half of the 20.50-ft-long track length.

$$L_1 = L_t/2 = \frac{20.50 \text{ ft}}{2} = 10.25 \text{ ft}$$

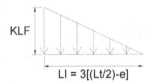

KLF

LI = 3[(Lt/2)-e]

FIGURE 12.21   Crane track loading over the front.

Another method is to calculate the moment that occurs during the maximum lift (39.9 k × 65 ft) and then to find the eccentricity of the moment and back into the length of track receiving the load:

$$e = \frac{M}{P}$$

where e = distance from the resultant of the triangle to the centerline of the loading diagram
M = moment generated from the maximum lift
P = total load between the crane and the lift (370 k)

$$M = 39.9 \text{ k} \times 65 \text{ ft} = 2594 \text{ ft-k}$$

$$e = \frac{2594 \text{ ft-k}}{370 \text{ k}} = 7.0 \text{ ft}$$

$$L_l = 3 \left[\left(\frac{L_t}{2} \right) - e \right]$$

where L_2 = loaded track length
         L_t = length of the track

$$L_2 = 3 \left[\left(\frac{20.50 \text{ ft}}{2} \right) - 7.0 \text{ ft} \right] = 9.75 \text{ ft}$$

The maximum load of the varying uniform diagram (W) is the total track load of 333 k (from above) divided by the loaded track length (L).

Method 1:

$$W = P/L_1, \quad \frac{330 \text{ k}}{10.25 \text{ ft}} = 32.2 \text{ klf} \quad \left(\frac{32.3 \text{ klf}}{8 \text{ ft}} = 4.0 \text{ ksf} \right)$$

Method 2:

$$W = P/L_2, \quad \frac{330 \text{ k}}{9.75 \text{ ft}} = 33.8 \text{ klf} \quad \left(\frac{33.8 \text{ klf}}{8 \text{ ft}} = 4.3 \text{ ksf} \right)$$

The total klf was divided by 8.0 ft, which represents two track widths. Since it was decided earlier to use method 1 (the simpler method), the gradient for the front loading would be 4.0 ksf at the high end and 0 ksf at the low end over a 10.25-ft length. This is illustrated in Figure 12.22.

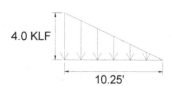

4.0 KLF

10.25'

FIGURE 12.22   Crane track loading method 1.

Since the side loading was going to be the determining factor, the rest of the problem will use 16.24 klf and 4.1 ksf, respectively. This was the side load value using 90% of total load.

Timber Decking

If timber decking is used, the decking should be checked for shear and bending resistance. Maximum shear allowable stress (F_v) is 140 psi and maximum bending allowable stress (F_b) is 1500 psi. The crane should be positioned in order to generate the maximum shear on the timber. As a rule, it was discussed earlier in this chapter to place loads 12 in from the reaction point and test the shear at this point. The contractor has a large number of 8 × 18 timbers available in the yard; therefore, these will be the most desirable timbers to use. The timbers will be full-sawn, which means their actual and nominal dimensions are the same. Check the shear on an 8 × 18.

8 × 18 properties (full sawn)

$A = 96$ in^2

$S_y = 128$ in^3 (weak direction)

16.24 klf/4 ft = 4.1 ksf × 18″/12″/ft = 6.2 k

$f_v = 1.5(6.2 \text{ k}/96 \text{ in}^2) = 0.0.96$ ksi or 96 psi < 140 psi   (OK)

Demand = 96 psi/140 psi = 0.69

Bending over a 5-ft span between stringers, the beam program is used to generate the maximum moment (see Figure 12.23). Note that the section modulus listed for the 8 × 18 timber is in the "weak" direction (S_y).

$$f_b = M/S_y, f_b = (12.3 \text{ ksi} \times 12″/\text{ft})/128 \text{ in}^3$$
$$= 1.153 \text{ ksi (or 1153 psi)} < 1500 \text{ psi}   \text{(OK)}$$

Demand = 1153/1500 = 0.77

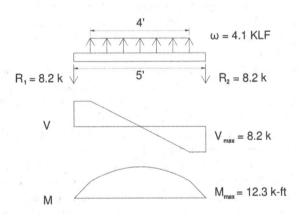

FIGURE 12.23   Shear and moment diagram of single track.

12.1.7   Stringer Design

We can assume one stringer takes the total track load or we can distribute a large percentage, say 75%, of the load to one beam. The more conservative approach is usually appreciated when dealing with human lives and equipment that costs hundreds of thousands of dollars. Place 100% of the track load onto the 30-ft stringer span and determine the maximum moment for this condition. The beam software has

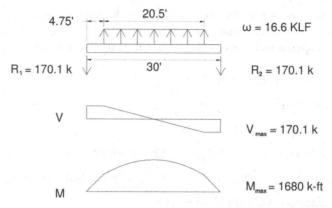

FIGURE 12.24 Shear and moment diagram.

generated a maximum moment of 1680 ft-k (see Figure 12.24). Add an impact factor to this moment.

$$IF = 50/125 + 30 = 0.32, \text{ use } 1.30 \text{ maximum}$$

$$1680 \text{ ft-k} \times 1.3 = 2184 \text{ ft-k}$$

Stringer Size To size the stringer, use the maximum bending moment determined by the shear and moment diagram. Stringer size depends on $S_{x,min} = (2184 \text{ ft-k} \times 12''/\text{ft})/21.6 \text{ ksi} = 1213 \text{ in}^3$ minimum section modulus without considering the compactness of the beam. The beam will be used in the strong direction. Table 12.5 lists potential stringers.

None of the beams listed require bracing. Therefore, the lightest of the three choices would be preferable. Remember, this is a conservative approach because the whole track load was placed over one beam. In reality, the second stringer would pick at least 25% of the load, thus reducing the moment by this same amount. In this case the stringer requirement could have been as low as 900–920 in^3 section moduli. This is a designer's decision and should be looked at closely.

As far as the bracing, this does not mean that braces will not be used at all in order to "team" the beams for placement purposes. As mentioned earlier, this is a technique contractors use so they can set two beams at one time, transport them together (which require fewer picks), weld in more favorable conditions, and avoid having to stabilize them temporarily during placement. This method is discussed in the trestle case study of the Golden State Bridge Ord Ferry bridge project.

TABLE 12.5 Potential A36 Stringers

Beam	S_y (in^3)	d/A_f (/in)	L_c (ft)	Bracing (ea)
W30 × 391	1250	0.87	48	None
W33 × 354	1230	1.06	39.7	None
W36 × 328	1210 (close)	1.21	34.8	None

Example 12.6   Truck or Rough Terrain (RT) Hydraulic Crane Trestle
Design the decking, stringers, cap beams and pile for the information given below.
A plan view and side view have been provided in Figure 12.25.

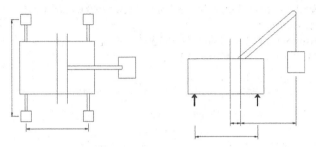

FIGURE 12.25   Plan and side view of hydraulic crane.

A truck or RT crane gets its support from outriggers. Outriggers are hydraulic
extensions that widen the footprint of the cranes base and lift the crane from the
ground so that only the outriggers make contact with the ground. At this point, the
tipping potential for the crane has been reduced because the base dimensions have
increased. Outriggers are typically founded on round or square distribution plates that
spread the concentrated load over a 1- to 2-ft dimension. It is very common and almost
mandatory that these outrigger pads set on wood cribbing in order to widen the contact
area even more so that the supporting surface (ground or trestle) does not experience
high point load pressure (P/A). Decent ground conditions can support from 2000 to
4000 psf pressure on average, and trestle decking can handle 140–160 psi in shear,
500–650 psi in perpendicular crushing, and 1500–1800 psi in bending.
   Given the following information:

Weight of crane (without load) = 250,000 lb (50 k counterweight, 200 k
   carrier weight)
Max pick = 80,000 lb
Max radius with max pick = 40 ft
Boom works, rigging and wire rope = 20,000 lb @ 20-ft
   radius center of gravity
Stringers are spaced at 5 ft OC and one outrigger (4 × 4) is supported in between
   the two.
12 × 12 wood decking (F_v = 140 psi, F_b = 1500 psi)
A36 steel, F_b = 21.6 ksi, F_v = 14.4 ksi
Outrigger pad = 4′ × 4′
Space stringers 5 ft apart and span 34 ft

   To start, the reactions under each outrigger need to be determined at the point when
the heaviest load is being hoisted.
   Add the moments about a given point. In this case, the rear outrigger location will
be fine.

$(200 \text{ k} \times 2 \text{ ft}) + (20 \text{ k} \times 20 \text{ ft}) + (80 \text{ k} \times 50 \text{ ft})/20 \text{ ft (outrigger spread)} = 240 \text{ k}$ on front outrigger (FOR)

Add the vertical forces.

Rear outrigger (ROR) = +240 k – 50 k – 200 k – 20 k – 80 k = 110 k

Since there are two outriggers in the front and two in the rear, divide these reactions in half for one outrigger.

Front outriggers (FOR) = 240 k/2 = 120 k (Figure 12.25)

ROR = 110 k/2 = 55 k (Figure 12.26)

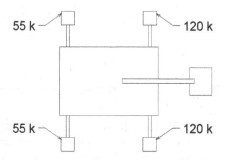

FIGURE 12.26   Diagram of crane showing outriggers.

The heaviest outrigger load (120 k) should be placed on the wood decking so the decking material must support the outrigger load in between the stringers. The FBD would look like Figure 12.27 with the uniform load being 120 k/4 ft wide pad = 30 klf, the width of the pad = 4 ft and the stringer spacing = 5 ft. See Figure 12.27 for the loading, shear and moment diagram.

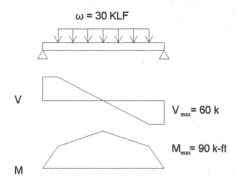

FIGURE 12.27   Loading diagram of outrigger.

In order to check the decking material for stress in shear and bending, calculate the maximum shear and bending moment, then check the stresses compared to wood decking that is good for $F_v = 140$ psi and $F_b = 1500$ psi as given above.

Shear and moment diagram:

Maximum shear = (30 klf × 4 ft)/2 = 60 k

Maximum moment = area under the shear diagram from the left support to the center of the outrigger.

$$M_{\text{max}} = (60 \text{ k} \times 0.5 \text{ ft}) + (60 \text{ k} \times 2 \text{ ft}/2) = 90 \text{ ft-k}$$

Check shear over 4 ft of timber (use 11.5″ as actual dimension)

Cross-section area of a $12 \times 12 = 11.5″ \times 11.5″ = 132.3 \text{ in}^2 \times 4 = 529.2 \text{ in}^2$

$$f_v = 1.5 \, (60 \text{ k}/(4)132.3 \text{ in}^2) = 0.170 \text{ ksi or } 170 \text{ psi} > 140 \text{ psi}$$

$$(\text{demand} = 170/140 = 1.21) \quad (\text{No Good})$$

Check bending over 4 ft of timber (use 11.5″ as actual dimension)
Section modulus of the same $12 \times 12 = (11.5″)(11.5″)^2/6 = 253.5 \text{ in}^3 \times 4 = 1014 \text{ in}^3$

$$f_b = (90 \text{ ft-k})(12)/1014 \text{ in}^3 = 1065 \text{ psi} < 1500 \text{ psi} \quad (\text{OK})$$

The calculations indicate that the wood decking is failing in shear (170 psi), even though it is fine in terms of bending. Therefore, a change has to be made in order for both checks to work. The options are:

Use a larger timber. This option is difficult because the most common (and economical) timber is a 12×12.

Reduce the spacing of the stringer. This was set as a given piece of information. If the stringer spacing changes, this will change the stringer design. This is a potential option.

Widen the outrigger pad to extend over the 5-ft beam spacing using additional wood cribbing. This option is least expensive and allows the beams to remain at a 5-ft spacing. The amount of cross-sectional area of timber needed to lower the shear stress has to be enough to lower the stress by 30 psi.

$A = 1.5 \, (60 \text{ k})/0.140 \text{ ksi} = 643 \text{ in}^2 - (4 \times 132.3 \text{ in}^2) = 114 \text{ in}^2$. If we divide this by the 4-ft (48-in) outrigger width, then we can back into the thickness of 2.39 in. This is the same thickness and same type of wood that would have to be added to the decking in this location. A 4× pad, the same width of the outrigger, would be adequate since it measures 3.5 in in thickness as shown in Figure 12.28.

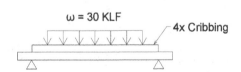

FIGURE 12.28 Outrigger on 4× wood cribbing.

The next step would be to position the crane on a set of stringers to generate the maximum bending moment for the stringer. Some assumptions can be made, such as the outrigger load (centered on two stringers) will be divided in half. Apply half the load to one stringer and half the load to the other. Draw a sketch of how the crane is positioned on the trestle and the location of the stringers. Also, the maximum outrigger loading on one set of stringers (either left or right) is when the crane swings around and picks the load off the side of the trestle. Place the two 120-k loads on one pair of stringers as shown in Figure 12.29.

The next question should be: Where do we place these two 120-k loads to generate the largest bending moment on the stringer? Like Example 12.4, which had unequal

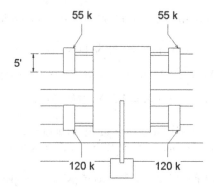

FIGURE 12.29 Plan view of crane outrigger pattern.

wheel loads, this will place the *centerline* of the stringer in between *one outrigger* and the *resultant*, which is the center of the outriggers. The outrigger spacing from front to back is 20 ft.

A beam program was used to check the outcome of this crane positioning and determined the M_{max} moment to be 508.2 ft-k (Figure 12.30). Since the loads were already divided by two for one individual stringer, this moment can be used to size one single stringer.

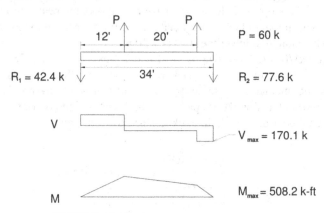

FIGURE 12.30 Shear and moment diagram.

Impact Factor

Previously, an impact factor was calculated for Example 12.4 involving a truck trestle on which trucks stopping and starting hard is very likely. If we look at the same possibility for a crawler or hydraulic crane, it should be obvious that the likelihood is much less. Some engineers would not use an impact factor at all in this case. Others may use 5 or 10%. Let's calculate the impact factor the way we learned in Example 12.2 and 12.3 but use a much lower factor.

IF $= 50/125 + 34' = 0.314$. An IF of 1.3 would be the maximum. However, an acceptable IF would be 10% for this condition because the truck and hydraulic cranes are still going to be driving on their rubber tires while positioning.

$$M_{max} = 508.2 \text{ ft-k} \times 1.10 = 559 \text{ ft-k}$$

$$S_{x,min} = (559 \text{ ft-k} \times 12''/\text{ft})/21.6 \text{ ksi} = 310.6 \text{ in}^3$$

Table 12.6 summarizes beam options with a flange buckling check.

TABLE 12.6 Beam Summary with A36 Steel[a]

Beam	$S_x(\text{in}^3)$	d/A_f	L_c	Bracing ($L_u = 34'$)
W18 × 175	344	1.11	455.5″/12 = 37 ft	None
W21 × 147	329	1.53	330.4″/12 = 27 ft	One in center
W24 × 131	**329**	**1.98**	**255.3″/12 = 21 ft**	**One in center**

[a]Best beam is in boldface.

The best stringer appears to be the W24 × 131 without doing a cost analysis between stringer weight and lateral bracing labor and materials. The beam is much lighter than the W18 × 175 and requires the same number of braces as the heavier W21 × 147 (one in center). Remember that this beam was designed based on the two outside stringers taking the entire load. The other two outside stringers would be in the same condition if the crane went to that side of the trestle to make a similar lift. The center girder, however, is never going to see the same loads, other than some wheel loading from trucks, forklifts, and the like. If that is the case, this stringer can be much lighter than the four outside stringers.

Cap Beam and Pile Design

Position the crane to generate the largest end reaction on the stringers so that the cap beam experiences the maximum load. At the same time, swing the crane 90° so that the maximum load is being lifted over the front or rear, and not over the side as shown in Figure 12.31.

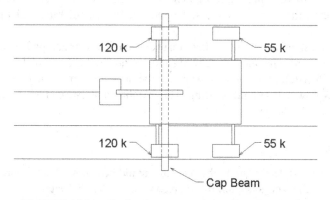

FIGURE 12.31 Hydraulic crane positioned over cap beam.

The pile should be located under the cap beam at a spacing that minimizes the bending moment in the cap beam and also maintains a manageable pile load. The beam program will be used to find the best pile location in order to achieve this best economy. It will be assumed that two piles are sufficient and that bracing can be added if the piles are unstable due to water flow and seismic shifting at 2% dead load.

The beam program results that are most desirable are shown in Figure 12.32 with an accompanying sketch. The first beam results represent the track loads (120 k/ea) over the five stringers, both loads more concentrated over stringers 1, 2, 4, and 5. The track loads are actually creating an "uplift" reaction at the center stringer (3) at the time the crane's heaviest outriggers are over the cap. The four outside stringers are supporting the complete crane load. The piles are eventually going to carry the complete 240-k crane load, 120 k per pile.

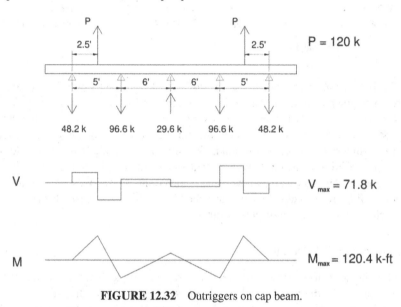

FIGURE 12.32   Outriggers on cap beam.

The beam shear and moment diagram above calculated the stringer loads. The calculated loads from this diagram can be used on the second diagram (Figure 12.33), which requires the stringer loads and the pile locations in order to check the cap beam loading.

This beam shear and moment diagram shows the five stringer loads with the center (3) acting upward. The piles are showing a force equal to the crane pad forces at 120 k. The pile will require this force in axial load, while at the same time resisting bending from water flow and seismic shifting. This combined stress analysis will be done in the next step of pile design.

Pile Design

The total live and dead load on the pile cap should be accumulated in order to determine how much force acts laterally when considering a 2–5% lateral force on the total weight over a single pile. The percent of dead load used varies between falsework

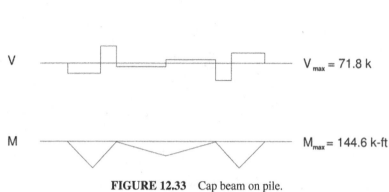

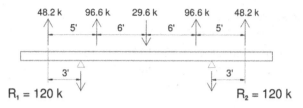

FIGURE 12.33 Cap beam on pile.

and equipment trestles in the industry. The minimum is 2%, and actually 3% can be achieved from the dynamics of the equipment using published manufacturers' equipment data coming from speeds, start and stop forces, and swinging. For trestles, many engineering firms use forces from 3 to 5% of dead load to take this into account, plus add an additional small factor of safety. In this text, 2% will be used.

The weight is the total of the crane weight with the load it is carrying plus the dead load of the timber, steel stringers, cap beam, and any other miscellaneous dead load weight that the pile will have to support. Figure 12.34 shows this condition as a combined stress load.

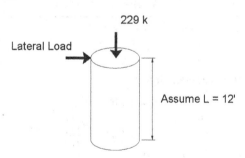

FIGURE 12.34 Pile combined loading.

Pile Load

Live crane load—all four outrigger loads = 350,000 lb

Live personnel load—$28' \times 34' \times 50$ psf = 47,600 lb

Stringers—$5 \times 38' \times 131$ lb/ft = 24,890 lb

Decking—28' × 34' × 35 pcf = 34,320 lb

Total live and dead load = 456,810 lb

Distributed to two pile (divide by 2) = 228,405 lb per pile

Lateral seismic load at 2% of DL + LL = 4568 lb (see Figure 12.35) (see also Table 12.7)

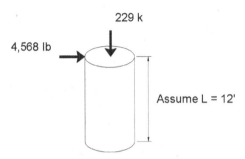

229 k

4,568 lb

Assume L = 12'

FIGURE 12.35   Seismic force at top of pile.

TABLE 12.7   Common Pipe Used as Trestle Pile

Pipe Size	OD (in)	ID (in)	Area (in²)	Section Modulus (in³)	Moment of Inertia (in⁴)	r (in)
18" × 0.25"	18	17.5	13.94	61.0	549.1	6.28
18" × 0.50"	18	17	27.49	117.0	1053.2	6.19
20" × 0.25"	20	19.5	15.51	75.6	756.4	6.98
20" × 0.375"	20	19.25	23.12	111.3	1113.5	6.94
24" × 0.25"	24	23.5	18.65	109.6	1315.3	8.40
24" × 0.50"	24	23	36.91	212.4	2549.4	8.31

OD = outside diameter and ID = inside diameter.

Lateral Load due to River Velocity

$$\text{Depth} = 12 \text{ ft}$$

$$P = wC_d \times \gamma \, \frac{(v)^2}{2(g)}$$

Units in psf

where P = pressure in plf of pile

w = width of pile including potential debris buildup

γ = unit weight of water (pcf)

v = velocity of flowing water (ft/s)

g = acceleration of gravity at Earth's atmosphere = 32.2 ft/s²

C_d = drag coefficient for shape of pile (assume 1.3 for round)

$$P = (2 \text{ ft}) (1.3) \, 62.4 \text{ pcf} \, \frac{[(9 \text{ ft/s})^2]}{2(32.2 \text{ ft/s}^2)} = 204 \text{ plf of pile}$$

$$R = 204 \text{ plf} \times 12 \text{ ft} = 2448 \text{ lb (Figure 12.36)}$$

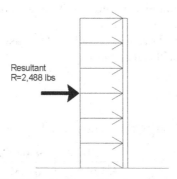

Resultant
R=2,488 lbs

FIGURE 12.36 Resultant force from water velocity.

Another method for stream velocity from the AASHTO (American Association of State Highway and Transportation Officials) version is

$$P = Kv^2$$

where $K = 0.7$ for round pile
$K = 1.4$ for HP pile
v = velocity of flowing water (ft/s)

With this formula, a velocity of 12 ft/s against an H pile would generate a 202 psf force.

$$P = (1.4)(12 \text{ ft/s}^2) = 202 \text{ psf}$$

Pile can be sized for several conditions. Typically, the size would be determined from bending or buckling since these are more critical than normal compression stresses.

Bending from lateral 2% load and river velocity is

$$M_{\text{max}} = (4568 \text{ lb} \times 12 \text{ ft}) + (2448 \text{ lb} \times 6 \text{ ft}) = 69{,}504 \text{ ft-lb}$$

The required section modulus = 69,504 ft-lb × 12″/ft/21,600 psi = 38.6 in³

$$F_b = 0.60 F_y \quad 0.60(36 \text{ ksi}) = 21.6 \text{ ksi}$$

The smallest pipe on Table 12.8 would be adequate for this bending condition (61 in³, 18″ × 0.25″). This extra section modulus may be needed when combined stress is checked.

Buckling from axial load would then be checked.

$$P = 229 \text{ k} \quad \text{Try } 18'' \text{ diameter} \times 0.25 \text{ thick pipe} \quad (\text{see Table 12.8})$$

$$F = P/A \quad F = 0.90F_y, \quad P_{\text{all}} = 0.90(36 \text{ ksi}) \times 13.94 \text{ in}^2 = 451{,}656 \text{ lb}$$

Due to the pile being braced, a k value of 1.0 was assumed. See Chapter 2 for end connections.

A combined stress analysis should be looked at to make sure the pile does not fail due to the bending and buckling combined.

Combined Stress Check

$$f_{\text{cs}} = \text{load ratio} + \text{section modulus ratio} \leq 1.0$$

$$\frac{228{,}405 \text{ lb}}{451{,}656 \text{ lb}} + \frac{38.6 \text{ in}^3}{61 \text{ in}^3} \leq 1.0$$

$0.5057 + 0.643 \leq 1.149$, since the combined stress ratio is greater than 1.0, this combined condition is exceeded by 14.9%. The next largest pile size should be considered.

The Golden State Bridge case study is a good example of a typical trestle used over a river.

CASE STUDY: GOLDEN STATE BRIDGE—ORD FERRY BRIDGE RETROFIT PROJECT

Project Overview

In 2011, the Golden State Bridge (GSB) of Benicia, California, was awarded a bridge retrofit project for the County of Butte. The bridge was the Ord Ferry Bridge over a very sensitive portion of the Sacramento River in Ord Bend, California. This project involved increasing the foundation capacity by widening and deepening the existing foundations in the river and in the floodplain. Additional piles were added in the widened footing area. In order to accomplish this foundation work, cofferdams were constructed around the existing footing, just outside the new work limits and outside the existing seal course. There was other work required to the superstructure; however, the focus of this case study is the work access from a trestle over the river. The construction season was short so GBS had to retrofit the east side in the first year, pull out of the river, and then retrofit the west side the second season (Figure 12.CS4).

FIGURE 12.CS4   First season trestle—south side.

The work trestle was shown in the contract plans, but GBS was responsible for their own design. They selected a trestle design they had used on previous projects and were able to draw from this experience and do prefabrication of materials in one of their nearby yards. This proved to be the right choice. The trestle design was a fairly typical design utilizing timber decking, wide-flange(WF) stringers teamed together with steel cross bracing, and steel WF cap beams on "star pile" (combination WF beam and T section). The river in this area was not very deep (some areas only 4–6 ft deep). However, regardless of the depth, the contractor was still working over a flowing, sensitive river, and floating equipment for access was not an option.

Crane Capacity: 110-Ton Crane, Heaviest Pick 45 k During Trestle Construction

The trestle was designed to support a 110-ton capacity crawler crane while hoisting a maximum load of almost 23 tons. The heaviest load the crane would lift was not during the cofferdam construction or concrete placement. The heaviest lift would actually be during the construction of the trestle itself. The weight of the required boom length, rigging, hammer, leads, and pile proved to be the most weight the crane would experience during the life of the trestle.

(continued)

CASE STUDY: GOLDEN STATE BRIDGE—ORD FERRY BRIDGE RETROFIT PROJECT (Continued)

Star Pile: Noise and Drivability in Hard Driving and Cobbles

The star pile consisted of two T sections welded to the middle of the WF beam (Figure 12.CS5). This created a symmetrical section that had two benefits. The first benefit was the reduction in noise and vibration. Since the Sacramento River is habitat to a variety of fish wildlife, keeping the noise and underwater vibrations to a minimum was critical according to the special provisions and project permitting through the Department of Fish and Game. The second benefit the star pile offered was the stout section made it easier to penetrate hard soils and maneuver around large cobbles present in the river subsurface.

FIGURE 12.CS5   Star pile.

Teamed Beams

The stringers for the trestle arrived on the project with the cross bracing (flange buckling resistance) and stiffeners (web yielding resistance) already installed. This welding and other fabrication was performed in their yard, where the condition for this type of work was ideal (Figure 12.CS6).

Prefabricated Cap Beams with Stiffeners and Pile Cap Cans and Pins

Also, prefabricated in the yard were the cap beams with stiffeners and pile cap cans (Figure 12.CS7). The stiffeners were located over the pile, and the cans were used to go over the star pile so that quick placement of the caps could be achieved. A shear pin was placed through the can and the star pile for uplift control. By using ideas like the prefabricated cans, welding and cutting is eliminated, thus reducing hundreds, if not thousands, of man-hours.

FIGURE 12.CS6   Teamed stringers.

FIGURE 12.CS7   Prefabricated pile caps.

(continued)

CASE STUDY: GOLDEN STATE BRIDGE—ORD FERRY BRIDGE RETROFIT PROJECT (*Continued*)

The trestle used on this project was very common except for the use of the star pile. To determine if the star pile concept is economical for a specific project, one would have to do an analysis on production labor, equipment, and materials. This same trestle concept has been used successfully throughout the nation on various projects with pipe, H, and WF piles. The project characteristics will dictate the pile types necessary to provide safe and economical access.

12.1.8 Star Pile Design and Properties

After seeing this project and the use of the star pile, this author was curious about the new properties of this star pile shape. In other words, for bearing and buckling resistance, what was the cross-sectional area and new radius of gyration? Figure 12.37 shows the cross section of the star pile. What axis was now the weak axis? What was the shape's moment of inertia in both the x and y axes?

TABLE 12.8 Properties of Sample Star Pile

Section	Area (in^2)	d (in)	Ad_2 (in^3)	Moment of Inertia (in^4)	ΣAd_2 (in^4)	$\Sigma I + Ad_2$ (in^4)
x **Axis (Strong)**						
W18 × 97	28.5	0	0	1750	0	1750
WT6 × 26.5	7.78	0	0	47.9	0	47.9
WT6 × 26.5	7.78	0	0	47.9	0	47.9
Totals					0	1845.8 in^4
y **Axis (Strong)**						
W18 × 97	28.5	0	0	201		201
WT6 × 26.5	7.78	5.28	216.1	17.7	216.1	233.8
WT6 × 26.5	7.78	5.28	216.1	17.7	216.1	233.8
Totals						668.6 in^4

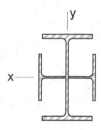

FIGURE 12.37 Cross section of star pile.

Beam W18 × 97, $A = 28.5$ in^2, $I_x = 1750$, $I_y = 201$

T's WT6 × 26.5, $A = 7.78$ in^2, $t_w = 0.345''$, $d = 6.03''$, $b_f = 10''$, $t_f = 0.575$, $I_x = 17.7$, $I_y = 47.9$

Calculations determined the properties of a sample star pile shown in Table 12.8.

$$\sqrt{r_x} = \text{square root of } \frac{1845.8 \text{ in}^4}{28.5 + 2(7.78)} = 6.47 \text{ in}$$

$$\sqrt{r_y} = \text{square root of } \frac{668.6 \text{ in}^4}{28.5 + 2(7.78)} = 3.90 \text{ in}$$

If these values are compared to a W18 × 97 by itself, the differences are interesting.

W18 × 97 before $r_x = 7.82$ in, $r_y = 2.65$ in

W18 × 97 after $r_x = 6.47$, $r_y = 3.90$ in

The radius of gyration improved in the weak direction and was lowered in the strong direction.

12.2 OTHER PROJECTS UTILIZING METHODS OF ACCESS

The following projects are other examples of work access situations that were solved using various methods.

The Galena Creek Project in Carson City, Nevada, shown in Figure 12.38, used an earthen fill to raise the valley floor to an elevation that allowed the use of more standard falsework. The cost of moving the material compacted in place and removing the material was compared to the cost to use an alternate method of support. Alternate methods were to cast the arch segmentally or install falsework the additional height to the bottom of the valley floor, which was more than 200 ft.

FIGURE 12.38 Galena Creek bridge.

FIGURE 12.39 Trans Bay Terminal project.

The bridge was a combination of a flat deck supported by an arch structure. The arch was constructed on pipe falsework and the flat deck was supported by falsework to the completed arch.

The TransBay Terminal project in San Francisco, California, is a public transportation project that is creating a terminal in downtown to merge all the major public transportations into one terminal. The excavation takes up almost three city blocks. Figure 12.39 shows strutting and a road crossing. The excavation included over 3 million cubic yards of earth removal.

This aerial photo (Figure 12.40) from Google Earth shows the excavation support system consisting of deep soil mixing and pipe strutting. Strutting was utilized despite

FIGURE 12.40 Aerial view of TransBay Terminal.

the excavation being over 150 ft wide. Included in the system were trestles for both public traffic (1st St. and Fremont St.) and crane access (longitudinally) throughout the excavation.

The very tight deflection requirements set by the engineer of record were up to 1/1000, which is 3 times most normal limits of temporary structure design. These requirements were enforced due to the close proximity of traffic and buildings to project site.

12.3 CONCLUSION

Access can be the difference between a successful project and an unsuccessful one. If the contractor was worried about spending money on good access, he may not realize the dividends that may pay off in the end. Sometimes it is hard to put a number on good access and how the supporting operations may benefit. However, terrible access will definitely not give the contractor the opportunity to maximize the production units of all the operations relying on the access.

As with most temporary structures, without a detailed analysis (estimate), one cannot make the most economical decisions that benefit all operations including the subcontractors.

CHAPTER 13

SUPPORT OF EXISTING STRUCTURES

Temporary supports are used in construction when an existing structure requires temporary support during demolition, construction, or retrofitting. The owner of the structure could be concerned about the structure settling, shifting, or completely failing. Sometimes the structures continue to be occupied, such as a building or bridge; and sometimes the facility is completely put out of service during construction. Regardless of the situation, the condition of the structure and safety of the public is most important.

During the construction of temporary supports, and while the support is in service, settlement monitoring is commonly performed to detect any unwanted settlement or lateral movement in excess of the specifications and allowable stresses to the structure.

During the planning stages for a temporary support design, the contractor must consider, at a minimum, the following:

1. How stiff is the existing structure compared to the stiffness of the temporary support? Can the existing structure allow any deflection or must the temporary support be designed for no deflection?
2. How much does the existing structure weigh?
3. Which portion of the structure footprint will require support?
4. Where can the loads be transferred?
5. Is there sufficient support capability adjacent to the structure?
6. Are temporary piles required?
7. What materials are available?

8. Do the specifications dictate the testing methods for welding and material usage?
9. Is the structure itself, in part, going to be part of the support? And if so, will it be able to stay within its allowable stresses?
10. Does the temporary support system have to be preloaded with a jacking device to eliminate or minimize deflection and stress? If so, how much preload is required?

Temporary supports could be necessary on a variety of project types, including the following:

1. State and federal highway projects during retrofit work for seismic upgrades.
2. Building construction in congested areas to prevent undermining and settlement.
3. Support of utilities during crossing utility installation or junction structure construction.

These are just a few situations that could require temporary supports.

13.1 BASIC BUILDING MATERIALS

Structures that may need to be supported usually contain basic building materials that a contractor utilizes every day, including the following:

Concrete (150 pcf)
Reinforced concrete (160 pcf)
Steel (490 pcf or 0.2836 lb per cubic inch)
Wood with 19% moisture content (33–40 pcf)
Water (62.4 pcf)

Other materials that a contractor may need to acquire unit weights for are:

Drywall
Roofing
Siding
Fiberglass

Material weights are a function of the volume of the material and its unit weight. If a material is a mixture or composite, the conservative approach would be to use the unit weight of the heaviest material and multiply that times the volume. This almost always results in a heavier load than actual, but this is the safest approach. If the volumes of the two different materials can be determined, then a more accurate weight can be determined. Another way to be more accurate and not too conservative is to weigh the material and a "new" unit weight can be used.

Table 13.1 lists example calculations of three different materials.

TABLE 13.1 Sample Calculations for Different Materials

Item	L	W	H	Volume	Unit Weight	Weight (lb)
Unit of 16' long 2 × 4	16'	4'	3'	192 CF	35.0	6750
Steel plate	12	6	1½"	15, 552 in³	0.2836	4,401
Concrete deadman	4	3	3	36 CF	150.0	5,400

13.1.1 Example 13.1 Pipe Unit Weight

Not all material volumes are perfect rectangles. Many times, the engineer will have to use averages, interpolate curves, round up, or subtract out voids. Pipe calculations are a good example of having to subtract voids. Generally, the unit weights of pipes can be obtained from the manufacturer's data sheets. However, when this information is not available, an engineer must draw from the basics in geometry and physics. How would the weight of a steel pipe be figured? Let's look at how this might be calculated.

Pipe: 48" × 0.50" steel pipe. Large steel pipe is designated by its outside diameter, for example, OD = 48".

Contents: ½ full of water.

Step 1: Calculate the outside area in inches squared because the unit weight of steel is in cubic inches.
$$A = \pi D^2/4$$
$$\pi \times (48")^2/4 = 1810 \text{ in}^2$$

Step 2: Calculate the inside area of the pipe the same way.
$$\pi \times (47")^2/4 = 1735 \text{ in}^2$$

Step 3: Determine the steel area by subtracting the OD from the ID as in Figure 13.1:
$$1810 - 1735 = 75 \text{ in}^2$$

Step 4: Calculate the weight of the pipe per linear foot:
$$75 \text{ in}^2 \times 12"(\text{length of one foot}) \times 0.2836 \text{ lb/in}^3 = 255 \text{ plf of pipe}$$

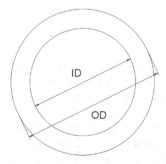

FIGURE 13.1 Steel pipe example.

Step 5: Use the inside diameter in step 2 and determine the water volume if the pipe is half full.

$$\frac{1735 \text{ in}^2}{144 \text{ in}^2/\text{ft}^2} = 12.05 \text{ ft}^2 \times 1 \text{ ft} \times \frac{1}{2} \text{ pipe} = 6.02 \text{ ft}^3$$

Step 6: Determine the water weight per linear foot of pipe.

$$6.02 \text{ ft}^3 \times 62.4 \text{ pcf} = 376 \text{ plf}$$

Step 7: What is the total weight per linear foot of this pipe, half full of water?

$$255 \text{ plf} + 376 \text{ plf} = 631 \text{ plf}$$

These steps can be followed for almost any volume that involves an outer and an inner portion. The unit weights of the material must be known and the dimensions should be accessible and rounded to ensure the safest scenario.

The rest of this chapter provides examples and case studies of temporary support of existing structures. The examples discussed may not be everyday occurrences but will set the foundation for an engineer/designer for approaching almost any temporary support condition, analyzing it, and making sound engineering decisions that leave the owner confident with the plan. The examples and case studies involve a water treatment plant building support adjacent to an excavation and a pipe support for the construction of a junction structure.

13.1.2 Example 13.2 Existing Water Treatment Plant

A water treatment project was undertaken to increase the plant's capacity. Part of the scope of the work was to add a tank next to an existing building/structure. The close proximity and depth of the new structure made it necessary to excavate adjacent to the existing structure and risk undermining and subsequently damaging the structure. Figure 13.2 shows the temporary structure in place. The materials being used are steel wide-flange beams (W shapes), 25-k per leg shoring towers (legs are doubled for 50-k capacity), double steel C-channels, high-strength coil rod tension anchors, and associated plates and nuts. The designers, for these cases, must make sound judgment decisions when determining how much the soil underneath the structure will help support or if the soil support should be completely neglected during the design of the temporary support. Many structures within water and wastewater treatment plants have to stay in service during the installation and life of the temporary support. This means water may have to remain within the structure and its weight added to the structure weight.

The steps necessary for temporary structure supports can follow this simple list:

1. Determine (calculate) the weight of the structure being supported, including any anticipated live loads as well. In this case, part of the structure could be filled with water. See Figures 13.3 and 13.4 for schematics of the structure requiring support.

FIGURE 13.2   Temporary support of an existing structure.

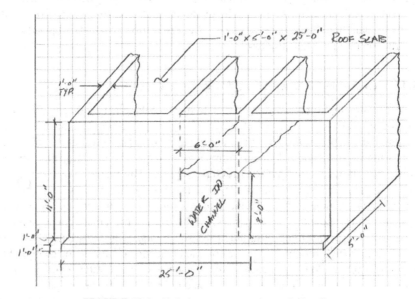

FIGURE 13.3   Existing structure schematic drawing.

2. Determine where the structure can be supported. Often these locations can experience high concentrated loads; therefore, the designer must determine what has to happen to the structure to be able to handle the load concentration. Figure 13.3 contains dimensions to assist in weight calculations and locations for support members. This example does not measure the stiffness of the

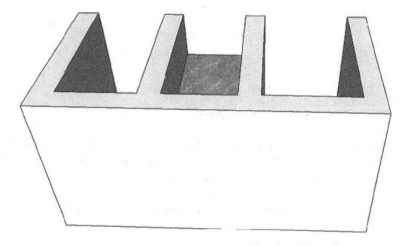

FIGURE 13.4 Water occupying the center cell.

reinforced concrete structure. It is assumed that the structure will be supported at enough points that stiffness is not a concern.

3. Determine what beams can be used and what are the longest spans. The project has W24 × 131's, HP 14 × 89's, and some random, smaller beams available.

Determining Structure Weight **Step 1:** Determine the weight of the existing structure and associated live loads.

Front Wall: $(25 \times 11 \times 12'') \times 160$ pcf $= 44,000$ lb

Wall 1 Ext (2 ea): $2(7 \times 11 \times 12'') \times 160$ pcf $= 24,640$ lb

Wall 2 Int (2 ea): $2(7 \times 11 \times 2'') \times 160$ pcf $= 24,640$ lb

Slab on grade: $(25 \times 7 \times 12'') \times 160$ pcf $= 44,000$ lb

Water in center cell: $6 \times 7 \times 8 \times 62.4$ pcf $= 20,966$ lb

Total structure weight $= 158,246$ lb (160 k)

Weight per linear foot of building (average) 158.3 kips/25 ft $= 6.4$ klf

The layout was conducive to using double stringers. These stringers would be supported by shoring towers on the ground on one side and the existing portion of the structure founded on a lower, stable level on the other side. Figure 13.5 shows the plan view of this temporary support layout.

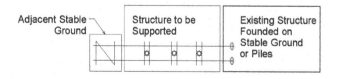

FIGURE 13.5 Plan view of temporary support layout.

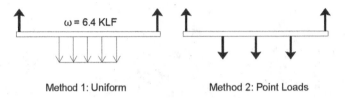

Method 1: Uniform   Method 2: Point Loads

FIGURE 13.6   Two loading condition options.

The load can be figured as a linear load (6.4 klf) or as three concentrated loads representing each cell (see Figure 13.6). Regardless of the method, the maximum moment figured should be the worst-case scenario.

Method 1: 6.4 klf (see previous calculation).

Method 2: Four concentrated loads, one from each wall of the three cells. The center cell is heavier due to the water inside the channel.

Method 1 from Beam Software:

$$M_{max} = 1140 \text{ ft-k}$$
$$V_{max} = 80 \text{ k}$$

Method 2 from Beam Software:

$$M_{max} = 1087 \text{ ft-k}$$
$$V_{max} = 82 \text{ k (loads off center slightly)}$$

Stringer Design   Using the most conservative method would mean using the moment and shear from method 1 shown in Figure 13.7. With these values we can check if two W24 × 131 work side by side. For clarity, the weight of the actual beams will not be added to the live loads. In practice, these weights would not be ignored.

Allowable stresses for A36 steel:

$$F_b = 0.60F_y = 21.6 \text{ ksi}$$
$$Fv = 0.40F_y = 14.4 \text{ ksi}$$

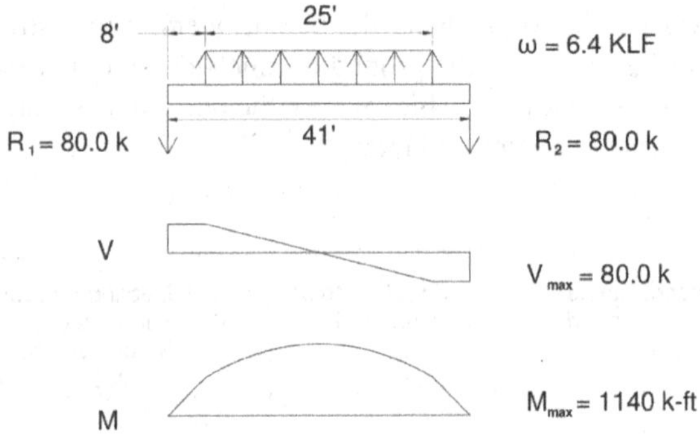

8' |← 25' →| ω = 6.4 KLF

R_1 = 80.0 k   41' R_2 = 80.0 k

V V_{max} = 80.0 k

M M_{max} = 1140 k-ft

FIGURE 13.7   Shear and moment diagram method 1.

S_x for a single W24 × 131 = 329 in^3, two beams S_x = 329 in^3 × 2 = 658 in^3
Using method 1:

$$f_b = (1140 \text{ ft-k} \times 12''/\text{ft})/658 \text{ in}^3 = 20.8 \text{ ksi} < 21.6 \text{ ksi} \quad (\text{OK})$$

$$f_v = 82 \text{ k}/2(24.48'' \times 0.605'') = 2.77 \text{ ksi} < 14.4 \text{ ksi} \quad (\text{OK})$$

Check the stability of the beam with $d/Af = 1.98$.

20,000/(36 × 1.98)/12″/ft = 24.0 < 41 ft, brace one beam to the other at least in the middle. Two braces would allow the contractor to set the beams as a team and offer more stability.

Cap Beam Design   The reaction at the tower from the previous beam design is 82 k. This is the force from the double stringer to the HP14 × 89 cap beam. Figure 13.8 shows the tower arrangement.

The design of the HP14 × 89 cap beam has a double point load from the two W24 × 131's on the HP14 × 89 spanning from one side of the tower to the other side. The two point loads can be combined as one 82-k load in order to simplify the problem and still end up with the same moment. The tower is made up of double frames on each side. They are generally 25 k per leg frames, but, when doubled up, become 50 k per leg towers. The span of the HP14 × 89 is approximately 6 ft. Therefore, using the 82-k load and a 6-ft span, calculate the maximum bending moment in the cap beam as shown in Figure 13.9. As a rule, the shear should also be checked.

$$M = PL/4$$

$$M = 82 \text{ k} (6 \text{ ft})/4 = 123 \text{ ft-k}$$

$$F_b = 123 \text{ ft-k} \times 12''/\text{ft}/131 \text{ in}^3 = 11.27 \text{ ksi} < 21.6 \text{ ksi} \quad (\text{OK})$$

$$F_v = 41 \text{ k}/(13.83'' \times 0.615'') = 4.82 \text{ ksi} < 14.4 \text{ ksi} \quad (\text{OK})$$

Subcap Beam Design   The subcaps run the short distance of the tower (4 ft), and there are two on each side in order to distribute the cap beam load equally.

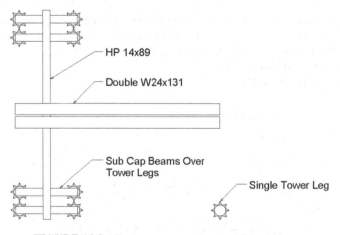

FIGURE 13.8   Tower arrangement supporting stringers.

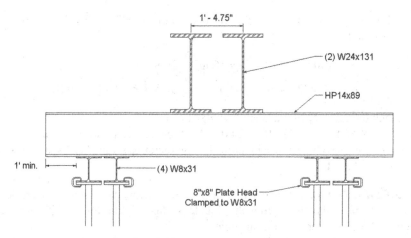

FIGURE 13.9 Cap beam layout, end view section.

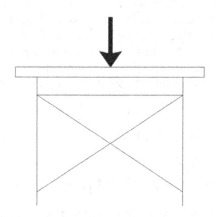

FIGURE 13.10 Subcap beam layout.

The company has available some short W8 × 31 × 8′ beams. Determine if this size beam will be sufficient. Apply the force from the cap beam to the center of the subcap beam similar to Figure 13.10. There are double subcaps on each side so be sure to divide the total 82-k force by 4.

Properties of a W8 × 31:

$$S_x = 27.5 \text{ in}^3$$

$$d = 8.0''$$

$$t_w = 0.285''$$

$$M = \frac{PL}{4} \quad M = \frac{(82 \text{ k}/4 \text{ ea})(4 \text{ ft})}{4} = 20.5 \text{ ft-k}$$

$$f_b = \frac{(20.5 \text{ ft-k})(12''/\text{ft})}{27.5 \text{ in}^3} = 8.95 \text{ ksi} < 21.6 \text{ ksi}$$

This is more than half the allowable stress. Would one subcap beam work?

$$f_b = \frac{41 \text{ ft-k} \times 12''/\text{ft}}{27.5 \text{ in}^3} = 17.9 \text{ ksi} < 21.6 \text{ ksi} \quad \text{(OK)}$$

A single subcap beam would work. However, the reason a contractor may have used the double tower ($25 \times 2 = 50$ k) is to be sure the tower capacity is adequate. The next few steps will address the towers.

The tower leg loads come from the total 82-k force to the whole tower. If this 82-k force was distributed equally to the eight tower legs, the force on each leg would be slightly over 10 k. In this case, this falsework system is slightly underused.

$$\frac{82 \text{ k}}{8 \text{ legs}} = 10.25 \text{ k/leg} < 25 \text{ k} \quad (\text{ OK})$$

The foundation for this system could be a series of timber pads distributing the tower leg forces to the soil. The pads must be a safe distance from the top of slope into the excavation. Figure 13.11 shows the location of the tower supports and the importance of this distance. A pair of two legs are supporting 20.5 k in this example. Figure 13.11 shows four 4×6's supporting a 4×6 corbel. If the 4×6 corbel projects a 1 :1 load path to the supporting soil, what would the soil load be?

$$F = 20.5 \text{ k}/\{(4 \times 5.5'') \times [5.5'' + (2 \times 3.5)]\}/144 \text{ in}^2 = 10.73 \text{ ksf or } 10,730 \text{ psf}$$

Besides some rare occasions, soil pressure can support anywhere from 2000 psf to 6000 psf. Therefore, this soil either needs to be supported on imported aggregate fill compacted to 95% or the base dimensions on the corbels and pads need to be widened.

FIGURE 13.11 Elevation view of structure.

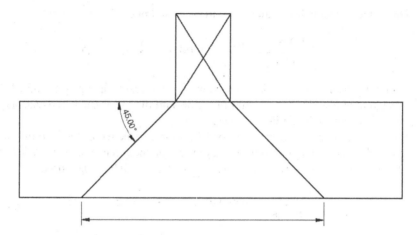

FIGURE 13.12   Pad load distribution.

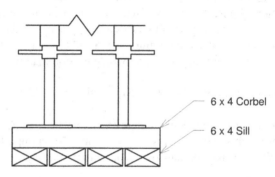

FIGURE 13.13   Corbel and pad (sill) arrangement.

The projection of the load through the timber pads is illustrated in Figure 13.12. Figure 13.13 shows the actual design condition of the corbels and pads (sills).

The two beams are attached to the structure with coil rods, plates, and nuts (see Figure 13.14). The coil rod capacity should be adequate to resist the weight of the portion of structure associated with the rod location. In this case, four rods are proposed. If we go back to the method 2 case for the stringer design, the two center rods were estimated to hold 30 k at the two outside rods and 50 k at the two inside rods. If the system was designed to support the 50-k load, then the other rods would be more than adequate. Of course, the two scenarios could have different rod diameters, but then the project staff would have to inventory these rods, and the chance of placing the wrong rod in the wrong location would increase.

Appendix 8 shows coil rod capacities. The Williams coil rod that has minimum yield strength of 58.1 k is the $1\frac{1}{4}''$ diameter coil rod.

13.1.3   Example 13.3 Temporary Pipe Supports

The temporary support example that will be illustrated next is the support of an existing pipe that will be undermined in order to build a complete new structure around it

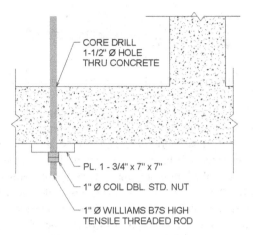

CORE DRILL
1-1/2" Ø HOLE
THRU CONCRETE

PL. 1 - 3/4" x 7" x 7"

1" Ø COIL DBL. STD. NUT

1" Ø WILLIAMS B7S HIGH
TENSILE THREADED ROD

FIGURE 13.14   Coil rod attachment.

FIGURE 13.15   Pipe support.

(see Figure 13.15). This condition is fairly common when a water facility desires to direct water to multiple locations. By adding a junction box, the operators can direct flow to one, two, or three directions or stop the flow completely. The pipe in this case may have to remain in service so the plant can continue water delivery. It is not until the junction structure is complete and the pipe inside demolished that the new structure will operate with its new configuration.

The pipe is assumed to be 50% filled with water during the life of the temporary support. The support system should be incorporated with the shoring of the excavation. If there is no shoring and the excavation is sloped (open-cut), then the

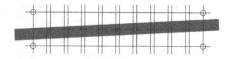

FIGURE 13.16   Pipe and support structure.

increased plan dimension of the excavation should be considered. An increased plan dimension would significantly lengthen the required beams.

The pipe that needs to be supported is an 84-in (inside diameter) reinforced concrete pipe (RCP). The water weight inside the pipe will be assumed at 62.4 pcf. Pipe weights can usually be looked up in a manufacturer's table. However, this pipe has been in the ground for 18 years and the specifications could not be found. Therefore, the unit weight of water, concrete, and steel with the dimensions of the pipe will be used to determine the weight per linear foot of pipe. The excavation is 52 ft across from one shoring wall to a steel frame support at the other side. Three stringers spanning the 52-ft excavation will be used due to a slight curve in the pipe. The general arrangement of the pipe and support structure are shown in Figure 13.16. The pipe is slightly skewed to the new structure design. Unlike Example 13.2, the weight of the components will be carried to the next step as the problem is solved.

The load path for supporting the RCP will follow this path:

Pipe will be saddled by wire rope.

The wire rope will terminate through a double C-channel support over the three stringers using coil rod, eye nuts, and wire rope clips.

The stringers will span 52 ft from one side of the excavation to the other.

The stringers will be supported at each end with the shoring wall or a short cap beam.

This example will follow the described load path using a series of steps. Standard A36 steel will be assumed using a $0.60F_y$ factor of safety.

Weight of Existing Conditions   **Step 1**: Determine the weight of permanent pipe containing 50% water.

Reinforced Concrete Pipe   RCP sizes are designated by the inside diameter (ID). Therefore, an 84″ RCP measures 84″ inside, and the outside dimension (OD) is the sum of the ID plus twice the pipe's thickness. The 84″ RCP is approximately 8″ thick, which means the OD is $84″ + (2 \times 8″) = 100″$. The process of calculating pipe weight at the beginning of this chapter holds true for concrete as well. The unit weight of the material becomes the variable:

$$OD = \frac{\pi(100″/12″/ft)^2}{4} = 54.5 \text{ SF}$$

$$ID = \frac{\pi(84″/12″/ft)^2}{4} = 38.5 \text{ SF}$$

Concrete area $= 54.5 - 38.5 = 16 \text{ SF} \times 1 \text{ ft} = 16 \text{ CF}$

Pipe weight $= 16 \text{ SF} \times 160 \text{ pcf} = 2560 \text{ lb pipe weight per linear foot}$

Water The pipe can only be half full at its maximum operating capacity due to the hydraulics of the plant. This must be confirmed so that, during the life of this temporary support, the pipe will never contain more than 50% water.

Using the ID from the pipe above, the weight of 50% water can be obtained.

ID = 38.5 SF

Water volume = 38.5 SF × 1 ft = 38.5 CF

Water weight = 38.5 CF × 62.4 pcf = 2402 lb × 50% = 1201 lb water weight per LF

Total weight of pipe and water per liner foot = 3761 lb

Step 2: Determine the size required for the wire rope.

At this point, it should be decided how far apart the wire rope saddles and double channels will be spaced. If there was a particular cable and channel size available, the engineer could use reverse calculations to obtain the proper spacing to match the cable and channel available. Another approach would be to choose spacing for the wire rope and channels, then size the cable and choose the channel. Moving forward, the spacing will be assumed to be 6 ft on center from saddle to saddle. Using this spacing, determine the weight per saddle and size the wire rope. The wire rope as a saddle is considered a two-parted line similar to a two-parted crane line that goes from the boom tip, through a "headache ball," and back to the boom with a wedge and socket (becket). In other words, the load that is distributed from the pipe to the wire rope is shared by the two legs of the wire rope.

The load associated with each saddle support is the weight of the pipe and water times 6 ft of pipe (6 ft on center).

P = 3761 lb × 6 ft = 22,566 lb/2 parts of line = 11,283 lb on a single wire rope as shown in Figure 13.17

At this time since this figure will dictate the rest of the design, it is appropriate to round this value to the nearest 100 lb and into kip units, say 11.3 k.

The factor of safety on wire rope can be between 2.0 and 6.0 (Table 13.2), depending on the application. For this application, the engineer should consider the risk in

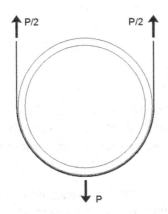

FIGURE 13.17   Wire rope saddle support.

TABLE 13.2   Wire Rope Safe Working Loads (SWL)

Minimum Breaking Force and Allowable Loads for 6 × 19 IWRC-XIPS Wire Rope with
Wire Rope Reduction of 80% < 1″ and 90% 1″ and Greater for Using Wire Rope Clips

Wire Rope Diameter (in)	Breaking Strength (tons)	SWL w/ 2.0 FOS & 80–90% (tons)	SWL w/ 3.0 FOS & 80–90% (tons)	SWL w/ 4.0 FOS & 80–90% (tons)	SWL w/ 5.0 FOS & 80–90% (tons)	SWL w/ 6.0 FOS & 80–90% (tons)
$\frac{3}{8}$—80%	7.6	3.0	2.0	1.5	1.2	1.0
$\frac{1}{2}$—80%	13.3	5.3	3.5	2.7	2.1	1.8
$\frac{5}{8}$—80%	20.6	8.2	5.5	4.1	3.3	2.7
$\frac{3}{4}$—80%	29.4	11.8	7.8	5.9	4.7	3.9
$\frac{7}{8}$—80%	39.8	15.9	10.6	8.0	6.4	5.3
1—90%	51.7	23.3	15.5	11.6	9.3	7.8

this pipe moving, shifting, or failing. The consequences, if any of these misfortunes occur, could be very costly to the contractor and probably shut down the existing plant, thus costing the owner lost revenue. With this said, the contractor will use a 4:1 factor of safety from the ultimate strength of the wire rope capacity.

$$P \times 4.0 = \text{safe working load (SWL)}$$

$$11.3 \text{ k} = 5.7 \text{ tons}$$

Referencing the wire rope chart shown in Table 13.2 with a 4:1 FOS and using wire rope clips, the minimum wire rope size that can be used is $\frac{3}{4}$″ 6 × 19 IWRC-XIPS.

To transition from wire rope to the tops of the double channel, the following hardware is required:

Wire rope clips
Eye bolt for the wire rope to loop through
Thimble to soften the radius of the wire rope bent
Coil rod threaded into the wire rope and going through the double channel
Plate washer (5 × 5)
$\frac{3}{4}$″ nut and washer

The hardware charts in Appendix 8 will indicate the SWL for each component. All the hardware associated with $\frac{3}{4}$″ diameter coil rod has an SWL of 19,000 lb, which includes a manufacturer recommended 2:1 FOS. Therefore, the $\frac{3}{4}$″ hardware with a $5 \times 5 \times \frac{1}{2}$ plate is adequate.

Double Channel Selection   Step 3: Extend the wire rope saddles through the double channels and determine the most economical channels necessary to support the intended load.

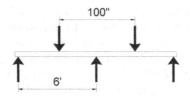

FIGURE 13.18 Double channels loading stringers.

The 11.3-k load will be applied to the channel in two locations and the channels will be supported by three stringers. A free-body diagram should be drawn similar to Figure 13.18. The three reactions at the bottom represent the three stringers, and the two forces coming from above represent the wire rope forces connected to the double channels. The wire rope connections are exactly 100 in apart because that is the outside diameter of the pipe and the cable wraps around the pipe. The stringers are placed 6 ft apart from each other.

Since the pipe is skewed, the engineer is looking for the worst-case scenario that causes the largest bending moment to the double channel. As it turns out, the most symmetrical arrangement also causes the largest bending moment. The beam program is used in Figure 13.19 because of the indeterminate loading condition. See the results below for the double channel.

$$M_{max} = 11.5 \text{ ft-k}$$

$$V_{max} = 6.29 \text{ k}$$

Stringer reactions:

Left: 6.3 k

Center: 10.0 k

Right: 6.3 k

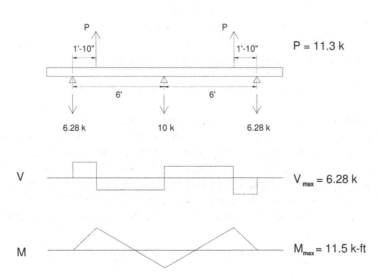

FIGURE 13.19 Shear and moment diagram of double channel.

With this information, the double channel size can be determined. These reactions will become the point loads on the stringers for the sizing of the stringers as well. The minimum size channel that can be used to resist a maximum bending moment requires an $S_x = 11.5$ ft-k $\times 12''/$ft $/21.6$ ksi $= 6.4$ in$^3/2$ channels $= 3.2$ in^3 for a single channel. A C5 $\times$ 9 has the closest section modulus that works; however, a C6 $\times$ 8.2 is lighter (9 lb vs. 8.2 lb). Either of these will work, but the 6'' channel would be more economical.

The shear can be checked on a channel in the same way shear is checked on a wide-flange beam. The depth of the beam and the thickness of the web make up the area resisting shear, and this is divided into the maximum shear value as such. Appendix 1 is used for d and t_w.

$$f_v = 6.29 \text{ k}/2 \text{ channels} \times (6'' \times 0.20'') = 2.62 \text{ ksi} < 14.4 \text{ ksi} \quad \text{(OK)}$$

Stringer Selection **Step 4:** Determine the size of all three stringers.

Draw a free-body diagram of the double channels loading on each of the three 52-ft-long stringers. There are two beam cases, the outside stringers, and the middle stringer. The shear and moment diagram in Figure 13.19 indicates the loads on the stringer without the weight of the channels from the three reaction values shown: 6.3, 10, and 6.3 k respectively. The downward forces represent the double channels loading the stringers, and the reactions acting upward represent the cap beam supports on each side of the excavation. The force from the channel now includes the weight for the channels and the wire rope hardware. With the addition of 300 lb pounds per side, the 10-k load becomes 10.3 k and the 6.3-k load on the outside beams is now 6.6 k. Figure 13.20 shows nine double channel sets loading a stringer.

The middle beam was put into the beam program and generated the following results (shown in Figure 13.21):

$$M_{\max} = 567 \text{ ft-k}$$

$$V_{\max} = 46.2 \text{ k}$$

$$\text{Reactions} = 46.2 \text{ k}$$

From this information, we can determine the beam size:

$$S_x = 568 \text{ ft-k} \times 12''/\text{ft} /21.6 \text{ ksi} = 312 \text{ in}^3$$

The W24 $\times$ 131 beam is adequate for bending, as is the W18 $\times$ 158.

The two outside beams can be much smaller. However, it is advisable to maintain the same beam height so that the cap beam and the double channels are provided two level surfaces on which to rest. Therefore, W24 wide flanges can be selected for the outsides as well (much lighter, of course). If this is not desired because the W24 becomes too light, then the center beam can be reduced to a W18 $\times$ 158 and the outside beam would be an 18'' beam as well. Lateral bracing should also be analyzed.

FIGURE 13.20   Double channels loading stringer.

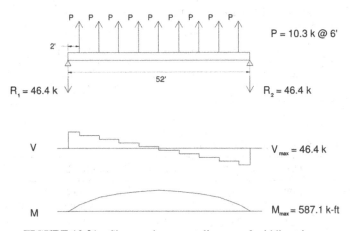

FIGURE 13.21 Shear and moment diagram of middle stringer.

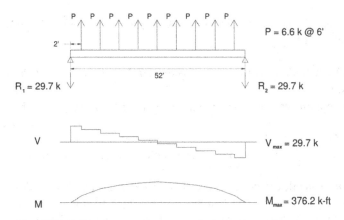

FIGURE 13.22 Shear and moment diagram of outside stringers.

This beam arrangement is favorable to bracing because the beams are fairly close together (6 ft apart) and parallel to each other.

Let's determine the size of the outside beams using the same free-body diagram, but including the lighter channel loads of 6.6 k. A new shear and moment diagram is necessary, as shown in Figure 13.22.

$M_{max} = 376.2$ ft-k

$V_{max} = 29.7$ k

Reaction $= 29.7$ k

$S_x = 376.2$ ft-k $\times 12''/$ft $/21.6$ ksi $= 209$ in^3

W24 $\times$ 94 and W18 $\times$ 119 are adequate for bending.

TABLE 13.3   Cost Comparison between W18s and W24s

Nominal Depth	Inside Stringer (1)	Outside Stringers (2)	Weight (52′ long) lb	Total $ (at $0.65)
W24 option	W24 × 131	W24 × 94	16,588	$10,782
W18 option	W18 × 158	W18 × 119	20,592	$13,385

TABLE 13.4   Lateral Bracing Requirements

Beam	d/A_f	L_c (ft)	Braces	Cost per Brace	Total Cost
W24 × 131	1.98	21.3	2	$1000	N/A
W24 × 94	3.06	13.8	3	$1000	$3000
W18 × 158	1.21	34.8	1	$1000	N/A
W18 × 119	1.59	26.5	1	$1000	$1000

The options for the three stringers (looking at only 18″ and 24″ WF beams) are both combinations:

1. W24 option: W24 × 131 on the inside and W24 × 94's (two each) on the outside, or
2. W18 option: W18 × 158 on the inside and W18 × 119's (two each) on the outside.

Table 13.3 shows the cost analysis and Table 13.4 the bracing analysis.

Pile Caps   The pile cap design is greatly affected by the number of piles supporting the stringers. In this example, the one side is supported on a continuous sheet pile wall, and the other side is supported on a cap beam supported by individual pile. The latter will be analyzed for the pile cap and pile loading and sizing.

Since the piles have to be outside the limits of the pipe being supported, the beam arrangement would be represented by the sketch shown in Figure 13.23. The loads from the three stringers can be found from the shear and moment diagrams (shown earlier in Figure 13.22). The stringers load the cap beam with a 29.7-k load from the two outsides and a 46.2-k load from the inside stringer. A beam weight should be added to both sides for the inside and outside stringers. An assumed weight of 6 k for the outside beams and 8 k for the inside beam will be adequate. Half of each of these additional weights will be added to the loads at the ends of the stringers.

$$29.7 + \left(\frac{1}{2} \text{ of } 6 \text{ k}\right) = 32.7 \text{ k}$$

$$46.2 + \left(\frac{1}{2} \text{ of } 8 \text{ k}\right) = 50.2 \text{ k}$$

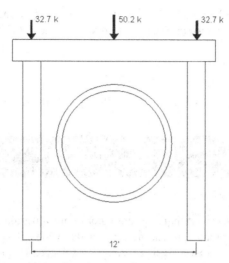

FIGURE 13.23 Cap beam loaded by stringers.

The piles are 12-ft apart on the opposite side of the shoring wall. Figure 13.23 lays out the loads on the cap beam supporting the stringers.

Maximum bending moment from the beam program (not shown) is 150.6 ft-k.

The beam used for this would require an 82.3-in^3 section modulus. Since HP shapes are good to use as cap beams, the lightest HP shape available is an HP 13 × 73 or an HP 14 × 73 (both weighing the same). There is a high concentrated load from the pile to the cap beam; therefore, the cap beam will need to be checked for web yielding. The pile size will not be determined so in order to check web yielding, it will be assumed that the pile width will be 12 in along the beam. The shear and moment diagram for these piles indicate a right and left reaction value of 57.8 k. This reaction is what the pile will have to support without a safety factor. This will also be the concentrated load used for web yielding. A web yielding sketch has been provided in Figure 13.24.

The values needed in order to check web yielding are the pile width, which was just established as 12 in, the k dimension of the cap beam $\left(1\frac{1}{8}''\right)$, and the thickness of the cap beam web (0.46 in).

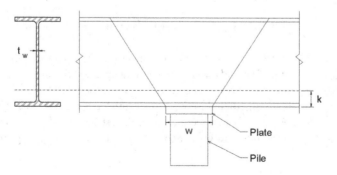

FIGURE 13.24 Web yielding at pile.

The concentrated load of 57.8 k is distributed at a 2.5 :1 angle, and the yielding stress occurs where it intersects the line at k distance. The HP 13 × 73 will be checked since it has a thicker web:

$$f_{wy} = P/[t_w \times (w + 5k)]$$

$$f_{wy} = 57.8 \text{ k}/\{(0.565'') \times [12'' + (5 \times 1.25'')]\} = 5.61 \text{ ksi} < 23.78 \text{ ksi} \quad \text{(OK)}$$

CASE STUDY: RETROFIT PROJECT IN SAN FRANCISCO, CALIFORNIA

Project Overview

After the Loma Prieta earthquake shook northern California in 1989, a decade of projects were initiated, including projects that widened and reinforced footings, wrapped columns in steel jackets, and replaced complete substructures. Such projects were something of the norm from 1990 to almost 2000. Many heavy civil contractors who did not want to miss out on an opportunity to contract in a new kind of work began to bid and build many of these retrofit projects. Since this work was fairly new, it brought not only new, exciting projects to the workplace but also a significant amount of uncertainty. What used to be considered normal ways to build bridges and their components was being changed. Even the California Department of Transportation had to develop standards and best practices for contractors to follow. The welding code previously used for temporary structures was changed to include welding procedures normally used for permanent structures. Figure 13.CS1 shows a schematic of a typical viaduct structure with two levels.

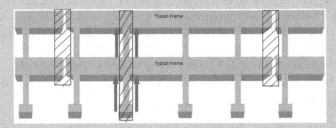

FIGURE 13.CS1   Viaduct schematic—two levels.

During this period, one structure of concern (following the collapse of a two-story bridge in Oakland) was a viaduct that spanned Silver Ave and Highway 101 and merged onto the Highway 280 corridor in San Francisco. This viaduct, originally constructed in the early 1960s, had sections that were constant but also had sections that were unique. Figure 13.CS2 shows a portion of Highway 280 going over Highway 101. The main purpose of this project was to support the existing reinforced concrete box-girder bridge; remove the footings, columns, and portions of the bent caps; install new piles, columns, and bent caps; and construct a longitudinal edge beam along each side of the lower deck portion. The project was sequenced (by specification) so that no adjacent bents could be worked on within a single frame. A frame on this project (hinge to hinge) consisted of approximately three to four bent caps. The bridge in most cases was two levels, and the lanes on the two levels were mostly closed to traffic for the duration of the project. In addition to the retrofitted bent caps, several hinges were removed completely (edge of deck to edge of deck) and rebuilt with new, state-of-the-art spherical Teflon bearings. While the hinge was removed, the bridge had to be supported back to the closest bent cap (the short span) while supporting the long span.

FIGURE 13.CS2   Highway 280 split going over Highway 101 in San Francisco.

Bent Cap Replacement

There were 5 frames with a total of 23 bents (approximately 4 to 5 bents per frame). As mentioned earlier, only one bent per frame could be retrofitted at once. In other words, only 5 bents could be worked on at the same time. Also, 2 adjacent bents could not be worked on simultaneously. In addition to these constraints, other specification constraints made it so there were actually only 4 bents being worked on simultaneously. As the project moved forward, partnering efforts allowed more work to be accomplished at the same time but never enough to satisfy an aggressive schedule.

(*continued*)

CASE STUDY: RETROFIT PROJECT IN SAN FRANCISCO, CALIFORNIA (*Continued*)

The image shown in Figure 13.CS3 was a typical bent cap retrofit. The bridge was supported by temporary supports on each side of the bent cap. The temporary supports were spaced far enough apart to allow access to the demolition and structures crew while at the same time not increasing the tributary load that had to be supported. The supports consisted of WF posts, WF beams, T shape bracing, and was founded on temporary piling in most cases. Hydraulic jacks and short beams were used between the bottom of the bridge and the top of the main beam. The system was jacked to a prescribed load, and then beam spacers were placed until the system was de-stressed and removed.

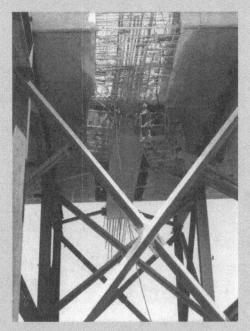

FIGURE 13.CS3   Typical bent cap replacement.

The demolition consisted of concrete removal where shown on the drawings and in almost all cases required the reinforcing steel to remain. The limits of box-girder removal were limited to 2 ft on each side of the bent cap, which would be replaced by additional new bent cap concrete (8 ft). The original bent caps were 4 ft wide. The reinforcing bar placement was difficult and couplers were used extensively to join the new bars to the old, remaining bars. In new construction, the ironworkers can determine the order of bar placement in the most economical fashion. In retrofit construction, where existing bars remain, the order of placement is somewhat dictated by the remaining bars. It should also be pointed out that every operation had to work around the temporary supports on each side of the bent.

Welding was critical on this project. The owner specified very strict welding requirements that would normally be used for new construction, and welder certifications were time consuming and expensive. Most of the welding was performed on-site so it took a while to get the on-site welding operations to run as smoothly as a shop might perform. In most cases, this goal was not reached.

Hinge Replacement

During the hinge replacement, the complete hinge, including concrete and reinforcing steel (except approximately 4 ft of reinforcing on each side), was removed. The bridge was approximately 40 ft wide and about 20 ft of hinge was removed longitudinally. The temporary support system selected doubled as a demolition platform and as support for the new concrete to be placed. The above deck support beams spanned from one bent to close to another bent since the removal took away any possibility that the bridge itself could offer any support. The work on the new hinge took approximately 2 months until the temporary support could be removed. Figure 13.CS4 shows a hinge support in place and the existing hinge concrete removed.

FIGURE 13.CS4 Hinge support.

The above deck beams were W36 beams and were capped at the locations where the support rods went through the deck and were also diagonally braced.

(continued)

CASE STUDY: RETROFIT PROJECT IN SAN FRANCISCO, CALIFORNIA (*Continued*)

The support rods went through holes that were drilled into the top and bottom deck of the box girder and terminated under the bridge with plates and nuts. The rods were $1\frac{1}{4}''$ and $1\frac{1}{2}''$ Dywidag type with 150-ksi high-strength capacity.

The falsework deck/demo platform was also supported by these rods. The new hinge was built in four stages of concrete placements and included the new bearing installation. The demolition was done either with small machines or by hand to preserve the remaining concrete and reinforcing steel. The specifications were very strict on not damaging the reinforcing steel that was to remain; and if damage did occur, the bars had to be fill-welded with welding rod. Figure 13.CS5 is a sketch of a typical hinge support and how the loads were transferred to each adjacent bridge bent.

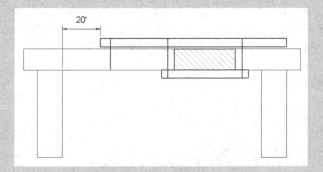

FIGURE 13.CS5   Sketch of hinge replacement.

The three examples provided in this chapter were from the author's experiences and represent more challenging and unusual temporary support situations. The case study discussed may be more representative of temporary support situations that one may encounter in heavy civil construction. Whatever the situation, contractors should procure the services of a qualified, registered, licensed engineer to assist in a safe and economical design and to distribute some unnecessary risk.

The student of temporary structure design should be proficient in calculating weights of all materials knowing the overall dimensions and unit weights of the materials. With these basic skills, one should be able to assist in any temporary structure supporting existing structures, buildings, and utilities.

APPENDIXES

APPENDIX 1

STEEL BEAMS (AISC)

Wide-Flange Beams

STEEL BEAMS (AISC)

Designation	Area A (in²)	Depth d (in)	Web Thickness t_w (in)	Flange Width b_f (in)	Flange Thickness t_f (in)	k (in)	Lateral Buckling d/A_f	Axis x-x I_x (in⁴)	Axis x-x S_x (in³)	Axis x-x r_x (in)	Axis y-y I_y (in⁴)	Axis y-y S_y (in³)	Axis y-y r_y (in)
W44x335	98.3	44.02	1.02	15.95	1.77	2 9/16	1.559	31100	1410	17.8	1200	150	3.49
W44x290	85.8	43.62	0.87	15.83	1.58	2 3/8	1.744	27100	1240	17.8	1050	133	3.5
W44x262	77.2	43.31	0.79	15.75	1.42	2 3/16	1.937	24200	1120	17.7	927	118	3.46
W44x230	67.7	42.91	0.71	15.75	1.22	2	2.233	20800	969	17.5	796	101	3.43
W40x593	174	42.99	1.79	16.69	3.23	4 7/16	0.797	50400	2340	17	2520	302	3.81
W40x503	148	42.05	1.54	16.42	2.76	3 15/16	0.928	41700	1980	16.8	2050	250	3.72
W40x431	127	41.26	1.34	16.22	2.36	3 9/16	1.078	34800	1690	16.6	1690	208	3.65
W40x372	109	40.63	1.16	16.06	2.05	3 1/4	1.234	29600	1460	16.4	1420	177	3.6
W40x321	94.1	40.08	1	15.91	1.77	2 15/16	1.423	25100	1250	16.3	1190	150	3.56
W40x297	87.4	39.84	0.93	15.825	1.65	3 1/16	1.526	23200	1170	16.3	1090	138	3.54
W40x277	81.3	39.69	0.83	15.83	1.575	2 3/4	1.592	21900	1100	16.4	1040	132	3.58
W40x249	73.3	39.38	0.75	15.75	1.42	2 5/8	1.761	19500	992	16.3	926	118	3.56
W40x215	63.3	38.98	0.65	15.75	1.22	2 3/8	2.029	16700	858	16.2	796	101	3.54
W40x199	58.4	38.67	0.65	15.75	1.065	2 1/4	2.305	14900	769	16	695	88.2	3.45
W40x174	51.1	38.2	0.65	15.75	0.83	2	2.922	12200	639	15.5	541	68.8	3.26
W40x466	137	42.44	1.67	12.64	2.95	4 1/8	1.138	36300	1710	16.3	1010	160	2.72

W40×392	115	41.57	1.42	12.36	2.52	3 11/16	1.335	29900	1440	16.1	803	130	2.64
W40×331	97.6	40.79	1.22	12.17	2.13	3 5/16	1.574	24700	1210	15.9	646	106	2.57
W40×278	81.8	40.16	1.02	11.97	1.81	3	1.854	20500	1020	15.8	521	87.1	2.52
W40×264	77.6	40	0.96	11.93	1.73	2 15/16	1.938	19400	971	15.8	493	82.6	2.52
W40×235	68.9	39.69	0.83	11.89	1.575	2 3/4	2.119	17400	874	15.9	444	74.6	2.54
W40×211	62	39.37	0.75	11.81	1.415	2 5/8	2.356	15500	785	15.8	390	66.1	2.51
W40×183	53.7	38.98	0.65	11.81	1.22	2 3/8	2.705	13300	682	15.7	336	56.9	2.5
W40×167	49.1	38.59	0.65	11.81	1.025	2 3/16	3.188	11600	599	15.3	283	47.9	2.4
W40×149	43.8	38.2	0.63	11.81	0.83	2	3.897	9780	512	14.9	229	38.8	2.29
W36×848	249	42.45	2.52	18.13	4.53	5 11/16	0.517	67400	3170	16.4	4550	501	4.27
W36×798	234	41.97	2.38	17.99	4.29	5 7/16	0.544	62600	2980	16.4	4200	467	4.24
W36×650	190	40.47	1.97	17.575	3.54	4 11/16	0.650	48900	2420	16	3230	367	4.12
W36×527	154	39.21	1.61	17.22	2.91	4 1/16	0.782	38300	1950	15.8	2490	289	4.02
W36×439	128	38.26	1.36	16.965	2.44	3 9/16	0.924	31000	1620	15.6	1990	235	3.95
W36×393	115	37.8	1.22	16.83	2.2	3 5/16	1.021	27500	1450	15.5	1750	208	3.9
W36×359	105	37.4	1.12	16.73	2.01	3 1/8	1.112	24800	1320	15.4	1570	188	3.87
W36×328	96.4	37.09	1.02	16.63	1.85	3	1.206	22500	1210	15.3	1420	171	3.84
W36×300	88.3	36.74	0.945	16.655	1.68	2 13/16	1.313	20300	1110	15.2	1300	156	3.83
W36×280	82.4	36.52	0.885	16.595	1.57	2 11/16	1.402	18900	1030	15.1	1200	144	3.81

(continued)

Designation	Area A (in^2)	Depth d (in)	Web Thickness t_w (in)	Flange Width b_f (in)	Flange Thickness t_f (in)	k (in)	Lateral Buckling d/A_f	Axis x-x I_x (in^4)	S_x (in^3)	r_x (in)	Axis y-y I_y (in^4)	S_y (in^3)	r_y (in)
W36×260	76.5	36.26	0.84	16.55	1.44	2 9/16	1.521	17300	953	15	1090	132	3.78
W36×245	72.1	36.08	0.8	16.51	1.35	2 1/2	1.619	16100	895	15	1010	123	3.75
W36×230	67.6	35.9	0.76	16.47	1.26	2 3/8	1.730	15000	837	14.9	940	114	3.73
W36×256	75.4	37.43	0.96	12.215	1.73	2 5/8	1.771	16800	895	14.9	528	86.5	2.65
W36×232	68.1	37.12	0.87	12.12	1.57	2 1/2	1.951	15000	809	14.8	468	77.2	2.62
W36×210	61.8	36.69	0.83	12.18	1.36	2 5/16	2.215	13200	719	14.6	411	67.5	2.58
W36×194	57	36.49	0.765	12.115	1.26	2 3/16	2.390	12100	664	14.6	375	61.9	2.56
W36×182	53.6	36.33	0.725	12.075	1.18	2 1/8	2.550	11300	623	14.5	347	57.6	2.55
W36×170	50	36.17	0.68	12.03	1.1	2	2.733	10500	580	14.5	320	53.2	2.53
W36×160	47	36.01	0.65	12	1.02	1 15/16	2.942	9750	542	14.4	295	49.1	2.5
W36×150	44.2	35.85	0.625	11.975	0.94	1 7/8	3.185	9040	504	14.3	270	45.1	2.47
W36×135	39.7	35.55	0.6	11.95	0.79	1 11/16	3.766	7800	439	14	225	37.7	2.38
W33×354	104	35.55	1.16	16.1	2.09	2 7/8	1.056	21900	1230	14.5	1460	181	3.74
W33×318	93.5	35.16	1.04	15.985	1.89	2 11/16	1.164	19500	1110	14.4	1290	161	3.71
W33×291	85.6	34.84	0.96	15.905	1.73	2 9/16	1.266	17700	1010	14.4	1160	146	3.69

W33×263	77.4	34.53	0.87	15.805	1.57	2 3/8	1.392	15800	917	14.3	1030	131	3.66
W33×241	70.9	34.18	0.83	15.86	1.4	2 3/16	1.539	14200	829	14.1	932	118	3.63
W33×221	65	33.93	0.775	15.805	1.275	2 1/16	1.684	12800	757	14.1	840	106	3.59
W33×201	59.1	33.68	0.715	15.745	1.15	1 15/16	1.860	11500	684	14	749	95.2	3.56
W33×169	49.5	33.82	0.67	11.5	1.22	2 1/16	2.411	9290	549	13.7	310	53.9	2.5
W33×152	44.7	33.49	0.635	11.565	1.055	1 7/8	2.745	8160	487	13.5	273	47.2	2.47
W33×141	41.6	33.3	0.605	11.535	0.96	1 3/4	3.007	7450	448	13.4	246	42.7	2.43
W33×130	38.3	33.09	0.58	11.51	0.855	1 11/16	3.362	6710	406	13.2	218	37.9	2.39
W33×118	34.7	32.86	0.55	11.48	0.74	1 9/16	3.868	5900	359	13	187	32.6	2.32
W30×477	140	34.21	1.63	15.865	2.95	3 3/4	0.731	26100	1530	13.7	1970	249	3.75
W30×391	114	33.19	1.36	15.59	2.44	3 1/4	0.873	20700	1250	13.5	1550	198	3.68
W30×326	95.7	32.4	1.14	15.37	2.05	2 13/16	1.028	16800	1030	13.2	1240	162	3.61
W30×292	85.7	32.01	1.02	15.255	1.85	2 5/8	1.134	14900	928	13.2	1100	144	3.58
W30×261	76.7	31.61	0.93	15.155	1.65	2 7/16	1.264	13100	827	13.1	959	127	3.54
W30×235	69	31.3	0.83	15.055	1.5	2 1/4	1.386	11700	746	13	855	114	3.52
W30×211	62	30.94	0.775	15.105	1.315	2 1/8	1.558	10300	663	12.9	757	100	3.49
W30×191	56.1	30.68	0.71	15.04	1.185	1 15/16	1.721	9170	598	12.8	673	89.5	3.46
W30×173	50.8	30.44	0.655	14.985	1.065	1 7/8	1.907	8200	539	12.7	598	79.8	3.43
W30×148	43.5	30.67	0.65	10.48	1.18	2	2.480	6680	436	12.4	227	43.3	2.28

(continued)

Designation	Area A (in^2)	Depth d (in)	Web Thickness t_w (in)	Flange Width b_f (in)	Flange Thickness t_f (in)	k (in)	Lateral Buckling d/A_f	Axis x-x I_x (in^4)	S_x (in^3)	r_x (in)	Axis y-y I_y (in^4)	S_y (in^3)	r_y (in)
W30×132	38.9	30.31	0.615	10.545	1	1 3/4	2.874	5770	380	12.2	196	37.2	2.25
W30×124	36.5	30.17	0.585	10.515	0.93	1 11/16	3.085	5360	355	12.1	181	34.4	2.23
W30×116	34.2	30.01	0.565	10.495	0.85	1 5/8	3.364	4930	329	12	164	31.3	2.19
W30×108	31.7	29.83	0.545	10.475	0.76	1 9/16	3.747	4470	299	11.9	146	27.9	2.15
W30×99	29.1	29.65	0.52	10.45	0.67	1 7/16	4.235	3990	269	11.7	128	24.5	2.1
W30×90	26.4	29.53	0.47	10.4	0.61	1 5/16	4.655	3620	245	11.7	115	22.1	2.09
W27×539	158	32.52	1.97	15.255	3.54	4 1/4	0.602	25500	1570	12.7	2110	277	3.66
W27×448	131	31.42	1.65	14.94	2.99	3 11/16	0.703	20400	1300	12.5	1670	224	3.57
W27×368	108	30.39	1.38	14.665	2.48	3 3/16	0.836	16100	1060	12.2	1310	179	3.48
W27×307	90.2	29.61	1.16	14.445	2.09	2 13/16	0.981	13100	884	12	1050	146	3.42
W27×258	75.7	28.98	0.98	14.27	1.77	2 1/2	1.147	10800	742	11.9	859	120	3.37
W27×235	69.1	28.66	0.91	14.19	1.61	2 5/16	1.254	9660	674	11.8	768	108	3.33
W27×217	63.8	28.43	0.83	14.115	1.5	2 3/16	1.343	8870	624	11.8	704	99.8	3.32
W27×194	57	28.11	0.75	14.035	1.34	2 1/16	1.495	7820	556	11.7	618	88.1	3.29
W27×178	52.3	27.81	0.725	14.085	1.19	1 7/8	1.659	6990	502	11.6	555	78.8	3.26

W27×161	47.4	27.59	0.66	14.02	1.08	1 13/16	1.822	6280	455	11.5	497	70.9	3.24
W27×146	42.9	27.38	0.605	13.965	0.975	1 11/16	2.011	5630	411	11.4	443	63.5	3.21
W27×129	37.8	27.63	0.61	10.01	1.1	1 13/16	2.509	4760	345	11.2	184	36.8	2.21
W27×114	33.5	27.29	0.57	10.07	0.93	1 5/8	2.914	4090	299	11	159	31.5	2.18
W27×102	30	27.09	0.515	10.015	0.83	1 9/16	3.259	3620	267	11	139	27.8	2.15
W27×94	27.7	26.92	0.49	9.99	0.745	1 7/16	3.617	3270	243	10.9	124	24.8	2.12
W27×84	24.8	26.71	0.46	9.96	0.64	1 3/8	4.190	2850	213	10.7	106	21.2	2.07
W24×492	144	29.65	1.97	14.115	3.54	4 5/16	0.593	19100	1290	11.5	1670	237	3.41
W24×408	119	28.54	1.65	13.8	2.99	3 3/4	0.692	15100	1060	11.3	1320	191	3.33
W24×335	98.4	27.52	1.38	13.52	2.48	3 1/4	0.821	11900	864	11	1030	152	3.23
W24×279	82	26.73	1.16	13.305	2.09	2 7/8	0.961	9600	718	10.8	823	124	3.17
W24×250	73.5	26.34	1.04	13.185	1.89	2 11/16	1.057	8490	644	10.7	724	110	3.14
W24×229	67.2	26.02	0.96	13.11	1.73	2 1/2	1.147	7650	588	10.7	651	99.4	3.11
W24×207	60.7	25.71	0.87	13.01	1.57	2 3/8	1.259	6820	531	10.6	578	88.8	3.08
W24×192	56.3	25.47	0.81	12.95	1.46	2 1/4	1.347	6260	491	10.5	530	81.8	3.07
W24×176	51.7	25.24	0.75	12.89	1.34	2 1/8	1.461	5680	450	10.5	479	74.3	3.04
W24×162	47.7	25	0.705	12.955	1.22	2	1.582	5170	414	10.4	443	68.4	3.05
W24×146	43	24.74	0.65	12.9	1.09	1 7/8	1.759	4580	371	10.3	391	60.5	3.01
W24×131	38.5	24.48	0.605	12.855	0.96	1 3/4	1.984	4020	329	10.2	340	53	2.97

(continued)

Designation	Area A (in^2)	Depth d (in)	Web Thickness t_w (in)	Flange Width b_f (in)	Flange Thickness t_f (in)	k (in)	Lateral Buckling d/A_f	Axis x-x I_x (in^4)	S_x (in^3)	r_x (in)	Axis y-y I_y (in^4)	S_y (in^3)	r_y (in)
W24×117	34.4	24.26	0.55	12.8	0.85	1 5/8	2.230	3540	291	10.1	297	46.5	2.94
W24×104	30.6	24.06	0.5	12.75	0.75	1 1/2	2.516	3100	258	10.1	259	40.7	2.91
W24×103	30.3	24.53	0.55	9	0.98	1 3/4	2.781	3000	245	9.96	119	26.5	1.99
W24×94	27.7	24.31	0.515	9.065	0.875	1 5/8	3.065	2700	222	9.87	109	24	1.98
W24×84	24.7	24.1	0.47	9.02	0.77	19/16	3.470	2370	196	9.79	94.4	20.9	1.95
W24×76	22.4	23.92	0.44	8.99	0.68	17/16	3.913	2100	176	9.69	82.5	18.4	1.92
W24×68	20.1	23.73	0.415	8.965	0.585	1 3/8	4.525	1830	154	9.55	70.4	15.7	1.87
W24×62	18.2	23.74	0.43	7.04	0.59	1 3/8	5.716	1550	131	9.23	34.5	9.8	1.38
W24×55	16.2	23.57	0.395	7.005	0.505	1 5/16	6.663	1350	114	9.11	29.1	8.3	1.34
W21×201	59.2	23.03	0.91	12.575	1.63	2 3/8	1.124	5310	461	9.47	542	86.1	3.02
W21×182	53.6	22.72	0.83	12.5	1.48	2 1/4	1.228	4730	417	9.4	483	77.2	3
W21×166	48.8	22.48	0.75	12.42	1.36	2 1/8	1.331	4280	380	9.36	435	70.1	2.98
W21×147	43.2	22.06	0.72	12.51	1.15	1 7/8	1.533	3630	329	9.17	376	60.1	2.95
W21×132	38.8	21.83	0.65	12.44	1.035	1 13/16	1.695	3220	295	9.12	333	53.5	2.93
W21×122	35.9	21.68	0.6	12.39	0.96	1 11/16	1.823	2960	273	9.09	305	49.2	2.92

Designation												
W21×111	32.7	21.51	0.55	12.34	0.875	1 5/8	1.992	2670	249	9.05	274	44.5	2.9
W21×101	29.8	21.36	0.5	12.29	0.8	1 9/16	2.172	2420	227	9.02	248	40.3	2.89
W21×93	27.3	21.62	0.58	8.42	0.93	1 11/16	2.761	2070	192	8.7	92.9	22.1	1.84
W21×83	24.3	21.43	0.515	8.355	0.835	1 9/16	3.072	1830	171	8.67	81.4	19.5	1.83
W21×73	21.5	21.24	0.455	8.295	0.74	1 1/2	3.460	1600	151	8.64	70.6	17	1.81
W21×68	20	21.13	0.43	8.27	0.685	1 7/16	3.730	1480	140	8.6	64.7	15.7	1.8
W21×62	18.3	20.99	0.4	8.24	0.615	1 3/8	4.142	1330	127	8.54	57.5	13.9	1.77
W21×57	16.7	21.06	0.405	6.555	0.65	1 3/8	4.943	1170	111	8.36	30.6	9.35	1.35
W21×50	14.7	20.83	0.38	6.53	0.535	1 5/16	5.962	984	94.5	8.18	24.9	7.64	1.3
W21×44	13	20.66	0.35	6.5	0.45	1 3/16	7.063	843	81.6	8.06	20.7	6.36	1.26
W18×311	91.5	22.32	1.52	12.005	2.74	3 7/16	0.679	6960	624	8.72	795	132	2.95
W18×283	83.2	21.85	1.4	11.89	2.5	3 3/16	0.735	6160	564	8.61	704	118	2.91
W18×258	75.9	21.46	1.28	11.77	2.3	3	0.793	5510	514	8.53	628	107	2.88
W18×234	68.8	21.06	1.16	11.65	2.11	2 3/4	0.857	4900	466	8.44	558	95.8	2.85
W18×211	62.1	20.67	1.06	11.555	1.91	2 9/16	0.937	4330	419	8.35	493	85.3	2.82
W18×192	56.4	20.35	0.96	11.455	1.75	2 7/16	1.015	3870	380	8.28	440	76.8	2.79
W18×175	51.3	20.04	0.89	11.375	1.59	2 1/4	1.108	3450	344	8.2	391	68.8	2.76
W18×158	46.3	19.72	0.81	11.3	1.44	2 1/8	1.212	3060	310	8.12	347	61.4	2.74
W18×143	42.1	19.49	0.73	11.22	1.32	2	1.316	2750	282	8.09	311	55.5	2.72

(continued)

Designation	Area A (in²)	Depth d (in)	Web Thickness t_w (in)	Flange Width b_f (in)	Flange Thickness t_f (in)	k (in)	Lateral Buckling d/A_f	Axis x-x I_x (in⁴)	S_x (in³)	r_x (in)	Axis y-y I_y (in⁴)	S_y (in³)	r_y (in)
W18×130	38.2	19.25	0.67	11.16	1.2	1 7/8	1.437	2460	256	8.03	278	49.9	2.7
W18×119	35.1	18.97	0.655	11.265	1.06	1 3/4	1.589	2190	231	7.9	253	44.9	2.69
W18×106	31.1	18.73	0.59	11.2	0.94	1 5/8	1.779	1910	204	7.84	220	39.4	2.66
W18×97	28.5	18.59	0.535	11.145	0.87	1 9/16	1.917	1750	188	7.82	201	36.1	2.65
W18×86	25.3	18.39	0.48	11.09	0.77	1 7/16	2.154	1530	166	7.77	175	31.6	2.63
W18×76	22.3	18.21	0.425	11.035	0.68	1 3/8	2.427	1330	146	7.73	152	27.6	2.61
W18×71	20.8	18.47	0.495	7.635	0.81	1 1/2	2.987	1170	127	7.5	60.3	15.8	1.7
W18×65	19.1	18.35	0.45	7.59	0.75	1 7/16	3.224	1070	117	7.49	54.8	14.4	1.69
W18×60	17.6	18.24	0.415	7.555	0.695	1 3/8	3.474	984	108	7.47	50.1	13.3	1.69
W18×55	16.2	18.11	0.39	7.53	0.63	1 5/16	3.818	890	98.3	7.41	44.9	11.9	1.67
W18×50	14.7	17.99	0.355	7.495	0.57	1 1/4	4.211	800	88.9	7.38	40.1	10.7	1.65
W18×46	13.5	18.06	0.36	6.06	0.605	1 1/4	4.926	712	78.8	7.25	22.5	7.43	1.29
W18×40	11.8	17.9	0.315	6.015	0.525	1 3/16	5.668	612	68.4	7.21	19.1	6.35	1.27
W18×35	10.3	17.7	0.3	6	0.425	1 1/8	6.941	510	57.6	7.04	15.3	5.12	1.22

W16x100	29.4	16.97	0.585	10.425	0.985	1 11/16	1.653	1490	175	7.1	186	35.7	2.51
W16x89	26.2	16.75	0.525	10.365	0.875	1 9/16	1.847	1300	155	7.05	163	31.4	2.49
W16x77	22.6	16.52	0.455	10.295	0.76	1 7/16	2.111	1110	134	7	138	26.9	2.47
W16x67	19.7	16.33	0.395	10.235	0.665	1 3/8	2.399	954	117	6.96	119	23.2	2.46
W16x57	16.8	16.43	0.43	7.12	0.715	1 3/8	3.227	758	92.2	6.72	43.1	12.1	1.6
W16x50	14.7	16.26	0.38	7.07	0.63	1 5/16	3.651	659	81	6.68	37.2	10.5	1.59
W16x45	13.3	16.13	0.345	7.035	0.565	1 1/4	4.058	586	72.7	6.65	32.8	9.34	1.57
W16x40	11.8	16.01	0.305	6.995	0.505	1 3/16	4.532	518	64.7	6.63	28.9	8.25	1.57
W16x36	10.6	15.86	0.295	6.985	0.43	1 1/8	5.280	448	56.5	6.51	24.5	7	1.52
W16x31	9.12	15.88	0.275	5.525	0.44	1 1/8	6.532	375	47.2	6.41	12.4	4.49	1.17
W16x26	7.68	15.69	0.25	5.5	0.345	1 1/16	8.269	301	38.4	6.26	9.59	3.49	1.12
W14x808	237	22.84	3.74	18.56	5.12	5 13/16	0.240	16000	1400	8.21	5510	594	4.82
W14x730	215	22.42	3.07	17.89	4.91	5 9/16	0.255	14300	1280	8.17	4720	527	4.69
W14x665	196	21.64	2.83	17.65	4.52	5 3/16	0.271	12400	1150	7.98	4170	472	4.62
W14x605	178	20.92	2.595	17.415	4.16	4 13/16	0.289	10800	1040	7.8	3680	423	4.55
W14x550	162	20.24	2.38	17.2	3.82	4 1/2	0.308	9430	931	7.63	3250	378	4.49
W14x500	147	19.6	2.19	17.01	3.5	4 3/16	0.329	8210	838	7.48	2880	339	4.43
W14x455	134	19.02	2.015	16.835	3.21	3 7/8	0.352	7190	756	7.33	2560	304	4.38
W14x426	125	18.67	1.875	16.695	3.035	3 11/16	0.368	6600	707	7.26	2360	283	4.34

(continued)

Designation	Area A	Depth d	Web Thickness t_w	Flange Width b_f	Flange Thickness t_f	k	Lateral Buckling d/A_f	Axis x-x I_x	S_x	r_x	Axis y-y I_y	S_y	r_y
	(in^2)	(in)	(in)	(in)	(in)	(in)		(in^4)	(in^3)	(in)	(in^4)	(in^3)	(in)
W14×398	117	18.29	1.77	16.59	2.845	3 1/2	0.388	6000	656	7.16	2170	262	4.31
W14×370	109	17.92	1.655	16.475	2.66	3 5/16	0.409	5440	607	7.07	1990	241	4.27
W14×342	101	17.54	1.54	16.36	2.47	3 1/8	0.434	4900	559	6.98	1810	221	4.24
W14×311	91.4	17.12	1.41	16.23	2.26	2 15/16	0.467	4330	506	6.88	1610	199	4.2
W14×283	83.3	16.74	1.29	16.11	2.07	2 3/4	0.502	3840	459	6.79	1440	179	4.17
W14×257	75.6	16.38	1.175	15.995	1.89	2 9/16	0.542	3400	415	6.71	1290	161	4.13
W14×233	68.5	16.04	1.07	15.89	1.72	2 3/8	0.587	3010	375	6.63	1150	145	4.1
W14×211	62	15.72	0.98	15.8	1.56	2 1/4	0.638	2660	338	6.55	1030	130	4.07
W14×193	56.8	15.48	0.89	15.71	1.44	2 1/8	0.684	2400	310	6.5	931	119	4.05
W14×176	51.8	15.22	0.83	15.65	1.31	2	0.742	2140	281	6.43	838	107	4.02
W14×159	46.7	14.98	0.745	15.565	1.19	1 7/8	0.809	1900	254	6.38	748	96.2	4
W14×145	42.7	14.78	0.68	15.5	1.09	1 3/4	0.875	1710	232	6.33	677	87.3	3.98
W14×132	38.8	14.66	0.645	14.725	1.03	1 11/16	0.967	1530	209	6.28	548	74.5	3.76
W14×120	35.3	14.48	0.59	14.67	0.94	1 5/8	1.050	1380	190	6.24	495	67.5	3.74
W14×109	32	14.32	0.525	14.605	0.86	1 9/16	1.140	1240	173	6.22	447	61.2	3.73
W14×99	29.1	14.16	0.485	14.565	0.78	1 7/16	1.246	1110	157	6.17	402	55.2	3.71

STEEL BEAMS (AISC) **383**

W14×90	26.5	14.02	0.44	14.52	0.71	1 3/8	1.360	999	143	6.14	362	49.9	3.7
W14×82	24.1	14.31	0.51	10.13	0.855	1 5/8	1.652	882	123	6.05	148	29.3	2.48
W14×74	21.8	14.17	0.45	10.07	0.785	1 9/16	1.793	796	112	6.04	134	26.6	2.48
W14×68	20	14.04	0.415	10.035	0.72	1 1/2	1.943	723	103	6.01	121	24.2	2.46
W14×61	17.9	13.89	0.375	9.995	0.645	1 7/16	2.155	640	92.2	5.98	107	21.5	2.45
W14×53	15.6	13.92	0.37	8.06	0.66	1 7/16	2.617	541	77.8	5.89	57.7	14.3	1.92
W14×48	14.1	13.79	0.34	8.03	0.595	1 3/8	2.886	485	70.3	5.85	51.4	12.8	1.91
W14×43	12.6	13.66	0.305	7.995	0.53	1 5/16	3.224	428	62.7	5.82	45.2	11.3	1.89
W14×38	11.2	14.1	0.31	6.77	0.515	1 1/16	4.044	385	54.6	5.87	26.7	7.88	1.55
W14×34	10	13.98	0.285	6.745	0.455	1	4.555	340	48.6	5.83	23.3	6.91	1.53
W14×30	8.85	13.84	0.27	6.73	0.385	15/16	5.341	291	42	5.73	19.6	5.82	1.49
W14×26	7.69	13.91	0.255	5.025	0.42	15/16	6.591	245	35.3	5.65	8.91	3.54	1.08
W14×22	6.49	13.74	0.23	5	0.335	7/8	8.203	199	29	5.54	7	2.8	1.04
W12×336	98.8	16.82	1.775	13.385	2.955	3 11/16	0.425	4060	483	6.41	1190	177	3.47
W12×305	89.6	16.32	1.625	13.235	2.705	3 7/16	0.456	3550	435	6.29	1050	159	3.42
W12×279	81.9	15.85	1.53	13.14	2.47	3 3/16	0.488	3110	393	6.16	937	143	3.38
W12×252	74.1	15.41	1.395	13.005	2.25	2 15/16	0.527	2720	353	6.06	828	127	3.34
W12×230	67.7	15.05	1.285	12.895	2.07	2 3/4	0.564	2420	321	5.97	742	115	3.31
W12×210	61.8	14.71	1.18	12.79	1.9	2 5/8	0.605	2140	292	5.89	664	104	3.28

(continued)

Designation	Area A (in^2)	Depth d (in)	Web Thickness t_w (in)	Flange Width b_f (in)	Flange Thickness t_f (in)	k (in)	Lateral Buckling d/A_f	Axis x-x I_x (in^4)	S_x (in^3)	r_x (in)	Axis y-y I_y (in^4)	S_y (in^3)	r_y (in)
W12×190	55.8	14.38	1.06	12.67	1.735	2 7/16	0.654	1890	263	5.82	589	93	3.25
W12×170	50	14.03	0.96	12.57	1.56	2 1/4	0.715	1650	235	5.74	517	82.3	3.22
W12×152	44.7	13.71	0.87	12.48	1.4	2 1/8	0.785	1430	209	5.66	454	72.8	3.19
W12×136	39.9	13.41	0.79	12.4	1.25	1 15/16	0.865	1240	186	5.58	398	64.2	3.16
W12×120	35.3	13.12	0.71	12.32	1.105	1 13/16	0.964	1070	163	5.51	345	56	3.13
W12×106	31.2	12.89	0.61	12.22	0.99	1 11/16	1.065	933	145	5.47	301	49.3	3.11
W12×96	28.2	12.71	0.55	12.16	0.9	1 5/8	1.161	833	131	5.44	270	44.4	3.09
W12×87	25.6	12.53	0.515	12.125	0.81	1 1/2	1.276	740	118	5.38	241	39.7	3.07
W12×79	23.2	12.38	0.47	12.08	0.735	1 7/16	1.394	662	107	5.34	216	35.8	3.05
W12×72	21.1	12.25	0.43	12.04	0.67	1 3/8	1.519	597	97.4	5.31	195	32.4	3.04
W12×65	19.1	12.12	0.39	12	0.605	1 5/16	1.669	533	87.9	5.28	174	29.1	3.02
W12×58	17	12.19	0.36	10.01	0.64	1 3/8	1.903	475	78	5.28	107	21.4	2.51
W12×53	15.6	12.06	0.345	9.995	0.575	1 1/4	2.098	425	70.6	5.23	95.8	19.2	2.48
W12×50	14.7	12.19	0.37	8.08	0.64	1 3/8	2.357	394	64.7	5.18	56.3	13.9	1.96
W12×45	13.2	12.06	0.335	8.045	0.575	1 1/4	2.607	350	58.1	5.15	50	12.4	1.94
W12×40	11.8	11.94	0.295	8.005	0.515	1 1/4	2.896	310	51.9	5.13	44.1	11	1.93

W12x35	10.3	12.5	0.3	6.56	0.52	1	3.664	285	45.6	5.25	24.5	7.47	1.54
W12x30	8.79	12.34	0.26	6.52	0.44	15/16	4.301	238	38.6	5.21	20.3	6.24	1.52
W12x26	7.65	12.22	0.23	6.49	0.38	7/8	4.955	204	33.4	5.17	17.3	5.34	1.51
W12x22	6.48	12.31	0.26	4.03	0.425	7/8	7.187	156	25.4	4.91	4.66	2.31	0.847
W12x19	5.57	12.16	0.235	4.005	0.35	13/16	8.675	130	21.3	4.82	3.76	1.88	0.822
W12x16	4.71	11.99	0.22	3.99	0.265	3/4	11.340	103	17.1	4.67	2.82	1.41	0.773
W12x14	4.16	11.91	0.2	3.97	0.225	11/16	13.333	88.6	14.9	4.62	2.36	1.19	0.753
W10x112	32.9	11.36	0.755	10.415	1.25	1 7/8	0.873	716	126	4.66	236	45.3	2.68
W10x100	29.4	11.1	0.68	10.34	1.12	1 3/4	0.958	623	112	4.6	207	40	2.65
W10x88	25.9	10.84	0.605	10.265	0.99	1 5/8	1.067	534	98.5	4.54	179	34.8	2.63
W10x77	22.6	10.6	0.53	10.19	0.87	1 1/2	1.196	455	85.9	4.49	154	30.1	2.6
W10x68	20	10.4	0.47	10.13	0.77	1 3/8	1.333	394	75.7	4.44	134	26.4	2.59
W10x60	17.6	10.22	0.42	10.08	0.68	1 5/16	1.491	341	66.7	4.39	116	23	2.57
W10x54	15.8	10.09	0.37	10.03	0.615	1 1/4	1.636	303	60	4.37	103	20.6	2.56
W10x49	14.4	9.98	0.34	10	0.56	1 3/16	1.782	272	54.6	4.35	93.4	18.7	2.54
W10x45	13.3	10.1	0.35	8.02	0.62	1 1/4	2.031	248	49.1	4.32	53.4	13.3	2.01
W10x39	11.5	9.92	0.315	7.985	0.53	1 1/8	2.344	209	42.1	4.27	45	11.3	1.98
W10x33	9.71	9.73	0.29	7.96	0.435	1 1/16	2.810	170	35	4.19	36.6	9.2	1.94
W10x30	8.84	10.47	0.3	5.81	0.51	15/16	3.533	170	32.4	4.38	16.7	5.75	1.37

(continued)

Designation	Area A (in²)	Depth d (in)	Web Thickness t_w (in)	Flange Width b_f (in)	Flange Thickness t_f (in)	k (in)	Lateral Buckling d/A_f	Axis x-x I_x (in⁴)	S_x (in³)	r_x (in)	Axis y-y I_y (in⁴)	S_y (in³)	r_y (in)
W10×26	7.61	10.33	0.26	5.77	0.44	7/8	4.069	144	27.9	4.35	14.1	4.89	1.36
W10×22	6.49	10.17	0.24	5.75	0.36	3/4	4.913	118	23.2	4.27	11.4	3.97	1.33
W10×19	5.62	10.24	0.25	4.02	0.395	13/16	6.449	96.3	18.8	4.14	4.29	2.14	0.874
W10×17	4.99	10.11	0.24	4.01	0.33	3/4	7.640	81.9	16.2	4.05	3.56	1.78	0.844
W10×15	4.41	9.99	0.23	4	0.27	11/16	9.250	68.9	13.8	3.95	2.89	1.45	0.81
W10×12	3.54	9.87	0.19	3.96	0.21	5/8	11.869	53.8	10.9	3.9	2.18	1.1	0.785
W8×67	19.7	9	0.57	8.28	0.935	1 7/16	1.163	272	60.4	3.72	88.6	21.4	2.12
W8×58	17.1	8.75	0.51	8.22	0.81	1 5/16	1.314	228	52	3.65	75.1	18.3	2.1
W8×48	14.1	8.5	0.4	8.11	0.685	1 3/16	1.530	184	43.3	3.61	60.9	15	2.08
W8×40	11.7	8.25	0.36	8.07	0.56	1 1/16	1.826	146	35.5	3.53	49.1	12.2	2.04
W8×35	10.3	8.12	0.31	8.02	0.495	1	2.045	127	31.2	3.51	42.6	10.6	2.03
W8×31	9.13	8	0.285	7.995	0.435	15/16	2.300	110	27.5	3.47	37.1	9.27	2.02
W8×28	8.25	8.06	0.285	6.535	0.465	15/16	2.652	98	24.3	3.45	21.7	6.63	1.62
W8×24	7.08	7.93	0.245	6.495	0.4	7/8	3.052	82.8	20.9	3.42	18.3	5.63	1.61
W8×21	6.16	8.28	0.25	5.27	0.4	13/16	3.928	75.3	18.2	3.49	9.77	3.71	1.26

W8x18	5.26	8.14	0.23	5.25	0.33	3/4	4.698	61.9	15.2	3.43	7.97	3.04	1.23
W8x15	4.44	8.11	0.245	4.015	0.315	3/4	6.412	48	11.8	3.29	3.41	1.7	0.876
W8x13	3.84	7.99	0.23	4	0.255	11/16	7.833	39.6	9.91	3.21	2.73	1.37	0.843
W8x10	2.96	7.89	0.17	3.94	0.205	5/8	9.768	30.8	7.81	3.22	2.09	1.06	0.841
W6x25	7.34	6.38	0.32	6.08	0.455	13/16	2.306	53.4	16.7	2.7	17.1	5.61	1.52
W6x20	5.87	6.2	0.26	6.02	0.365	3/4	2.822	41.4	13.4	2.66	13.3	4.41	1.5
W6x15	4.43	5.99	0.23	5.99	0.26	5/8	3.846	29.1	9.72	2.56	9.32	3.11	1.46
W6x16	4.74	6.28	0.26	4.03	0.405	3/4	3.848	32.1	10.2	2.6	4.43	2.2	0.966
W6x12	3.55	6.03	0.23	4	0.28	5/8	5.384	22.1	7.31	2.49	2.99	1.5	0.918
W6x9	2.68	5.9	0.17	3.94	0.215	9/16	6.965	16.4	5.56	2.47	2.19	1.11	0.905
W5x19	5.54	5.15	0.27	5.03	0.43	13/16	2.381	26.2	10.2	2.17	9.13	3.63	1.28
W5x16	4.68	5.01	0.24	5	0.36	3/4	2.783	21.3	8.51	2.13	7.51	3	1.27
W4x13	3.83	4.16	0.28	4.06	0.345	11/16	2.970	11.3	5.46	1.72	3.86	1.9	1

American Standard C-Channel

Designation	Area A (in²)	Depth d (in)	Web Thickness t_w (in)	Flange Width b_f (in)	Flange Thickness t_f (in)	k (in)	Lateral Buckling d/A_f	Axis x-x			Axis y-y		
								I_x (in⁴)	S_x (in³)	r_x (in)	I_y (in⁴)	S_y (in³)	r_y (in)
C15×50	14.7	15.0	0.716	3.72	0.650	1.44	6.203	404	53.8	5.24	11.0	3.77	0.865
C15×40	11.8	15.0	0.520	3.52	0.650	1.44	6.556	348	46.5	5.43	9.17	3.34	0.883
C15×33.9	10.0	15.0	0.400	3.40	0.650	1.44	6.787	315	42.0	5.61	8.07	3.09	0.901
C12×30	8.81	12.0	0.510	3.17	0.501	1.13	7.556	162	27.0	4.29	5.12	2.05	0.762
C12×25	7.34	12.0	0.387	3.05	0.501	1.13	7.853	144	24.0	4.43	4.45	1.87	0.779
C12×20.7	6.08	12.0	0.282	2.94	0.501	1.13	8.147	129	21.5	4.61	3.86	1.72	0.797
C10×30	8.81	10.0	0.673	3.03	0.436	1.00	7.570	103	20.7	3.43	3.93	1.65	0.668
C10×25	7.35	10.0	0.526	2.89	0.436	1.00	7.936	91.1	18.2	3.52	3.34	1.47	0.675
C10×20	5.87	10.0	0.379	2.74	0.436	1.00	8.371	78.9	15.8	3.67	2.80	1.31	0.690
C10×15.3	4.48	10.0	0.240	2.60	0.436	1.00	8.821	67.3	13.5	3.88	2.27	1.15	0.711

C9×20	5.87	9.00	0.448	2.65	0.413	1.00	8.223	60.9	13.5	3.22	2.41	1.17	0.640
C9×15	4.40	9.00	0.285	2.49	0.413	1.00	8.752	51.0	11.3	3.40	1.91	1.01	0.659
C9×13.4	3.94	9.00	0.233	2.43	0.413	1.00	8.968	47.8	10.6	3.48	1.75	0.954	0.666
C8×18.75	5.51	8.00	0.487	2.53	0.390	0.938	8.108	43.9	11.0	2.82	1.97	1.01	0.598
C8×13.75	4.03	8.00	0.303	2.34	0.390	0.938	8.766	36.1	9.02	2.99	1.52	0.848	0.613
C8×11.5	3.37	8.00	0.220	2.26	0.390	0.938	9.076	32.5	8.14	3.11	1.31	0.775	0.623
C7×14.75	4.33	7.00	0.419	2.30	0.366	0.875	8.316	27.2	7.78	2.51	1.37	0.772	0.561
C7×12.25	3.59	7.00	0.314	2.19	0.366	0.875	8.733	24.2	6.92	2.59	1.16	0.696	0.568
C7×9.8	2.87	7.00	0.210	2.09	0.366	0.875	9.151	21.2	6.07	2.72	0.957	0.617	0.578
C6×13	3.82	6.00	0.437	2.16	0.343	0.813	8.098	17.3	5.78	2.13	1.05	0.638	0.524
C6×10.5	3.07	6.00	0.314	2.03	0.343	0.813	8.617	15.1	5.04	2.22	0.860	0.561	0.529
C6×8.2	2.39	6.00	0.200	1.92	0.343	0.813	9.111	13.1	4.35	2.34	0.687	0.488	0.536

(continued)

Designation	Area A (in²)	Depth d (in)	Web Thickness t_w (in)	Flange Width b_f (in)	Flange Thickness t_f (in)	k (in)	Lateral Buckling d/A_f	Axis x-x I_x (in⁴)	S_x (in³)	r_x (in)	Axis y-y I_y (in⁴)	S_y (in³)	r_y (in)
C5×9	2.64	5.00	0.325	1.89	0.320	0.750	8.267	8.89	3.56	1.84	0.624	0.444	0.486
C5×6.7	1.97	5.00	0.190	1.75	0.320	0.750	8.929	7.48	2.99	1.95	0.470	0.372	0.489
C4×7.25	2.13	4.00	0.321	1.72	0.296	0.750	7.857	4.58	2.29	1.47	0.425	0.337	0.447
C4×6.25	1.77	4.00	0.247	1.65	0.272	0.750	8.913	4.00	2.00	1.50	0.345	0.284	0.441
C4×5.4	1.58	4.00	0.184	1.58	0.296	0.750	8.553	3.85	1.92	1.56	0.312	0.277	0.444
C4×4.5	1.38	4.00	0.125	1.58	0.296	0.750	8.553	3.65	1.83	1.63	0.289	0.265	0.457
C3×6	1.76	3.00	0.356	1.60	0.273	0.688	6.868	2.07	1.38	1.09	0.300	0.263	0.413
C3×5	1.47	3.00	0.258	1.50	0.273	0.688	7.326	1.85	1.23	1.12	0.241	0.228	0.405
C3×4.1	1.20	3.00	0.170	1.41	0.273	0.688	7.794	1.65	1.10	1.18	0.191	0.196	0.398
C3×3.5	1.09	3.00	0.132	1.37	0.273	0.688	8.021	1.57	1.04	1.20	0.169	0.182	0.394

APPENDIX 2

STEEL PIPE

PIPE Dimensions and Properties									
Dimensions					Properties				
Nominal Diameter in.	Outside Diameter in.	Inside Diameter in.	Wall Thickness in.	Weight lb per ft	A in.^{2}	I in.^{4}	S in.^{3}	r in.	Schedule No.
Standard Weight									
1/2	0.840	0.622	0.109	0.85	0.250	0.017	0.041	0.261	40
3/4	1.050	0.824	0.113	1.13	0.333	0.037	0.071	0.334	40
1	1.315	1.049	0.133	1.68	0.494	0.087	0.133	0.421	40
1-1/4	1.660	1.380	0.140	2.27	0.669	0.195	0.235	0.540	40
1-1/2	1.900	1.610	0.145	2.72	0.799	0.310	0.326	0.623	40
2	2.375	2.067	0.154	3.65	1.075	0.666	0.561	0.787	40
2-1/2	2.875	2.469	0.203	5.79	1.704	1.530	1.064	0.947	40
3	3.500	3.068	0.216	7.58	2.228	3.017	1.724	1.164	40
3-1/2	4.000	3.548	0.226	9.11	2.680	4.788	2.394	1.337	40
4	4.500	4.026	0.237	10.79	3.174	7.233	3.214	1.510	40
5	5.563	5.047	0.258	14.62	4.300	15.16	5.451	1.878	40
6	6.625	6.065	0.280	18.97	5.581	28.14	8.496	2.245	40
8	8.625	7.981	0.322	28.55	8.399	72.49	16.81	2.938	40
10	10.75	10.02	0.365	40.48	11.91	160.7	29.90	3.674	40
12	12.75	12.00	0.375	49.56	14.58	279.3	43.82	4.377	—

PIPE Dimensions and Properties									
Dimensions					Properties				
Nominal Diameter in.	Outside Diameter in.	Inside Diameter in.	Wall Thickness in.	Weight lb per ft	A in.^{2}	I in.^{4}	S in.^{3}	r in.	Schedule No.
Extra Strong									
1/2	0.840	0.546	0.147	1.090	0.320	0.020	0.048	0.250	80
3/4	1.050	0.742	0.154	1.470	0.433	0.045	0.085	0.321	80
1	1.315	0.957	0.179	2.170	0.639	0.106	0.161	0.407	80
1-1/4	1.660	1.278	0.191	3.000	0.881	0.242	0.291	0.524	80
1-1/2	1.900	1.500	0.200	3.630	1.068	0.391	0.412	0.605	80
2	2.375	1.939	0.218	5.020	1.477	0.868	0.731	0.766	80
2-1/2	2.875	2.323	0.276	7.660	2.254	1.924	1.339	0.924	80
3	3.500	2.900	0.300	10.25	3.016	3.894	2.225	1.136	80
3-1/2	4.000	3.364	0.318	12.50	3.678	6.280	3.140	1.307	80
4	4.500	3.826	0.337	14.98	4.407	9.610	4.271	1.477	80
5	5.563	4.813	0.375	20.78	6.112	20.67	7.431	1.839	80
6	6.625	5.761	0.432	28.57	8.405	40.49	12.22	2.195	80
8	8.625	7.625	0.500	43.39	12.76	105.7	24.51	2.878	80
10	10.75	9.750	0.500	54.74	16.10	212.0	39.43	3.628	60
12	12.75	11.75	0.500	65.42	19.24	361.5	56.71	4.335	—
Double-Extra Strong									
2	2.375	1.503	0.436	9.03	2.66	1.31	1.10	0.703	—
2-1/2	2.875	1.771	0.552	13.69	4.03	2.87	2.00	0.844	—
3	3.500	2.300	0.600	18.58	5.47	5.99	3.42	1.047	—
4	4.500	3.152	0.674	27.54	8.10	15.3	6.79	1.374	—
5	5.563	4.063	0.750	38.55	11.34	33.6	12.09	1.722	—
6	6.625	4.897	0.864	53.16	15.64	66.3	20.02	2.060	—
8	8.625	6.875	0.875	72.42	21.30	162	37.56	2.757	—

APPENDIX 3

H PILE (AISC)

HP Shapes

Designation	Area A (in²)	Depth d (in)	Web Thickness t_w (in)	Flange Width b_f (in)	Flange Thickness t_f (in)	k (in)	Lateral Buckling d/A_f	Axis x-x I_x (in⁴)	S_x (in³)	r_x (in)	Axis y-y I_y (in⁴)	S_y (in³)	r_y (in)
HP14X117	34.4	14.21	0.805	14.885	0.805	1 1/16	1.186	1220	172	5.96	443	59.5	3.59
HP14X102	30	14.01	0.705	14.785	0.705	1	1.344	1050	150	5.92	380	51.4	3.56
HP14X89	26.1	13.83	0.615	14.695	0.615	15/16	1.530	904	131	5.88	326	44.3	3.53
HP14X73	21.4	13.61	0.505	14.585	0.505	7/8	1.848	729	107	5.84	261	35.8	3.49
HP13X100	29.4	13.15	0.765	13.205	0.765	1 7/16	1.302	886	135	5.49	294	44.5	3.16
HP13X87	25.5	12.95	0.665	13.105	0.665	1 3/8	1.486	755	117	5.45	250	38.1	3.13
HP13X73	21.6	12.75	0.565	13.005	0.565	1 1/4	1.735	630	98.8	5.4	207	31.9	3.1
HP13X60	17.5	12.54	0.46	12.9	0.46	1 1/8	2.113	503	80.3	5.36	165	25.5	3.07
HP12X84	24.6	12.28	0.685	12.295	0.685	1 3/8	1.458	650	106	5.14	213	34.6	2.94
HP12X85	21.8	12.13	0.605	12.215	0.605	1 5/16	1.641	569	92.8	5.11	186	30.4	2.92
HP12X86	18.4	11.94	0.515	12.125	0.515	1 1/4	1.912	472	79.1	5.06	153	25.3	2.88
HP12X87	15.5	11.78	0.435	12.045	0.435	1 1/8	2.248	393	66.8	5.03	127	21.1	2.86
HP10X57	16.8	9.99	0.656	10.225	0.656	1 3/16	1.489	294	58.8	4.18	101	19.7	2.45
HP10X42	12.4	9.7	0.415	10.075	0.415	1 1/16	2.320	210	43.4	4.13	71.7	14.2	2.41
HP8X36	10.6	8.02	0.445	8.155	0.445	15/16	2.210	119	29.8	3.36	40.3	9.88	1.95

APPENDIX 4

ALLOWABLE BUCKLING STRESS

AISC Allowable Compressive (Buckling) Stresses for A36 (36 KSI) Steel (Fbs)

kL/r	Fbs (ksi)	kL/r	Fbs (ksi)	kL/r	Fbs (ksi)	kL/r	Fbs (ksi)	kL/r	Fbs (ksi)
1	21.56	41	19.11	81	15.24	121	10.14	161	5.76
2	21.52	42	19.02	82	15.14	122	9.98	162	5.69
3	21.48	43	18.93	83	15.02	123	9.86	163	5.62
4	21.44	44	18.84	84	14.91	124	9.71	164	5.55
5	21.39	45	18.78	85	14.78	125	9.56	165	5.48
6	21.35	46	18.69	86	14.67	126	9.41	166	5.41
7	21.31	47	18.6	87	14.56	127	9.26	167	5.34
8	21.25	48	18.51	88	14.44	128	9.12	168	5.27
9	21.21	49	18.42	89	14.32	129	8.98	169	5.23
10	21.16	50	18.35	90	14.21	130	8.85	170	5.16
11	21.11	51	18.26	91	14.09	131	8.71	171	5.09
12	21.06	52	18.17	92	13.97	132	8.58	172	5.05
13	21.01	53	18.08	93	13.84	133	8.45	173	4.99
14	20.96	54	17.99	94	13.72	134	8.32	174	4.93
15	20.89	55	17.91	95	13.6	135	8.21	175	4.87
16	20.84	56	17.82	96	13.48	136	8.08	176	4.81

kL/r	Fbs (ksi)	kL/r	Fbs (ksi)	kL/r	Fbs (ksi)	kL/r	Fbs (ksi)	kL/r	Fbs (ksi)
17	20.79	57	17.73	97	13.35	137	7.97	177	4.77
18	20.73	58	17.64	98	13.23	138	7.85	178	4.72
19	20.67	59	17.55	99	13.1	139	7.74	179	4.67
20	20.61	60	17.43	100	12.98	140	7.63	180	4.62
21	20.54	61	17.34	101	12.85	141	7.52	181	4.57
22	20.48	62	17.25	102	12.72	142	7.42	182	4.52
23	20.42	63	17.16	103	12.58	143	7.31	183	4.47
24	20.36	64	17.07	104	12.48	144	7.21	184	4.42
25	20.29	65	16.94	105	12.34	145	7.11	185	4.38
26	20.23	66	16.85	106	12.21	146	7.02	186	4.33
27	20.16	67	16.74	107	12.08	147	6.92	187	4.28
28	20.08	68	16.64	108	11.94	148	6.83	188	4.24
29	19.95	69	16.53	109	11.81	149	6.75	189	4.19
30	19.88	70	16.43	110	11.67	150	6.65	190	4.15
31	19.82	71	16.33	111	11.54	151	6.56	191	4.11
32	19.75	72	16.22	112	11.41	152	6.47	192	4.06
33	19.68	73	16.12	113	11.27	153	6.39	193	4.02
34	19.61	74	16.01	114	11.14	154	6.31	194	3.98
35	19.58	75	15.91	115	10.99	155	6.23	195	3.94
36	19.51	76	15.79	116	10.86	156	6.14	196	3.89
37	19.42	77	15.69	117	10.72	157	6.06	197	3.86
38	19.35	78	15.58	118	10.57	158	5.99	198	3.81
39	19.28	79	15.47	119	10.42	159	5.92	199	3.78
40	19.19	80	15.35	120	10.27	160	5.85	200	3.74

APPENDIX 5

SHEET PILE (SKYLINE)

Section	Width (w) in (mm)	Height (h) in (mm)	Thickness Flange (tf) in (mm)	Thickness Web (tw) in (mm)	Cross-Sectional Area in²/ft (cm²/m)	Weight Pile lb/ft (kg/m)	Weight Wall lb/ft² (kg/m²)	Section Modulus Elastic in³/ft (cm³/m)	Moment of Inertia in⁴/ft (cm⁴/m)	Coation Area Both Sides ft²/ft of single (m²/m)	Coation Area Wall Surface ft²/ft² (m²/m²)
AZ 12-700	27.56	12.36	0.335	0.335	5.82	45.49	19.81	22.4	138.3	5.61	1.22
	700	314	8.5	8.5	123.2	67.7	96.7	1205	18880	1.71	1.22
AZ 13-700	27.56	12.4	0.375	0.375	6.36	49.72	21.65	24.3	150.4	5.61	1.22
	700	315	9.5	9.5	134.7	74	105.7	1305	20540	1.71	1.22
AZ 13-700-10/10	27.56	12.42	0.394	0.394	6.63	51.85	22.58	25.2	156.5	5.61	1.22
	700	316	10	10	140.4	77.2	110.2	1355	21370	1.71	1.22
AZ 14-700	27.56	12.44	0.413	0.413	6.9	53.96	23.5	26.1	162.5	5.61	1.22
	700	316	10.5	10.5	146.1	80.3	114.7	1405	22190	1.71	1.22
AZ 12-770	30.31	13.52	0.335	0.335	5.67	48.78	19.31	23.2	156.9	6.10	1.20
	770	343.5	8.5	8.5	120.1	72.6	94.3	1245	21430	1.86	1.20
AZ 13-770	30.31	13.54	0.354	0.354	5.94	51.14	20.24	24.2	163.7	6.10	1.20
	770	344	9	9	125.8	76.1	98.8	1300	22360	1.86	1.20
AZ 14-770	30.31	13.56	0.375	0.375	6.21	53.42	21.14	25.2	170.6	6.10	1.20
	770	344.5	9.5	9.5	131.5	79.5	103.2	1355	23300	1.86	1.2
AZ 14-770-10/10	30.31	13.58	0.394	0.394	6.48	55.71	22.06	26.1	177.5	6.07	1.2
	770	345	10	10	137.2	82.9	107.7	1405	24240	1.85	1.2

	1	2	3	4	5	6	7	8	9	10	11
AZ 18	24.8	14.96	0.375	0.375	7.11	49.99	24.19	33.5	250.4	5.64	1.35
	630	380	9.5	9.5	150.4	74.4	118.1	1800	34200	1.72	1.35
AZ 17-700	27.56	16.52	0.335	0.335	6.28	49.12	21.38	32.2	265.3	6.1	1.33
	700	419.5	8.5	8.5	133	73.1	104.4	1730	36230	1.86	1.33
AZ 18-700	27.56	16.54	0.354	0.354	6.58	51.41	22.39	33.5	276.8	6.1	1.33
	700	420	9	9	139.2	76.5	109.3	1800	37800	1.86	1.33
AZ 19-700	27.56	16.56	0.375	0.375	6.88	53.76	23.41	34.8	288.4	6.1	1.33
	700	420.5	9.5	9.5	145.6	80	114.3	1870	39380	1.86	1.33
AZ 20-700	27.56	16.58	0.394	0.394	7.18	56.11	24.43	36.2	299.9	6.1	1.33
	700	421	10	10	152	83.5	119.3	1945	40960	1.86	1.33
AZ 26	24.8	16.81	0.512	0.48	9.35	65.72	31.79	48.4	406.5	5.91	1.41
	630	427	13	12.2	198	97.8	155.2	2600	55510	1.8	1.41
AZ 24-700	27.56	18.07	0.441	0.441	8.23	64.3	28	45.2	408.8	6.33	1.38
	700	459	11.2	11.2	174.1	95.7	136.7	2430	55820	1.93	1.38
AZ 26-700	27.56	18.11	0.48	0.48	8.84	69.12	30.1	48.4	437.3	6.33	1.38
	700	460	12.2	12.2	187.2	102.9	146.9	2600	59720	1.93	1.38
AZ 28-700	27.56	18.15	0.52	0.52	9.46	73.93	32.19	51.3	465.9	6.33	1.38
	700	461	13.2	13.2	200.2	110	157.2	2760	63620	1.93	1.38
AZ 24-700N	27.56	18.07	0.492	0.354	7.71	60.28	26.26	45.3	409.3	6.3	1.37
	700	459	12.5	9	163.3	89.7	128.2	2435	55890	1.92	1.37
AZ 26-700N	27.56	18.11	0.531	0.394	8.33	65.11	28.37	48.4	437.8	6.3	1.37
	700	460	13.5	10	176.4	96.9	138.5	2600	59790	1.92	1.37

Section	Width (w) in (mm)	Height (h) in (mm)	Thickness Flange (tf) in (mm)	Thickness Web (tw) in (mm)	Cross-Sectional Area in²/ft (cm²/m)	Weight Pile lb/ft (kg/m)	Weight Wall lb/ft² (kg/m²)	Section Modulus Elastic in³/ft (cm³/m)	Moment of Inertia in⁴/ft (cm⁴/m)	Coation Area Both Sides ft²/ft of single (m²/m)	Coation Area Wall Surface ft²/ft² (m²/m²)
AZ 28-700N	27.56	18.15	0.571	0.433	8.95	69.95	30.46	51.4	466.5	6.3	1.37
	700	461	14.5	11	189.5	104.1	148.7	2765	63700	1.92	1.37
AZ 36-700N	27.56	19.65	0.591	0.441	10.2	79.7	34.61	66.8	656.2	6.76	1.47
	700	499	15	11.2	216	118.6	169	3590	89610	2.06	1.47
AZ 38-700N	27.56	19.69	0.63	0.48	10.87	84.94	37.07	70.6	694.5	6.76	1.47
	700	500	16	12.2	230	126.4	181	3795	94840	2.06	1.47
AZ 40-700N	27.56	19.72	0.669	0.52	11.53	90.18	39.32	74.3	732.9	6.76	1.47
	700	501	17	13.2	244	134.2	192	3995	100080	2.06	1.47
AZ 42-700N	27.56	19.65	0.709	0.551	12.22	95.49	41.57	78.2	766	6.76	1.47
	700	499	18	14	259	142.1	203	4205	104930	2.06	1.47
AZ 44-700N	27.56	19.69	0.748	0.591	12.89	100.73	43.83	81.9	804.1	6.76	1.47
	700	500	19	15	273	149.9	214	4405	110150	2.06	1.47
AZ 46-700N	27.56	19.72	0.787	0.63	13.55	105.97	46.08	85.7	842.2	6.76	1.47
	700	501	20	16	287	157.7	225	4605	115370	2.06	1.47
AZ 46	22.83	18.94	0.709	0.551	13.76	89.1	46.82	85.5	808.8	6.23	1.63
	580	481	18	14	291.2	132.6	228.6	4595	110450	1.9	1.63
AZ 48	22.83	18.98	0.748	0.591	14.48	93.81	49.28	89.3	847.1	6.23	1.63
	580	482	19	15	306.5	139.6	240.6	4800	115670	1.9	1.63
AZ 50	22.83	19.02	0.787	0.63	15.22	98.58	51.8	93.3	886.5	6.23	1.63
	580	483	20	16	322.2	146.7	252.9	5015	121060	1.9	1.63

APPENDIX 6

WOOD PROPERTIES

Rectangular (Wood) Properties

Nominal Size $b \times d$	Standard Dressed Size (S4S) $b \times d$ inches $\times$ inches	Area of Section A in^2	x-x AXIS (Strong)		y-y AXIS (Weak)		Approximate Weight in Pound per Linear Foot (lb/ft) of Piece When Density Equals 35 lb/ft^3
			Section Modulus S_{XX} in^3	Moment of Inertia I_{XX} in^4	Section Modulus S_{YY} in^3	Moment of Inertia I_{YY} in^4	
1×3	$3/4 \times 2\text{-}1/2$	1.875	0.781	0.977	0.234	0.088	0.456
1×4	$3/4 \times 3\text{-}1/2$	2.625	1.531	2.680	0.328	0.123	0.638
1×6	$3/4 \times 5\text{-}1/2$	4.125	3.781	10.40	0.516	0.193	1.003
1×8	$3/4 \times 7\text{-}1/4$	5.438	6.571	23.82	0.680	0.255	1.322
1×10	$3/4 \times 9\text{-}1/4$	6.938	10.70	49.47	0.867	0.325	1.686
1×12	$3/4 \times 11\text{-}1/4$	8.438	15.82	88.99	1.055	0.396	2.051
2×3	$1\text{-}1/2 \times 2\text{-}1/2$	3.750	1.563	1.953	0.938	0.703	0.911
2×4	$1\text{-}1/2 \times 3\text{-}1/2$	5.250	3.063	5.359	1.313	0.984	1.276
2×5	$1\text{-}1/2 \times 4\text{-}1/2$	6.750	5.063	11.39	1.688	1.266	1.641
2×6	$1\text{-}1/2 \times 5\text{-}1/2$	8.250	7.563	20.80	2.063	1.547	2.005
2×8	$1\text{-}1/2 \times 7\text{-}1/4$	10.88	13.15	47.66	2.720	2.040	2.644
2×10	$1\text{-}1/2 \times 9\text{-}1/4$	13.88	21.40	98.97	3.470	2.603	3.374
2×12	$1\text{-}1/2 \times 11\text{-}1/4$	16.88	31.65	178.0	4.220	3.165	4.103
2×14	$1\text{-}1/2 \times 13\text{-}1/4$	19.88	43.90	290.8	4.970	3.728	4.832

Nominal Size $b \times d$	Standard Dressed Size (S4S) $b \times d$ inches $\times$ inches	Area of Section A in^2	x-x AXIS (Strong)		y-y AXIS (Weak)		Approximate Weight in Pound per Linear Foot (lb/ft) of Piece When Density Equals 35 lb/ft^3
			Section Modulus S_{XX} in^3	Moment of Inertia I_{XX} in^4	Section Modulus S_{YY} in^3	Moment of Inertia I_{YY} in^4	
3×4	2-1/2 × 3-1/2	8.750	5.104	8.932	3.646	4.557	2.127
3×5	2-1/2 × 4-1/2	11.25	8.438	18.98	4.688	5.859	2.734
3×6	2-1/2 × 5-1/2	13.75	12.60	34.66	5.729	7.161	3.342
3×8	2-1/2 × 7-1/4	18.13	21.91	79.41	7.554	9.443	4.407
3×10	2-1/2 × 9-1/4	23.13	35.66	164.9	9.638	12.05	5.622
3×12	2-1/2 × 11-1/4	28.13	52.74	296.7	11.72	14.65	6.837
3×14	2-1/2 × 13-1/4	33.13	73.16	484.7	13.80	17.26	8.052
3×16	2-1/2 × 15-1/4	38.13	96.91	739.0	15.89	19.86	9.268
4×4	3-1/2 × 3-1/2	12.25	7.146	12.51	7.146	12.51	2.977
4×5	3-1/2 × 4-1/2	15.75	11.81	26.58	9.188	16.08	3.828
4×6	3-1/2 × 5-1/2	19.25	17.65	48.53	11.23	19.65	4.679
4×8	3-1/2 × 7-1/4	25.38	30.67	111.2	14.81	25.91	6.169
4×10	3-1/2 × 9-1/4	32.38	49.92	230.9	18.89	33.05	7.870
4×12	3-1/2 × 11-1/4	39.38	73.84	415.3	22.97	40.20	9.572
4×14	3-1/2 × 13-1/4	47.25	104.3	691.3	27.56	48.23	11.48
4×16	3-1/2 × 15-1/4	54.25	137.9	1051	31.65	55.38	13.19
5×5	4-1/2 × 4-1/2	20.25	15.19	34.17	15.19	34.17	4.922
6×6	5-1/2 × 5-1/2	30.25	27.73	76.26	27.73	76.26	7.352
6×8	5-1/2 × 7-1/2	41.25	51.56	193.4	37.81	104.0	10.03
6×10	5-1/2 × 9-1/2	52.25	82.73	393.0	47.90	131.7	12.70
6×12	5-1/2 × 11-1/2	63.25	121.2	697.1	57.98	159.4	15.37
6×14	5-1/2 × 13-1/2	74.25	167.1	1128	68.06	187.2	18.05
6×16	5-1/2 × 15-1/2	85.25	220.2	1707	78.15	214.9	20.72
6×18	5-1/2 × 17-1/2	96.25	280.7	2456	88.23	242.6	23.39
6×20	5-1/2 × 19-1/2	107.3	348.7	3400	98.36	270.5	26.08
6×22	5-1/2 × 21-1/2	118.3	423.9	4557	108.4	298.2	28.75
6×24	5-1/2 × 23-1/2	129.3	506.4	5950	118.5	325.9	31.43
8×8	7-1/2 × 7-1/2	56.25	70.31	263.7	70.31	263.7	13.67
8×10	7-1/2 × 9-1/2	71.25	112.8	535.9	89.06	334.0	17.32
8×12	7-1/2 × 11-1/2	86.25	165.3	950.5	107.8	404.3	20.96
8×14	7-1/2 × 13-1/2	101.3	227.9	1538	126.6	474.8	24.62
8×16	7-1/2 × 15-1/2	116.3	300.4	2328	145.4	545.2	28.27
8×18	7-1/2 × 17-1/2	131.3	383.0	3351	164.1	615.5	31.91
8×20	7-1/2 × 19-1/2	146.3	475.5	4636	182.9	685.8	35.56
8×22	7-1/2 × 21-1/2	161.3	578.0	6213	201.6	756.1	39.20
8×24	7-1/2 × 23-1/2	176.3	690.5	8113	220.4	826.4	42.85

Nominal Size $b \times d$	Standard Dressed Size (S4S) $b \times d$ inches $\times$ inches	Area of Section A in^2	x-x AXIS (Strong)		y-y AXIS (Weak)		Approximate Weight in Pound per Linear Foot (lb/ft) of Piece When Density Equals 35 lb/ft^3
			Section Modulus S_{XX} in^3	Moment of Inertia I_{XX} in^4	Section Modulus S_{YY} in^3	Moment of Inertia I_{YY} in^4	
10 × 10	9-1/2 × 9-1/2	90.25	142.9	678.8	142.9	678.8	21.94
10 × 12	9-1/2 × 11-1/2	109.3	209.5	1205	173.1	822.0	26.57
10 × 14	9-1/2 × 13-1/2	128.3	288.7	1949	203.1	964.9	31.18
10 × 16	9-1/2 × 15-1/2	147.3	380.5	2949	233.2	1108	35.80
10 × 18	9-1/2 × 17-1/2	166.3	485.0	4244	263.3	1251	40.42
10 × 20	9-1/2 × 19-1/2	185.3	602.2	5872	293.4	1394	45.04
10 × 22	9-1/2 × 21-1/2	204.3	732.1	7870	323.5	1537	49.66
10 × 24	9-1/2 × 23-1/2	223.3	874.6	10276	353.6	1679	54.27
12 × 12	11-1/2 × 11-1/2	132.3	253.6	1458	253.6	1458	32.16
12 × 14	11-1/2 × 13-1/2	155.3	349.4	2359	297.7	1712	37.75
12 × 16	11-1/2 × 15-1/2	178.3	460.6	3570	341.7	1965	43.34
12 × 18	11-1/2 × 17-1/2	201.3	587.1	5137	385.8	2218	48.93
12 × 20	11-1/2 × 19-1/2	224.3	729.0	7108	429.9	2472	54.52
12 × 22	11-1/2 × 21-1/2	247.3	886.2	9526	474.0	2725	60.11
12 × 24	11-1/2 × 23-1/2	270.3	1059	12439	518.1	2979	65.70
14 × 14	13-1/2 × 13-1/2	182.3	410.2	2769	410.2	2769	44.31
14 × 16	13-1/2 × 15-1/2	209.3	540.7	4190	470.9	3179	50.87
14 × 18	13-1/2 × 17-1/2	236.3	689.2	6031	531.7	3589	57.43
14 × 20	13-1/2 × 19-1/2	263.3	855.7	8343	592.4	3999	64.00
14 × 22	13-1/2 × 21-1/2	290.3	1040	11183	653.2	4409	70.56
14 × 24	13-1/2 × 23-1/2	317.3	1243	14602	713.9	4819	77.12
16 × 16	15-1/2 × 15-1/2	240.3	620.8	4811	620.8	4811	58.41
16 × 18	15-1/2 × 17-1/2	271.3	791.3	6924	700.9	5432	65.94
16 × 20	15-1/2 × 19-1/2	302.3	982.5	9579	780.9	6052	73.48
16 × 22	15-1/2 × 21-1/2	333.3	1194	12839	861.0	6673	81.01
16 × 24	15-1/2 × 23-1/2	364.3	1427	16765	941.1	7294	88.55
18 × 18	17-1/2 × 17-1/2	306.3	893.4	7817	893.4	7817	74.45
18 × 20	17-1/2 × 19-1/2	341.3	1109	10815	995.5	8710	82.95
18 × 22	17-1/2 × 21-1/2	376.3	1348	14495	1098	9603	91.46
18 × 24	17-1/2 × 23-1/2	411.3	1611	18928	1200	10497	99.97
20 × 20	19-1/2 × 19-1/2	380.3	1236	12051	1236	12051	92.43
20 × 22	19-1/2 × 21-1/2	419.3	1502	16152	1363	13287	101.9
20 × 24	19-1/2 × 23-1/2	458.3	1795	21091	1489	14522	111.4
22 × 22	21-1/2 × 21-1/2	462.3	1657	17808	1657	17808	112.4
22 × 24	21-1/2 × 23-1/2	505.3	1979	23254	1811	19465	122.8
24 × 24	23-1/2 × 23-1/2	552.3	2163	25417	2163	25417	134.2

APPENDIX 7

FORMWORK CHARTS (WILLIAMS)

Common Forming Lumber Properties

Properties of Structural Lumber

Nominal Size (in) b x h	American Standard Size (in) b x h S4S* 19% Maximum Moisture	Area of Section (in²) A = bh		Moment of Inertia (in) $I = (bh^3)/12$		Section Modulus (in) $S = (bh^2)/6$		Board Feet (per lineal ft of piece)	Approx. Weight (lbs per lineal ft)**
		Rough	S4S	Rough	S4S	Rough	S4S		
4 x 2	3-1/2 x 1-1/2	5.89	5.25	1.30	0.98	1.60	1.31	2/3	1.5
6 x 2	5-1/2 x 1-1/2	9.14	8.25	2.01	1.55	2.48	2.06	1	2.3
8 x 2	7-1/4 x 1-1/2	11.98	10.87	2.64	2.04	3.25	2.72	1-1/3	3.0
10 x 2	9-1/4 x 1-1/2	15.23	13.87	3.35	2.60	4.13	3.47	1-2/3	3.9
12 x 2	11-1/4 x 1-1/2	18.48	16.87	4.07	3.16	5.01	4.21	2	4.7
2 x 4	1-1/2 x 3-1/2	5.89	5.25	6.45	5.36	3.56	3.06	2/3	1.5
2 x 6	1-1/2 x 5-1/2	9.14	8.25	24.10	20.80	8.57	7.56	1	2.3
2 x 8	1-1/2 x 7-1/4	11.98	10.87	54.32	47.63	14.73	13.14	1-1/3	3.0
2 x 10	1-1/2 x 9-1/4	15.23	13.87	111.58	98.93	23.80	21.39	1-2/3	3.9
2 x 12	1-1/2 x 11-1/4	18.48	16.87	199.31	177.97	35.40	31.64	2	4.7
3 x 4	2-1/2 x 3-1/2	9.52	8.75	10.42	8.93	5.57	5.10	1	2.4
3 x 6	2-1/2 x 5-1/2	14.77	13.75	38.93	34.66	13.84	12.60	1-1/2	3.8
3 x 8	2-1/2 x 7-1/4	19.36	18.12	87.74	79.39	23.80	21.90	2	5.0
3 x 10	2-1/2 x 7-1/4	24.61	23.12	180.24	164.89	38.45	35.65	2-1/2	6.4
3 x 12	2-1/2 x 11-1/4	29.86	28.12	321.96	296.63	56.61	52.73	3	7.8
4 x 4	3-1/2 x 3-1/2	13.14	12.25	14.39	12.5	7.94	7.15	1-1/3	3.4
4 x 6	3-1/2 x 5-1/2	20.39	19.25	53.76	48.53	19.12	17.65	2	5.3
4 x 8	3-1/2 x 7-1/4	26.73	25.38	121.17	111.15	32.86	30.66	2-2/3	7.0
4 x 10	3-1/2 x 9-1/4	33.98	32.38	248.91	230.84	53.10	49.91	3-1/3	9.0

* Rough dry sizes are 1/8" larger, both dimensions.

** Based on a unit weight value of 40 lb. per cu. ft. Actual weights vary depending on species and moisture content.

Data supplied by the National Forest Products Association

Form Loading in Pounds/ Sq. Foot for Incremental Slab Thickness*

Concrete Weight (lbs per sq ft)	Slab Thickness							
	2 in.	4 in.	6 in.	8 in.	10 in.	12 in.	14 in.	16 in.
100	67	84	100	117	134	150	167	184
115	70	89	108	127	146	165	185	204
125	71	92	113	134	155	175	196	217
135	73	95	118	140	163	185	208	230
150	75	100	125	150	175	200	225	250

* Values above include 50 psf live load for construction loads. Formwork dead load is not included.

Safe Spacing (ℓ) in inches of supports for plywood sheathing continuous over four or more supports. Table based on APA rated plywood class 1.

$\Delta max = \ell / 360$, but not to exceed 1/16"
$F_s = 72$ psi
$F_b = 1930$ psi
$E_e = 1{,}500{,}000$ psi
$E = 1{,}650{,}000$ psi

Safe Spacing of Supports for Plywood Sheathing

Pressure or Load of Concrete (lbs per sq ft)	Sanded Thickness, Face Grain Parallel to Span				Sanded Thickness, Face Grain Perpendicular to Span			
	1/2 in.	5/8 in.	3/4 in.	1 in.	1/2 in.	5/8 in.	3/4 in.	1 in.
75	21	24	27	32	14	16	21	28
100	19	22	25	30	12	14	19	26
125	18	21	23	28	12	13	17	24
150	17	20	22	27	11	12	16	23
175	16	19	21	26	10	11	15	22
200	15	18	20	25	8	11	15	21
300	13	16	18	22	7	9	12	18
400	12	14	16	20	7	8	11	16
500	11	13	15	19	6	7	10	14
600	11	12	14	17	6	7	9	13
700	10	12	13	16	6	6	9	12
800	10	11	13	16	5	6	8	11
900	9	11	12	15	5	6	8	11
1000	9	10	12	14	5	5	7	10

Safe Spacing of Supports

Safe Spacing (ℓ) in inches of supports for joists, studs, etc. single span.

Δmax = ℓ / 360, but not to exceed 1/4"

E = 1,600,000 psi

Safe Spacing of Supports - Simple Span-Studs, Joists and Single Beams

Uniform Load (lbs per lineal ft)	Nominal Size of S4S Lumber							
	2x4	2x6	2x8	2x10	2x12	3x6	4x4	4x8
Fb (psi) =	1500	1250	1200	1050	975	1250	1500	1200
100	60	94	115	138	160	106	80	142
200	48	75	97	116	135	89	64	120
300	42	61	79	94	11	78	55	108
400	37	53	68	82	96	68	50	101
500	33	47	61	73	86	61	47	93
600	30	43	56	67	78	56	44	85
700	28	40	51	62	72	51	42	79
800	26	37	48	58	68	48	40	74
900	24	35	45	54	64	45	37	70
1000	23	33	43	51	60	43	35	66
1100	22	32	41	49	58	41	34	63
1200	21	30	39	47	55	39	32	60
1300	20	29	38	45	53	38	31	58
1400	19	28	36	43	51	36	30	56
1500	19	27	35	42	49	35	29	54
1600	18	26	34	41	48	34	28	52
1700	17	25	33	39	46	33	27	50
1800	17	25	32	38	45	32	26	49
1900	17	24	31	37	44	31	26	48
2000	16	23	30	36	43	30	25	46

Safe Spacing (ℓ) in inches of supports for joists, studs, etc. continuous over three or more supports.

$\Delta max = \ell / 360$, but not to exceed 1/4"

$E = 1,600,000$ psi

Safe Spacing of Supports - Multi-Span Studs, Joists & Single Beams

Uniform Load (lbs per lineal ft)	Nominal Size of S4S Lumber							
	2x4	2x6	2x8	2x10	2x12	3x6	4x4	4x8
Fb (psi) =	1500	1250	1200	1050	975	1250	1500	1200
100	75	110	135	162	188	125	97	167
200	58	84	108	129	152	105	79	140
300	47	68	88	105	124	88	69	127
400	41	59	76	91	107	76	62	117
500	37	53	68	82	96	68	56	105
600	33	48	62	74	87	62	51	95
700	31	44	58	69	81	58	47	88
800	29	42	54	64	76	54	44	83
900	27	39	51	61	71	51	42	78
1000	26	37	48	58	68	48	40	74
1100	25	35	46	55	64	46	38	70
1200	23	34	44	52	62	44	36	67
1300	23	33	42	50	59	42	35	65
1400	22	31	41	49	57	41	33	62
1500	21	30	39	47	55	39	32	60
1600	20	29	38	45	53	38	31	58
1700	20	28	37	44	52	37	30	56
1800	19	28	36	43	50	36	29	55
1900	19	27	35	42	49	35	29	53
2000	18	26	34	41	48	34	28	52

Values are based on NDS 2001 for S.Y.P. #2

All values below bold line indicate failure in bending.

Safe Spacing of Supports

Safe Spacing (ℓ) in inches of supports for double wales single span.

Δmax = ℓ / 360, but not to exceed 1/4"

E = 1,600,000 psi

Safe Spacing of Supports - Simple Span Double Walers

Uniform Load (lbs per lineal ft)	2x4	2x6	2x8	2x10	2x12	3x6	4x4	4x8
Nominal Size of S4S Lumber — Fb (psi) =	1500	1250	1200	1050	975	1250	1500	1200
100	76	111	137	164	190	126	98	169
200	60	94	115	138	160	106	80	142
300	53	83	104	125	145	96	70	129
400	48	75	97	116	135	89	64	120
500	44	67	86	103	121	83	59	113
600	42	61	79	94	111	78	55	108
700	39	56	73	87	102	73	53	104
800	37	53	68	82	96	68	50	101
900	34	50	64	77	90	64	48	98
1000	33	47	61	73	86	61	47	93
1100	31	45	58	70	82	58	45	89
1200	30	43	56	67	78	56	44	85
1300	29	41	53	6	75	53	43	82
1400	28	40	51	62	72	51	42	79
1500	27	38	50	59	70	50	41	76
1600	26	3	48	58	68	48	40	74
1700	25	36	47	56	65	47	38	72
1800	24	35	45	54	64	45	37	70
1900	24	34	44	53	62	44	36	68
2000	23	33	43	51	60	43	35	66

Safe Spacing (ℓ) in inches of supports for double wales continuous over three or more supports.

Δmax = ℓ / 360, but not to exceed 1/4"

E = 1,600,000 psi

Safe Spacing of Supports - Multi-Span Double Walers

Uniform Load (lbs per lineal ft)	Nominal Size of S4S Lumber	2x4	2x6	2x8	2x10	2x12	3x6	4x4	4x8
	Fb (psi) =	1500	1250	1200	1050	975	1250	1500	1200
100		93	130	161	193	223	148	115	198
200		75	110	135	162	188	425	97	167
300		65	97	122	146	170	113	86	151
400		58	84	108	129	152	105	79	140
500		52	75	97	116	136	97	73	133
600		47	68	88	105	124	88	69	127
700		44	63	82	98	114	82	65	122
800		41	59	76	91	107	76	62	117
900		39	56	72	86	101	72	59	110
1000		37	53	68	82	96	68	56	105
1100		35	50	65	78	91	65	54	100
1200		33	48	62	74	87	62	51	95
1300		32	46	60	71	84	60	49	92
1400		31	45	58	69	81	58	47	88
1500		30	43	56	67	78	56	46	85
1600		29	42	54	64	76	54	44	83
1700		28	40	52	62	72	52	43	80
1800		27	39	51	61	71	51	42	78
1900		26	38	49	59	69	49	41	76
2000		26	37	48	58	68	48	40	74

Values are based on NDS 2001 for S.Y.P. #2

All values below bold line indicate failure in bending.

APPENDIX 8

FORM HARDWARE VALUES (WILLIAMS)

Shebolt Tie-Rod Forming System

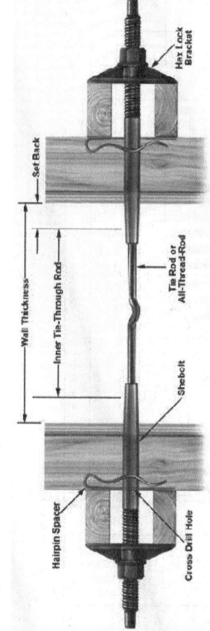

C5T Threaded Shebolt

Shebolt Diameter	Maximum Tap Size	Standard Lengths	Thread Lenghts	Weight	Wing Nuts	Vibra-Lock Bracket	Hex-Nut Bracket	Hex-Lock Bracket
9/16" (14 mm)	3/8" (9.5 mm)	16"; 18" (406; 457 mm)	5"; 8" (127; 203 mm)	0.94; 1.06 lbs. (0.43; 0.48 kg.)	W1	B2	BR5	H1
3/4"* (19 mm)	1/2" (13 mm)	20"; 24" (508; 610 mm)	9"; 12" (229; 305 mm)	2.04; 2.40 lbs. (0.93; 1.09 kg.)	W3	B3	BR10	H2
7/8"* (22 mm)	1/2" (13 mm)	20"; 24" (508; 610 mm)	9"; 12" (229; 305 mm)	2.72; 3.26 lbs. (1.23; 1.48 kg.)	W4	B5	BR15S	H3
1"* (25 mm)	5/8" (16 mm)	20"; 24" (508; 610 mm)	9"; 12" (229; 305 mm)	3.75; 4.63 lbs. (1.7; 2.1 kg.)	W5	B6	BR15S	H5
1-1/8" (29 mm)	3/4" (19 mm)	20"; 24" (508; 610 mm)	9"; 12" (229; 305 mm)	4.56; 5.69 lbs. (2.07; 2.58 kg.)	W6T	B8	BR15S	H5
1-1/4"* (32 mm)	3/4" (19 mm)	20"; 24" (508; 610 mm)	9"; 12" (229; 305 mm)	5.94; 7.13 lbs. (2.69; 3.23 kg.)	W6T	B8	BR21	H6
1-3/8" (35 mm)	7/8" (22 mm)	24"; 30" (610; 762 mm)	12" (305 mm)	8.56; 10.7 lbs. (3.88; 4.84 kg.)	W6T	B11	BR21	H11
1-1/2"* (38 mm)	1" (25 mm)	24"; 30" (610; 762 mm)	12" (305 mm)	9.88; 12.3 lbs. (4.48; 5.58 kg.)	W11	B11	(1)	H11
1-5/8" (41 mm)	1" (25 mm)	As Specified	As Specified	As Specified	W11	B11	(1)	H11
1-3/4" (45 mm)	1-1/8" (29 mm)	As Specified	As Specified	As Specified	W12	B12	(1)	H12
2" (51 mm)	1-1/4" (32 mm)	As Specified	As Specified	As Specified	(1)	AGB16	(1)	(1)

* Standard stock sizes - coil outer thread, coil tap.
(1) Uses a hex nut and steel plate assembly.
Shebolt diameters 9/16", 1-3/8", 1-5/8", 1-3/4", and 2" are available with acme outer thread.

B7G Mild Steel Coil Rod - ASTM C1018

Williams B7G Coil Rod is manufactured in mild steel. It is available all-threaded or with threaded ends. The product can be welded. Standard lengths are 12' and 20'.

Bar Diameter & Pitch	Minimum Net Area Thru Threads	Minimum Ultimate Strength	Minimum Yield Strength	Part Number
3/8" - 8 (10 mm)	0.078 in² (50.0 mm²)	5.43 kips (24.1 kN)	4.65 kips (20.7 kN)	B7G-03
1/2" - 6 (13 mm)	0.141 in² (90.9 mm²)	9.86 kips (43.9 kN)	8.45 kips (37.6 kN)	B7G-04
5/8" - 4-1/2 (16 mm)	0.221 in² (143 mm²)	15.5 kips (68.8 kN)	13.3 kips (59.0 kN)	B7G-05
3/4" - 4-1/2 (19 mm)	0.333 in² (215 mm²	23.3 kips (104 kN)	20.0 kips (88.9 kN)	B7G-06
7/8" - 4-1/2 (22 mm)	0.458 in² (296 mm²)	32.1 kips (143 kN)	27.5 kips (122 kN)	B7G-07
1" - 3-1/2 (25 mm)	0.590 in² (381 mm²)	41.3 kips (184 kN)	35.4 kips (158 kN)	B7G-08
1-1/8" - 3-1/2 (28 mm)	0.739 in² (477 mm²)	51.7 kips (230 kN)	44.3 kips (197 kN)	B7G-09
1-1/4" - 3-1/2 (32 mm)	0.969 in² (625 mm²)	67.8 kips (302 kN)	58.1 kips (259 kN)	B7G-10
1-1/2" - 3-1/2 (38 mm)	1.47 in² (951 mm²)	103 kips (459 kN)	88.4 kips (393 kN)	B7G-12

B8S UNC Threaded Tie Rods - ASTM C1045

Bar Diameter & Pitch	Minimum Net Area Thru Threads	Minimum Ultimate Strength	Minimum Yield Strength	Part Number
3/8" - 16 (10 mm)	0.078 in² (50.0 mm²)	9.30 kips (41.4 kN)	7.13 kips (31.7 kN)	B8S-03
1/2" - 13 (13 mm)	0.142 in² (91.6 mm²)	17.0 kips (75.7 kN)	13.1 kips (58.0 kN)	B8S-04
5/8" - 11 (16 mm)	0.226 in² (146 mm²)	27.1 kips (121 kN)	20.8 kips (92.5 kN)	B8S-05
3/4" - 10 (19 mm)	0.334 in² (216 mm²	40.0 kips (178 kN)	30.7 kips (137 kN)	B8S-06
7/8" - 9 (22 mm)	0.462 in² (298 mm²)	55.4 kips (247 kN)	42.5 kips (189 kN)	B8S-07
1" - 8 (25 mm)	0.606 in² (391 mm²)	72.7 kips (324 kN)	55.8 kips (248 kN)	B8S-08
1-1/8" - 7 (28 mm)	0.763 in² (492 mm²)	80.1 kips (356 kN)	61.8 kips (275 kN)	B8S-09
1-1/4" - 7 (32 mm)	0.969 in² (625 mm²)	102 kips (453 kN)	78.5 kips (349 kN)	B8S-10
1-3/8" (35 mm)	1.23 in² (794 mm²)	129 kips (575 kN)	99.6 kips (443 kN)	B8S-11
1-1/2" - 6 (38 mm)	1.41 in² (906 mm²)	148 kips (656 kN)	114 kips (506 kN)	B8S-12
1-3/4" (45 mm)	1.90 in² (1226 mm²)	200 kips (887 kN)	154 kips (685 kN)	B8S-14
2" - 6 (51 mm)	2.65 in² (1710 mm²)	278 kips (1238 kN)	215 kips (955 kN)	B8S-16

External Fasteners

Shebolt Diameter	Wing Nuts	Length	Diameter	Weight
9/16" (14 mm)	W1	5" (127 mm)	13/16" (20.6 mm)	0.31 lbs. (0.14 kg.)
3/4" (19 mm)	W3	4-3/4" (121 mm)	3/4" (19.0 mm)	0.44 lbs. (0.20 kg.)
7/8" (22 mm)	W4	5" (127 mm)	13/16" (20.6 mm)	0.50 lbs. (0.23 kg.)
1" (25 mm)	W5	5-3/8" (137 mm)	1" (25.4 mm)	0.69 lbs. (0.31 kg.)
1-1/8" (29 mm)	W6T	5-1/2" (140 mm)	1-1/8" (28.6 mm)	1.13 lbs. (0.51 kg.)
1-1/4" (32 mm)	W6T	5-1/2" (140 mm)	1-1/8" (28.6 mm)	1.13 lbs. (0.51 kg.)
1-3/8" (35 mm)	W6T	5-1/2" (140 mm)	1-1/8" (28.6 mm)	1.13 lbs. (0.51 kg.)
1-1/2" (38 mm)	W11	6-1/2" (165 mm)	1-25/32" (45.2 mm)	2.25 lbs. (1.02 kg.)
1-5/8" (41 mm)	W11	6-1/2" (165 mm)	1-25/32" (45.2 mm)	2.25 lbs. (1.02 kg.)
1-3/4" (45 mm)	W12	7" (178 mm)	2" (50.8 mm)	3.69 lbs. (1.68 kg.)

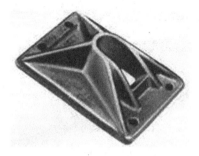

Shebolt Diameter	Vibra-Lock Bracket	Width	Length	Depth	Weight
9/16" (14 mm)	B2	3-1/2" (88.9 mm)	5" (127 mm)	1-3/8" (34.9 mm)	1.53 lbs. (0.69 kg.)
3/4" (19 mm)	B3	4" (102 mm)	5" (127 mm)	1-1/8" (28.6 mm)	2.00 lbs (0.90 kg.)
7/8" (22 mm)	B5	5" (127 mm)	6" (152 mm)	1-1/2" (38.1 mm)	2.44 lbs. (1.10 kg.)
1" (25 mm)	B6	5-1/4" (133 mm)	6-1/4" (159 mm)	1-7/8" (47.6 mm)	3.93 lbs. (1.76 kg.)
1-1/8" (29 mm)	B8	6" (152 mm)	7" (178 mm)	1-7/8" (47.6 mm)	4.72 lbs. (2.14 kg.)
1-1/4" (32 mm)	B8	6" (152 mm)	7" (178 mm)	1-7/8" (47.6 mm)	4.72 lbs. (2.14 kg.)
1-1/2" (38 mm)	B11	6-1/2" (165 mm)	7-1/2" (191 mm)	2" (50.8 mm)	7.50 lbs. (3.40 kg.)
1-5/8" (41 mm)	B11	6-1/2" (165 mm)	7-1/2" (191 mm)	2" (50.8 mm)	7.50 lbs. (3.40 kg.)
1-3/4" (45 mm)	B12	6-1/2" (165 mm)	7-1/2" (191 mm)	2-3/16" (55.6 mm)	8.75 lbs. (4.0 kg.)
2" (51 mm)	AGB16	7-1/4" (184 mm)	8-1/4" (210 mm)	2-3/16" (55.6 mm)	11.1 lbs. (5.0 kg.)

Shebolt Diameter	Hex-Lock Bracket	Width	Length	Depth	Weight
9/16" (14 mm)	H1	3-1/4" (82.6 mm)	4" (102 mm)	1-5/16" (33.3 mm)	1.09 lbs. (0.49 kg.)
3/4" (19 mm)	H2	4" (102 mm)	5" (127 mm)	1-11/16" (42.9 mm)	1.75 lbs. (0.79 kg.)
7/8" (22 mm)	H3	5" (127 mm)	6" (152 mm)	1-3/4" (44.5 mm)	2.81 lbs. (1.28 kg.)
1" (25 mm)	H5	5" (127 mm)	6" (152 mm)	2-1/16" (52.4 mm)	3.25 lbs. (1.48 kg.)
1-1/8" (29 mm)	H5	5" (127 mm)	6" (152 mm)	2-1/16" (52.4 mm)	3.25 lbs. (1.48 kg.)
1-1/4" (32 mm)	H6	6" (152 mm)	7" (178 mm)	2-3/4" (69.9 mm)	6.50 lbs. (2.95 kg.)
1-1/2" (38 mm)	H11	6-1/2" (165 mm)	7-1/2" (191 mm)	2-3/16" (55.6 mm)	9.25 lbs. (4.2 kg.)
1-5/8" (41 mm)	H11	6-1/2" (165 mm)	7-1/2" (191 mm)	2-3/16" (55.6 mm)	9.25 lbs. (4.2 kg.)
1-3/4" (45 mm)	H12	6-1/2" (165 mm)	7-1/2" (191 mm)	3-1/16" (79.4 mm)	12.9 lbs. (5.86 kg.)

Shebolt Diameter	Hex-Nut Bracket	Diameter	Depth	Weight	Weight
9/16" (14 mm)	BR5	3-3/8" (85.7 mm)	1-5/16" (33.3 mm)	0.97 lbs. (0.44 kg.)	0.97 lbs. (0.44 kg.)
3/4" (19 mm)	BR10	4-1/4" (108 mm)	1-5/16" (33.3 mm)	1.25 lbs. (0.57 kg.)	1.25 lbs. (0.57 kg.)
7/8" (22 mm)	BR15S	5" (127 mm)	1-23/32" (43.7 mm)	2.05 lbs. (0.93 kg.)	2.05 lbs. (0.93 kg.)
1" (25 mm)	BR15S	5" (127 mm)	1-23/32" (43.7 mm)	2.05 lbs. (0.93 kg.)	2.19 lbs. (1.0 kg.)
1-1/8" (29 mm)	BR15S	5" (127 mm)	1-23/32" (43.7 mm)	2.05 lbs. (0.93 kg.)	2.12 lbs. (0.96 kg.)
1-1/4" (32 mm)	BR21	5" (127 mm)	2-1/16" (52.4 mm)	3.56 lbs. (1.62 kg.)	3.56 lbs. (1.62 kg.)
1-3/8" (35 mm)	BR21	5" (127 mm)	2-1/16" (52.4 mm)	3.56 lbs. (1.62 kg.)	3.56 lbs. (1.62 kg.)

Taper Tie Forming System

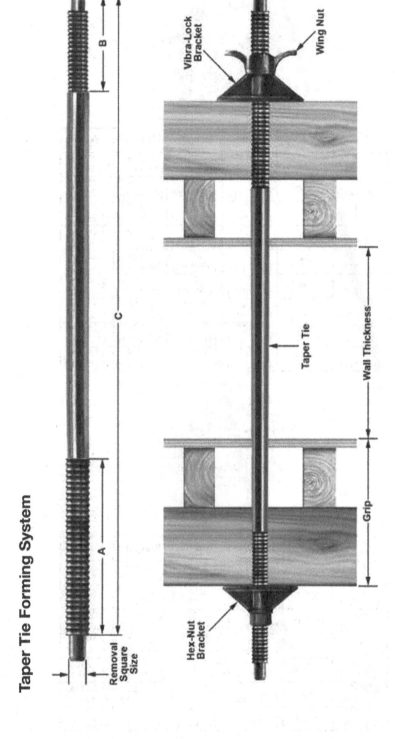

Vibra-Lock Bracket

Wing Nut

Taper Tie

Wall Thickness

Grip

Hex-Nut Bracket

Removal Square Size

C7T Taper Ties

Large End Diameter	Small End Diameter	Square Size	Working Load	Ultimate Strength	Wing Nuts		Vibra-Lock Bracket		Hex Nut Bracket		Hex-Lock Bracket	
					(A) End	(B) End	(A) End	(B) End	(A) End	(B) End	(A) End	(B) End
3/4" (20 mm)	1/2" (12.7 mm)	7/16" (11.1 mm)	7.5 kips (33.4 kN)	15 kips (66.7 kN)	W3	W1	B3	B2	BR10	BR5	H2	H1
1" (25 mm)	3/4" (19.0 mm)	5/8" (15.9 mm)	18 kips (80.0 kN)	36 kips (160 kN)	W5T	W5	B6	B3	BR15S	BR10	H5	H2
1-1/4" (32 mm)	1" (25.4 mm)	13/16" (20.6 mm)	34 kips (151 kN)	68 kips (303 kN)	W6T	W5T	B8	B6	BR21	BR15S	H6	H5
1-1/2" (38 mm)	1-1/4" (31.8 mm)	13/16" (20.6 mm)	50 kips (222 kN)	100 kips (445 kN)	W11	W6T	B11	B8	Hex Nut & Plate	BR21	H11	H6
1-5/8" (41 mm)	1-3/8" (34.9 mm)	1" (25.4 mm)	58 kips (258 kN)	116 kips (516 kN)	W11	W6T	B11	B11	Hex Nut & Plate	BR21	H11	H11

APPENDIX 9

ALUMINUM BEAMS (ALUMA)

Aluma Strongback Load Chart
Imperial*

Span (ft)	Allowable Deflection L/360 (in)	1 Span (lbs/ft)		2 Span (lbs/ft)		3 Span (lbs/ft)	
4.00	0.13	10006 M		7603 R		8640 R	
4.50	0.15	7906 M		6758 R		7680 R	
5.00	0.17	6404 M		6082 R		6912 R	
5.50	0.18	5101 D		5292 M		6283 R	
6.00	0.20	3929 D		4447 M		5559 M	
6.50	0.22	3090 D		3789 M		4737 M	
7.00	0.23	2474 D		3267 M		4084 M	
7.50	0.25	2012 D	*2012	2846 M		3558 M	
8.00	0.27	1657 D	*1554	2502 M		3127 M	*2932
8.50	0.28	1382 D	*1219	2216 M		2608 D	*2301
9.00	0.30	1164 D	*970	1977 M		2197 D	*1831
9.50	0.32	990 D	*781	1774 M		1868 D	*1475
10.00	0.33	849 D	*636	1601 M	*1533	1601 D	*1201
10.50	0.35	733 D	*524	1452 M	*1261	1383 D	*988
11.00	0.37	638 D	*435	1323 M	*1047	1203 D	*820
11.50	0.38	558 D	*364	1211 M	*877	1053 D	*687
12.00	0.40	491 D	*307	1112 M	*739	927 D	*579

Aluma Beam Load Chart
Imperial*

Span (ft)	Allowable Deflection L/360 (in)	1 Span (lbs/ft)		2 Span (lbs/ft)		3 Span (lbs/ft)	
4.00	0.13	3151 M**		2471 R		2808 R	
4.50	0.15	2490 M		2196 R		2496 R	
5.00	0.17	2017 M		1977 R		2246 R	
5.50	0.18	1537 D		1797 R		2042 R	
6.00	0.20	1184 D		1402 M		1753 M	
6.50	0.22	931 D		1193 M		1728 R	
7.00	0.23	745 D		1030 M		1288 M	
7.50	0.25	606 D	*606	896 M		1144 D	*1144
8.00	0.27	499 D	*468	788 M		942 D	*883
8.50	0.28	416 D	*367	676 M		786 D	*693
9.00	0.30	351 D	*292	622 M		662 D	*551
9.50	0.32	298 D	*235	558 M		563 D	*444
10.00	0.33	256 D	*192	509 M	*462	482 D	*362

This information is proprietary to Aluma Systems and is subject to change. It is intended to be used by technically skilled designers knowledgeable in the field and is to be used with other data.

INDEX